W. Kahle · H. Leonhardt · W. Platzer

Color Atlas and Textbook of Human Anatomy

in 3 Volumes

Volume 2:
Internal Organs

by Helmut Leonhardt

Translated by H. L. and A. D. Dayan

3rd revised edition

170 color plates with 584 drawings
by Gerhard Spitzer

1986
Georg Thieme Verlag   Thieme Inc.
Stuttgart · New York   New York

Prof. Dr. med. *Werner Kahle*
Neurologisches Institut
(Edinger Institut) der Univer-
sität Frankfurt/Main, FRG

Prof. Dr. med. *Helmut Leonhardt*
Direktor des Anatomischen
Instituts der Universität Kiel,
FRG

Univ.-Prof. Dr. med. univ.
Werner Platzer
Vorstand des Anatomischen
Instituts der Universität
Innsbruck, Austria

Gerhard Spitzer,
Frankfurt/Main, FRG

Hedi L. Dayan, M.B., and
Anthony D. Dayan, M.D.,
Beckenham, Kent, UK

1st German edition 1976	1st Dutch edition 1978
2nd German edition 1978	2nd Dutch edition 1981
3rd German edition 1979	1st French edition 1979
4th German edition 1984	1st Greek edition 1985
5th German edition 1986	1st Italian edition 1979
	1st Japanese edition 1979
1st English edition 1978	2nd Japanese edition 1981
2nd English edition 1984	3rd Japanese edition 1984
	1st Spanish edition 1977

**Library of Congress Cata-
loguing in Publication Data**

Kahle, W. (Werner)
Color atlas and textbook of
human anatomy.

Translation of: Ta-
schenatlas der Anatomie.
Includes bibliographies
and indexes.

Contents: v. 1. Locomotor
system / by Werner Platzer –
v. 2. Internal organs / by Helmut Leonhardt – – v. 3. Nervous system and sensory organs / Werner Kahle.

1. Anatomy, Human-Atlases.
I. Leonhardt, Helmut.
II. Platzer, Werner. III. Title.
[DNLM: 1. Anatomy-atlases.
QS 17 K12t]
QM25.K3413   1986
611'.022'2   86-5679

© 1978, 1986 Georg Thieme Verlag, Rüdigerstrasse 14,
D-7000 Stuttgart 30, FRG
Typesetting by Druckhaus Dörr, (Linotype System 5
[202])
Printed in West Germany by Druckhaus Dörr,
D-7140 Ludwigsburg

ISBN 3-13-533403-1 (Georg Thieme Verlag, Stuttgart)
ISBN 0-86577-250-9 (Thieme Inc., New York)
                          1 2 3 4 5 6

Foreword

This pocket atlas is designed to provide a plain and clear compendium of the essential facts of human anatomy for the student of medicine. It also demonstrates the basic knowledge of the subject for students of related disciplines and for the interested layman. For all students preparation for their examinations and practice requires repetition of visual experiences. Text and illustrations in this book have been deliberately juxtaposed to provide visual demonstration of the topics of anatomy.

The pocket atlas is divided according to organ systems into three volumes: Volume 1 deals with the locomotor system, Volume 2 with the internal organs and skin and Volume 3 with the nervous system and the organs of the special senses. The topographic relationships of the peripheral pathways of nerves and vessels are considered in Volume 1, in so far as they are closely related to the locomotor system; Volume 2 systematically describes the distribution of the vessels. The floor of the pelvis (pelvic cavity), which has a close functional relationship with the organs of the lesser pelvis, and the relevant topography are incorporated in Volume 2. The developmental anatomy (embryology) of the teeth is briefly mentioned in Volume 2 because it aids unterstanding of the eruption of the teeth. The common embryological origins of the male and female genital organs are also discussed because it helps to explain their structure in the adult, as well as their not infrequent variants and malformations. Certain problems connected with pregnancy and childbirth are mentioned in the chapter on the female reproductive organs. But these do not cover all the knowledge of embryology required by students. The notes on physiology and biochemistry are deliberately brief and only serve to provide better understanding of structural details. Reference should be made to textbooks of physiology and biochemistry. Finally, it must be emphasized that no pocket atlas can replace a major textbook or the opportunity to examine macroscopic dissections and microscopic preparations.

The reference list mentions textbooks and original papers as a guide to the more advanced literature, and it also cites clinical textbooks of relevance to the study of anatomy.

Those who require less detailed knowledge of the structure of the human body will find clear illustrations, too, of the anatomic bases of the more important methods of medical examination. To help the nonmedical reader, everyday English terms for the major organs and their parts have been supplied as far as feasible; these terms are also listed in the index.

Frankfurt/Main, Kiel, Innsbruck *The Editors*

Foreword to the 3rd English Edition of Volume 2

I owe thanks to many colleagues who have given valuable advie about facts and technical matters. I particularly wish to thank Prof. *Junzo Ochi,* Shiga University, Seta, Otsu City, who translated the book into Japanese, for his numerous improvements. Suggestions from students have led to better presentation of the material. I must thank Dr. h. c. *G. Hauff* and Dr. *D. Bremkamp,* the initiators of the work and their colleagues, the publishers, Georg Thieme Verlag, for the willingness with which they have overcome the technical problems.

Kiel, December 1985   *Helmut Leonhardt*

From the Foreword to 1st Edition of Volume 2

Changes in medical education have led to a shortening of the preclinical studies and to major changes in their emphasis. Knowledge of anatomy in particular now tends to be gained mostly from formal lecture courses and examinations place little reliance on practical work, such as dissection. Under these circumstances, there is a danger that the learning of anatomy will become just repetition of lessons without any visualisation of the facts. However, anatomical knowledge can only live through visual understanding.

This pocket book is brought to life mainly by its illustrations. I thank Herr *G. Spitzer* who has produced them. The book originated from knowledge gained by teaching students. It is meant principally for medical students and the interested lay reader, whose place has been taken by *Renate* and *Matthias*.

Homburg (Saar), October 1972   *Helmut Leonhardt*

Contents

X   Contents

Volume 1: Locomotor System by W. Platzer

Volume 3: Central Nervous System and Sensory Organs by W. Kahle

How to Use This Book

Each illustration is marked by a capital letter and the illustrations on each page have consecutive reference numbers. Duplicate pictures of the same structure or organ have the same number. References in the text to the illustrations show the corresponding letters and numbers. Thus, it is easy to refer from the picture to the corresponding text and from the text to look at the appropriate illustration.

Students can use the book as a synopsis and as an aid to recall facts acquired during lectures and courses in macroscopic and microscopic anatomy. In order to refresh the memory, e. g. when preparing for an examination, it is useful for two candidates to work together; one reads aloud a page of the text with its references and the other looks up the structures indicated on the page of illustrations. Subsequently, students should exchange roles and go through the same part of the text again. In this way the information is taken in by both the eye and the ear and is repeated in a convenient fashion. An advantage of the concise text is that it refreshes the memory of facts acquired in regular courses. If particular topics are not recalled, or if any doubts arise during revison, the two candidates should discuss the subject in question and look it up in one of the larger text books (see the list of references at the end of the book). Despite its brevity this book contains a sufficient number of repetitions as particularly important or complicated matters are often discussed again from another aspect on the same or a subsequent page.

Viscera

The human viscera, organs on whose activities the life of the entire organism depends, lie within the neck, chest, abdomen and pelvis. From a functional point of view they can be classified into "systems": **circulatory system** (heart and blood vessels), **blood and defense system** (blood, lymphatics, bone marrow) **endocrine system** (ductless glands and cells which produce hormones) **respiratory system** (nose, airways, larynx and lungs), **digestive system** (mouth, esophagus, gastrointestinal tract, liver and pancreas), **urogenital system** and urinary and genital or reproductive organs.

A Survey of viscera immediately adjacent to the anterior and posterior walls of the trunk. Black: organs of the circulatory system; blue: organs of the respiratory system; red: organs of the digestive system; yellow: organs of the urogenital system.

The following description of the viscera reviews their macroscopic structure, but their microscopic and sub-microscopic (electron microscopic) appearances are included, in so far as the details are essential for understanding the function of the organ. The local topography and interrelationships of the organs are demonstrated in illustrations of preparations and schematized sections through the body.

B Position of the schematized sections through the body (with reference to the page numbers).

The skeleton and the muscles of the wall of the trunk, which also play a part in the functions of certain viscera (e. g., respiration) are discussed in Vol. 1. On the other hand, the floor of the pelvis, which completes the body wall at the pelvic outlet, can only be understood in relation to the viscera, because of its involvement in the arrangement of the sphincters. It is discussed in the appropriate sections of this

volume. The skin, a multifunctional organ, is also described in this volume, even though it is not included amongst the viscera, and has vital sensory functions which are discussed in Vol. 3. The nervous system, i. e., brain, spinal cord, peripheral nerves and sense organs, originally regarded as belonging to the viscera, is contained in Vol. 3. The nervous system, together with the endocrine glands, has a great influence on the functions of the viscera. The systemic anatomical approach, which predominates in this atlas, makes it necessary to include with the viscera other visceral structures situated in the head, as well as the arrangement of the blood and lymphatic vessels situated in the locomotor system. The topographic localization of the peripheral neurovascular pathways (vessels and nerves) is described in Vol. 1.

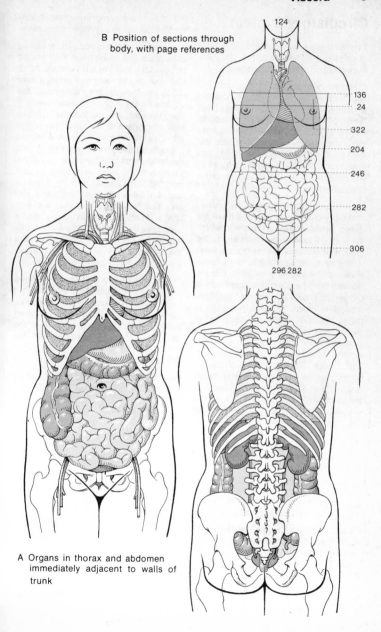

B Position of sections through
body, with page references

124

136
24

322

204

246

282

306

296 282

A Organs in thorax and abdomen
immediately adjacent to walls of
trunk

Circulatory System

The heart and the vessels are the organs of circulation. They contain blood (with the exception of lymphatic vessels).

Circulatory systems. In man and the higher mammals after birth it is possible to distinguish between the greater (systemic) circulation, that supplies blood to all the organs, and the *lesser (pulmonary) circulation,* which serves for the exchange of gases. Together they form a continuous circuit shaped like a figure 8. At its center is a suction and pressure pump, the heart.

Arteries, capillaries, veins. All vessels that *carry blood away from the heart* are called *arteries,* and all vessels which *carry blood toward the heart* are called *veins.* Between the arteries and the veins in the systemic and pulmonary circulations lies the *capillary network.* It is important to remember that blood rich in oxygen is known as "arterial" (oxygenated) and blood deficient in oxygen is called "venous" (deoxygenated). In the systemic circulation arteries carry oxygenated blood and veins carry deoxygenated blood. In the pulmonary circulation venous blood flows through the arteries into the lungs and arterial blood passes through the veins to the heart. In the pulmonary circulation there is only one organ the lungs, through which all the blood flows, and it is of uniform composition. In the systemic circulation there are various organs, e. g., the kidneys, intestines, endocrine glands, etc. The systemic blood flow can take different routes since the systemic circulation consists of numerous parallel subcircuits. The composition of the blood flowing through them is not uniform.

Heart. The heart is divided into *two parts.* Each part has an antechamber, the atrium, and a main chamber, the ventricle. The *"right heart"* (right atrium and ventricle) propels blood through the *pulmonary circulation,* and the *"left heart"* (left atrium and ventricle) is responsible for flow through the *systemic* circulation.

The course of the blood. *The systemic circulation.* The blood flows from the left atrium **AB1** into the left ventricle **AB2,** and thence via the main artery *(aorta* **AB3**) into the capillary networks of the organs. Figure **A** shows the capillaries of the intestines **A4** and **A5.** Blood flows from the legs and the lower half of the trunk via the *inferior vena cava* **AB6** into the right atrium **AB7;** it is returned from the head, arms and the upper half of the trunk via the *superior vena cava.* Deoxygenated blood passes from the intestine and the other unpaired abdominal viscera first into the *portal vein* **A8,** and finally it is returned again via the inferior vena cava into the right atrium. The capillary network of the liver **A9** is situated in the venous section of the stomach-intestine-spleen circuit (portal circulation).

The pulmonary circulation: Blood flows from the right atrium **AB7** into the right ventricle **AB10,** and then via the *pulmonary arteries* (the arteries of the lung) **AB11** into the capillary network of the lung **A12,** from where it flows back via the veins of the lungs *(pulmonary veins)* **AB13** into the left atrium.

Lymphatics. Tissue fluid in the capillary bed is partly drained via a special network of small **A14,15** and large **A16** lymphatic vessels, which return it to the venous side of the systemic circulation **A17.** In the lymph vessels biological filters, lymph nodes **A18,** are interpolated.

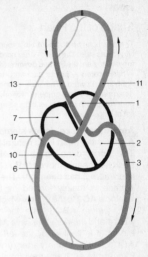

B Pulmonary circulation.
Systemic circulation
(schematic)

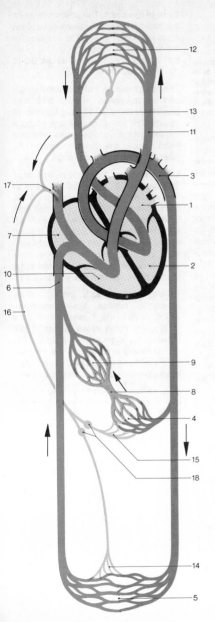

A Organs of circulation (schematic)

Heart

Shape of the Heart

The heart can be compared to a rounded cone lying on its side. The apex – *apex cordis* – points forward toward the left and downward; the base – *basis cordis* – points backward toward the right and upward. The size of the heart is determined by the amount of work it has to perform, and is at least as big as the clenched fist of its possessor.

Anterior surface, *facies sternocostalis.* **A** shows the heart in its natural position. The pericardium which envelops the heart has been removed (**AB 1** shows the line where it has been cut). The anterior surface is mainly formed by the *right ventricle* **AC 2** to the right of which lies the *right atrium* **ABC 3** with the *superior* **ABC 4** and *inferior vena cava* **ABC 5.** To the left it is flanked by the *left ventricle* **ABC 6.** The pulmonary artery (the *pulmonary trunk) AC 7* emerges from the right ventricle. The *aortic arch* **ABC 8** lies across the bifurcation of the pulmonary artery. It gives off the large vessels to the head and arms (the *brachiocephalic* or *innominate artery* **AB 9,** the *left common carotid artery* **AB 10,** and *left subclavian artery* **AB 11**). Between the pulmonary bifurcation and the aortic arch runs a ligament, the *ligamentum arteriosum (ligament of Botallo)* **A 12** (see p. 46). Each atrium has a pouch, the auricle. This helps to produce the rounding of the shape of the heart by filling in the niche between the great arteries and the base of the heart. The right auricle **A 13** is completely visible from the front, but only the apex of the left auricle **A 14** can be seen, because of the rotation of the heart to the left. A branch of the left coronary artery runs along the groove between the right and left ventricles – the *anterior interventricular sulcus* **A 15.** The right coronary artery runs in a groove between the right atrium and right ventricle – the *coronary sulcus* **A 16.** Both arteries, which are embedded in fatty tissue, supply blood to the heart muscle (myocardium).

Base of the heart. The posterior aspect, the base, shows the attachments of the great veins. Its right half is occupied by the essentially vertical right atrium, its left half by the horizontal left atrium **BC 17** (see "Venous Cross" p. 22).

The pulmonary trunk divides into the *right* and *left pulmonary arteries* **B 18.** The *pulmonary veins* **BC 19** open into the left atrium. The left coronary artery and a vessel for the venous return from the heart muscle, the *coronary sinus,* **B 20,** pass in the left *coronary sulcus.*

Lower surface, *facies diaphragmatica.* The lower or diaphragmatic surface of the heart rests on the diaphragm. It becomes visible if the heart is turned from its normal position (arrow). The undersurface is mainly occupied by the left ventricle. Between the left and right ventricle, in the *posterior interventricular sulcus,* runs the posterior interventricular branch of the right coronary artery **C 21.**

Beneath the perfectly smooth, shiny pericardial surface, the **epicardium,** which covers the heart muscle (see p. 22), are some areas of fatty tissue between the superficial muscle fascicles and coronary arteries and the myocardium ("building fat"). It helps to maintain the uniform rounded shape of the heart.

Since the heart borders the air filled lungs on both sides, it is readily visible on X-ray (see p. 28).

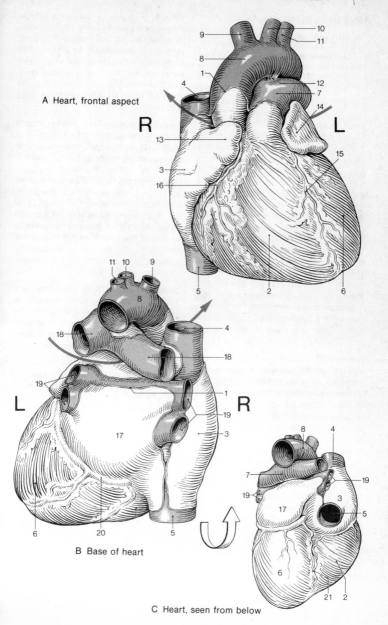

A Heart, frontal aspect

R
L

B Base of heart

L
R

C Heart, seen from below

Chambers of the Heart

The heart is cut open in two frontal planes, one behind the other; their position is marked in Fig. **C.** In **A** the anterior section passes through the right ventricle which lies furthest in front of the left rotation of the heart. The posterior section **B** has opened up both ventricles and atria.

Atria. The interior wall of the left **B 1** and right **AB 2** atria is mainly smooth. Only in the auricles **A 2, A 3** does the atrial muscle protrude in the shape of small, comblike ridges, the *pectinati muscles,* as shown in **B** in a section through the atrial wall. The *right atrium:* the *opening* of the superior vena cava **AB 4** widens as it passes into the atrium. The opening of the inferior vena cava **AB 5** is partly covered in front by a sickle-shaped ridge, the *'valve' of the inferior vena cava (valve of Eustachius)* **B 6.** During fetal life this valve directs the blood towards the membranous interatrial septum which at that time has an opening leading into the left atrium, the *foramen ovale* (see p. 318). Below the valve the venous return from the heart muscle enters via the *coronary sinus* **B 7.** It, too, is delimited by a semicircular ridge, *the valve of the coronary sinus (valvula of Thebesius).*

Ventricles. The interior wall of the right **A 8** and left **A 9** ventricles is very rugged, more so in the contracted (systole) than in the relaxed state of the heart (diastole). Strong muscular columns, the *trabeculae carneae cordis* **B 10,** project from the muscle walls, some of which, the *papillaris muscles,* are particularly prominent. The papillaris anterior muscle of the right ventricle **A 11** arises from the *septomarginalis muscle* and is the called the "moderator band." It forms with the *supraventricular crest* (cut of in **A**) an opening leading into the outflow tract (arrow in **A**). The ventricular septum is predominantly muscle, but on top, at the atrioventricular border, there is a zone of connective tissue where congenital defects of the septum are most frequently located.

Valves. The atrioventricular orifice is closed by a cusped valve (atrioventricular valve) **AB 12.** The outflow from each ventricle is also closed by a valve, that from the right ventricle into the *pulmonary trunk* **AB 13** by the pulmonary valve, and that from the left ventricle into the *aorta* **AB 14** by the semilunar aortic valve **B 15.**

Cusped valves: The *atrioventricular valve of the right heart* consists of 3 cusps; it is the *tricuspid valve* **AB 12.** The corresponding valve of the left heart is formed by 2 cusps: the *bicuspid "mitral" (atrioventricular) valve* **B 16.** It resembles the bilobed miter of a bishop. The cusps arise from a fibrous ring, the "skeleton" of the heart, which separates the muscles of the atrium from those of the ventricle (visible in the sectioned heart **B**). They are attached to the tips of the papillary muscles by their tendons, the *chordae tendinae.*

Semilunar valves: At the root of the aorta and of the pulmonary trunk are semilunar valves, the *aortic valve* **B 15** and the *pulmonary valve* (tip of the blue arrow in **A**). Each of the semilunar valves consists of 3 pockets with the base pointing towards the appropriate opening.

Arrows: Direction of flow of deoxygenated blood (blue) and oxygenated blood (red). Function of heart valves (see p. 11).

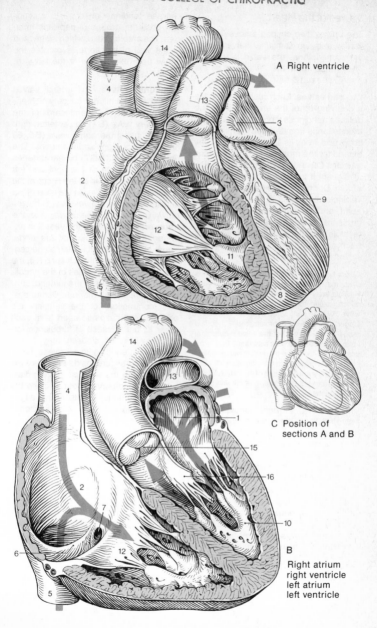

A Right ventricle

C Position of
sections A and B

B

Right atrium
right ventricle
left atrium
left ventricle

Valves of the Heart

The *valves,* two cusped atrioventricular valves and two semilunar valves, lie in the same plane, the *valve plane* **AB.** They arise from the *"skeleton of the heart"*, see p. 12.

Cusped valves. Each cusp (flap) **BC1** is a doubling up of the endocardium. The delicate tendinous cords, *chordae tendineae,* arise from the free margin of the cusp and from its lower surface. They extend for the most part to the papillaris muscles **C2** and hold the cusp in position to prevent it from flapping back into the atrium. During diastole (relaxation) of the ventricle the edges of the cusps move apart, and the valve opens **E.** When the pressure inside the ventricle increases (systole), the flaps spread out, approach each other and close the valve **D;** see Action of the Heart, p. 16.

Clinical tips: After inflammation of the cusps, scars may form along the margins of the valve **F.** This may cause narrowing of the valve orifice, called *stenosis.* If the margins of the valve become contracted by scars and can no longer be closely approximated, *"incompetence"* (insufficiency) of the valve results. Valvular incompetence may also result from gross dilatation of the heart in cardiac failure (failure of the heart muscle). In this condition the papillaris muscles and the base of the valve become so far separated that closure of the valve is no longer possible.

Tricuspid valve (right atrioventricular valve) **B3.** Of the three *cusps,* one is situated in front (anterior cusp), one behind (posterior cusp), and one in the middle *(septal cusp).* The anterior cusp is the largest; it is attached by the chordae tendinae to the powerful *papillaris anterior muscle* which arises from the *septomarginalis trabecula.* The posterior and the small septal cusps are anchored to smaller papillaris muscles. The septal cusp partly covers the membranous part of the interventricular septum.

Mitral (bicuspid) valve **B1.** We distinguish an anteromedial and a posterolateral cusp, which are kept in position by chordae tendinae attached to anterior and posterior groups of papillaris muscles. The anterior cusp, which is attached to the wall of the aorta, separates the inflow from the outflow of the left ventricle.

Semilunar valves. The arterial valves consist of 3 semilunar cusps which are also duplicates of the endocardium: the *pulmonary valve* **(B4)** has anterior, right and left cusps: the *aortic valve* **(B5, G)** has posterior, right and left cusps. The right **BG6** and left **BG7** coronary arteries arise in the depth of the right and left aortic sinuses formed by the cusps of the aortic semilunar valve. At the level of the semilunar valves the vessel wall bulges outward, *sinus of the aorta,* and the transverse diameter of the vessel is enlarged forming the *bulb of the aorta.* There is a nodular thickening at the center of the free edge of each cusp **H8,** the *nodule,* and on both sides of the nodule the edges of the cusps are thinned forming the *lunula* (half moons). During systole (contraction) of the ventricle the edges of the cusps lie apart from each other **K,** but because of turbulence formation within the sinuses, they do not become flattened against the vessel wall. When the pressure inside the artery exceeds that in the ventricle (during diastole) the cusps spread out to close the valve **I.** The nodules at the valve edges help to ensure tight closure; see Action of the Heart, p. 16. **B9** Coronary sinus.

Clinical tips: After inflammation of the valves, *stenosis* or *insufficiency* of the heart may occur, **L.**

A

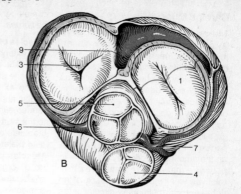

9
3
5
6
7
4
1

Valves of heart,
seen from above

B

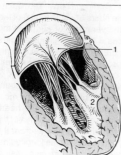

1

2

D Closed

E Open

F Pathologically
deformed

C Mitral valve (left atrioventricular valve)

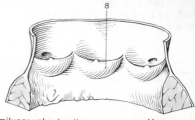

6

7

G

8

Semilunar valve (aortic
or pulmonary valve)

H

I Closed

K Open

L Pathologically
deformed

Cardiac Muscle

The *wall of the heart* consists of three layers: the *endocardium* (inner lining), *myocardium* (heart muscle) and *epicardium* (the outer fibrous connective tissue layer covering the heart). The thickness of the wall of the heart depends mainly on the myocardium. It is covered inside and outside by endocardium and epicardium as thin, smooth, skinlike layers. The degree to which the muscles of the various parts of the heart are developed depends on the amount of work they are made to perform – the walls of the atria are a thin, relatively weak muscle; the muscle of the right ventricle (pulmonary circulation) is thinner than that of the left (systemic circulation).

During fetal life, when the placenta is incorporated into the systemic circulation (see p. 324), the weight of the heart is about 0.6% of the body weight as against 0.4–0.5% during life after birth.

Clinical tips: In stenosis or insufficiency of an arterial valve, or of the mitral (biscuipid) valve, the ventricle that pumps the blood through that valve becomes hypertrophied and this can be seen on a radiograph.

Cardiac skeleton. The muscles of the atria are completely separated from the muscle of the ventricles by dense connective tissue, the *skeleton of the heart* **A1**. The latter is bridged only by the *impulse-conducting atrioventricular bundle (bundle of His)* **A2** (see p. 18). The cardiac skeleton serves as the place of origin and the area of attachment for the muscle of the atria and the ventricles. It is mainly formed by two rings of fibrous tissue, *anuli fibrosi,* from which the atrioventricular valves arise. They are connected to the adventitial fibrous structures of the aorta **AB3** and the pulmonary trunk **AB4**. The cardiac skeleton lies in the valve plane; it is marked on the outer cardiac surface by the *coronary groove* (see p. 6). At the borders between the anuli fibrosi and the adventitial connective tissue of the aorta arise very coarse connective tissue wedges, *trigona fibrosa.*

Muscles of the atria. The muscles of the atria extend partly as an arc from front to back **B5**, and partly diagonally across from one side of the atrium to the other. At the exits of the great vessels **B6**, anular and looped muscle bundles are formed, which follow the vessel orifices to the start of the pericardium. The muscle fibers are partly attached to the cardiac skeleton.

Muscles of the ventricles. Like other hollow organs, each ventricle is made up of an external layer of longitudinal fibers, a middle layer of circular and an internal layer of longitudinal fibers. The outermost muscle fibers (muscle tracts) of the longitudinal layer together surround the two ventricles **B7**. The longitudinal fibers arise from the cardiac skeleton, mainly from the fibrous trigones. They run anticlockwise in spirals toward the apex of the ventricle where individual fiber bundles enter the circular layer. Many muscle tracts go to the apex of the heart, where they form a whorl, the *vortex of the heart,* **C**. From this, as well as from the circular layer of fibers, some longitudinal fibers segregate and return clockwise in steep spirals to the cardiac skeleton. Among them are the *trabeculae carneae* and papillary muscles. The left ventricle and the outflow tracts of both chambers are surrounded by strong circular fibers. As seen in slow motion pictures, the inflow and outflow tracts contract one after the other.

B6 Pulmonary veins, **B8** superior vena cava, **B9** inferior vena cava, **A10** tricuspid valve, **A11** bicuspid valve.

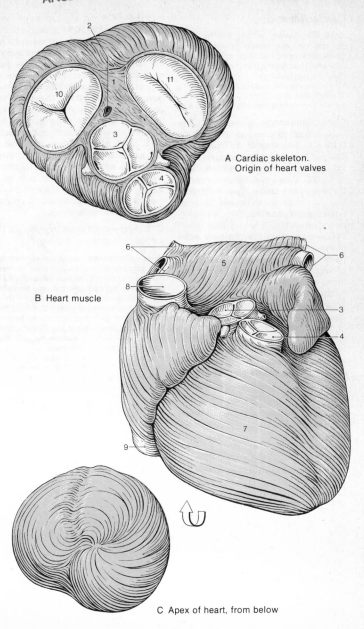

A Cardiac skeleton.
  Origin of heart valves

B Heart muscle

C Apex of heart, from below

Cardiac Muscle

A. van Leeuwenhoek, in 1692, was the first to describe the heart muscle as a lattice with slitlike interspaces. The lattice consists of single striated muscle cells joined end-on to each other. Cardiac muscle tissue is a special form of striated muscle (see Skeletal Muscles, Vol. 1). The striations of heart muscle correspond on the whole to those of skeletal muscle. See **A.**

Light microscope appearance. The oval, almost quadrangular cell nucleus **B 1** lies in the center of the muscle cell. The cell boundaries between neighboring cells are short lines, the *intercalated disks* **B 2.** Connective tissue surrounds the muscle cell. Under higher magnification the striations can be seen, even in unstained preparations. They are mainly bound up with the 0.5–1 μm thick *myofibrils* (muscle fibrils), which have been drawn greatly magnified in **B,** (right), the contractile elements of the muscle. We distinguish the light *I-band* (I = isotropic band) and dark *A-band* (A = anisotropic band). Within the I-band is the *Z-band* (intermediate disk) which traverses the muscle cell, and within the A-band there is the *H-zone* (bright, *Hensen* zone) crossed by a fine dark M-line (middle line). The bands form a regular repeating pattern of periodic intervals extending from one Z-band to the next. Each interval is about 2 μm long and is called a sarcomere.

Electron microscope appearance. Each myofibril (muscle fibril) **C 3** consists of *muscle filaments* – approximately 12 nm thick *myosin* filaments **CD 4,** and 6 nm thick *actin* filaments **CD 5.** The myosin filaments lie exactly side by side and form the A-band, the actin filaments make up the I-band of the myofibril in **B.** The actin filaments penetrate into the interspaces between the myosin filaments, but do not reach the actin filaments which intrude beyond the A-band. The H-zone in **B** is free of them. The M-band in **B** is formed by a protein network between the filaments **C 6.** The *plasmalemma* **C 7** is covered by a basement membrane **C 8** and with it forms periodically recurring folds **C 9** transverse to the course of the filaments. At the crest of each fold the actin filaments of neighboring sarcomeres join in such a way that each actin filament of one sarcomere is connected with several filaments of the next sarcomere **CD 10.** These bonds, and the pleats in the plasmalemma form the Z-line in **B.** The plasmalemmas of adjacent muscle cells form cell contacts **C 11,** amongst which there are regular gap junctions (nexus) **C 12,** which serve to spread excitatory stimuli. Together with the attachment of the actin filaments, they form the *intercalated disks,* the light bands in **B.** Myofibrils do not cross the intercellular space.

C 13 glycogen, **C 14** mitochondrion, **C 15** smooth endoplasmic reticulum.

Muscle contraction. During *isotonic* contraction (shortening of a muscle while its tone remains unchanged), which is initiated by ATP (adenosine triphosphate), the actin filaments **CD 5** slide deeper between the myosin fibrils **CD 4.** This causes the H-band to disappear, the I-bands grows extremely fine and the A-band becomes approximated to the Z-band **CD 10,** see **D.** In extreme contraction the actin filaments may overlap or even fold over each other to form a contraction band within the H-band (*Huxley's* sliding model). During this process ATP is reduced to ADP (adenosine diphosphate).

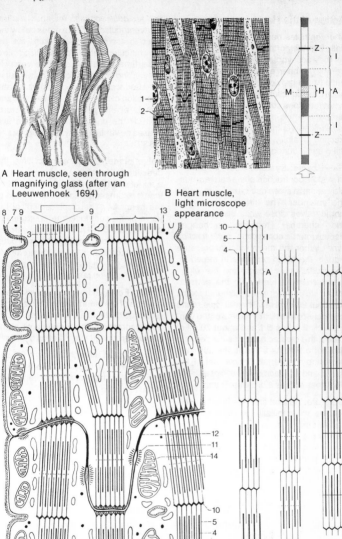

A Heart muscle, seen through magnifying glass (after van Leeuwenhoek 1694)

B Heart muscle, light microscope appearance

C Electron microscope appearance

D Shortening of actomyosin filaments

Action of the Heart

The ventricles pump the blood rhythmically into the aorta and pulmonary trunk. As they act synchronously and in a comparable manner, examination of one ventricle in **AB** provides us with information about the functioning of the heart. The heart functions continuously during life in a repetitive, two-phase cardiac cycle: the blood-filled ventricle is emptied by contraction *(systole),* after which it becomes filled again during relaxation *(diastole).*

Heart cycle. Systole: At the beginning of systole contraction of the cardiac muscle causes a steep rise in pressure inside the chamber. The atrioventricular and aortic valves close while the volume of the chamber remains unchanged (isovolumetric contraction = *contraction time).* When the intraventricular pressure equals the blood pressure in the aorta, the aortic valve opens and the blood pressure rises further (in the aorta to about 120 mmHg, in the pulmonary trunk to about 20 mmHg). The muscles of the ventricle contract, the volume of the ventricle diminishes **B1** and about 70 ml of blood, the *cardiac output,* is ejected into the aorta **B2.** As a result, the intraventricular pressure falls below that of the aorta and the *aortic valve closes.* The chamber acts as a pressure pump.

Diastole: Relaxation of the heart muscle follows, while initially the atrioventricular valve remains closed. The volume of the ventricular cavity remains unchanged (isovolumetric relaxation = relaxation time) and the residual content of the chamber stays at about 70 ml of blood,

the residual volume. As soon as the intraventricular pressure drops below the blood pressure in the atrium, the atrioventricular valve opens. Blood now flows from the atrium into the ventricle **A1** = *filling time.* The forces active in this process are the suction effect of the relaxation of the elastic chamber wall and systole of the atrium. The latter starts towards the end of the filling time and ceases as ventricular systole begins, see **AB3,C.**

Valve plane. During ejection time the direction of the valve changes toward the apex of the heart **B4,** and during filling time it returns again towards the base of the heart **A4.** The displacement of the valve plane contributes to enlargement of the atrium **B3.** It has a suction effect on the venous blood and together with other factors accelerates the venous return to the heart. During this phase the heart acts as a suction pump.

Timing. For the time taken by systole and diastole of the atrium and ventricle during a cardiac cycle, at a pulse rate of 75/min, see diagram **C.** The 8 segments of the circle correspond to a duration of 0.8 s.

Cross section through the right **E5** and left **E6** ventricle in systole; **D5,6** in diastole. **DE7** is the anterior interventricular sulcus (groove).

		Atrioventricular Valve	Aortic Valve
	Contraction time	closed	closed
Systole			
	Output time	closed	closed
Cardiac Cycle			
	Relaxation time	closed	closed
Diastole			
	Filling time	open	closed

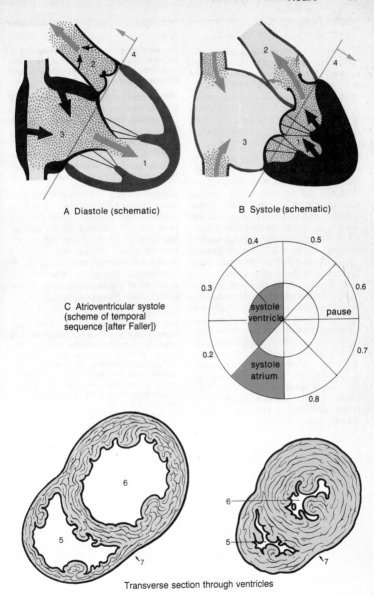

A Diastole (schematic)

B Systole (schematic)

C Atrioventricular systole
(scheme of temporal
sequence [after Faller])

0.4   0.5

0.3   0.6

0.2   0.7

0.8

systole
ventricle

pause

systole
atrium

Transverse section through ventricles

D Diastole

E Systole

Intrinsic Impulse-Conducting System and Cardiac Nerves

The stimuli which result in systole, i. e., contraction of the cardiac muscle, originate with the heart itself — *automatism of the heart*. The heart contains a specialized muscle tissue, the *impulse-conducting system*, which spontaneously generates rhythmical local impulses and which conducts them to stimulate the rest of the heart muscle to contract.

Sinus node, *sinuatrial node* **A1** *(Keith-Flack node)* (the pacemaker) is a network of muscle cells, in all about 2.5 cm long and 0.2 cm wide. It lies at the anterior margin of the orifice of the superior vena cava and radiates into the working muscle of the right atrium, from which the impulse reaches the **atrioventricular node A2** *(Aschoff-Tawara node),* situated in the right atrium near the orifice of the coronary sinus. The atrioventricular node continues into the *trunk* of the **bundle of His** *(truncus fasciculi atrioventricularis)* which pierces the cardiac skeleton (see p. 12). At the upper margin of the muscular part of the interventricular septum the trunk divides into two *branches.* These run along each side of the septum beneath the endocardium to the base of the papillaris muscles. The left branch **A3** fans out **A4,** the right branch **A5** runs mainly across the moderator band (see p. 8) to the papillaris anterior muscle **A6.**

The terminal fibers of the **bundles of His,** called **Purkinje fibers,** merge with the cardiac muscles. Impulses can originate from any part of the impulse-conducting system. The spontaneous *impulse frequency of the sinus node* (about 70/min sinus rhythm) is, however, greater than that of the *atrioventricular node* (about 50–60/min, *AV rhythm),* or that of the bundle of His (about 45–25/min, *ventricular rhythm).* As a rule, therefore, the heart works in a coordinated manner determined by the sinus node (pacemaker), because the subsequent centers remain silent.

The *electrocardiogram* (ECG) is a recording of potential changes in the electrical field of the heart. A block of the sinus node or *bundle of His* reveals a slower ventricular rhythm.

Microscopic structure. The specialized cardiac muscle cells **B5,** which make up the impulse-conducting system, have a greater diameter than the cells of the active muscles **B7,** and they contain more fluid and fewer fibrils. They are also able to generate energy by anaerobic metabolism. **B8** Endocardium.

Cardiac nerves. The nerves to the heart serve to adapt the automatic activity of the heart to the requirements of the body (see Vol. 3). Excitation of the cardiac branches of *sympathetic nerves* increases stroke volume, the speed of conduction of stimuli, cardiac excitability and pulse frequency. The cardiac branches of the *vagus nerve* parasympathetic have the opposite effect.

Both these nervous pathways also contain afferent fibers along which, among other messages, pass, e. g., pain stimuli (angina pectoris) which are projected into the left arm. The cardiac sympathetic nerves contain mainly postganglionic neurons, the parasympathetic (vagus) nerve branches mainly preganglionic ones. The postganglionic nerve cells, red dots in **C,** lie partly in the walls of the atria beneath the epicardium, and their nerve fibers run to the impulse-conducting system.

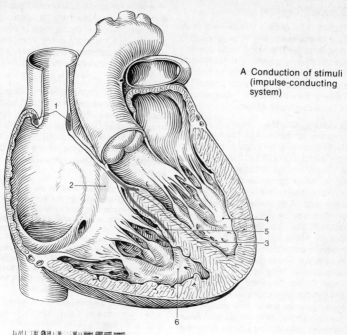

A Conduction of stimuli (impulse-conducting system)

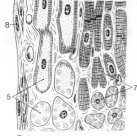

B Impulse-conducting modified cardiac muscle fibers, light microscope appearance

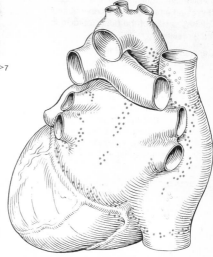

C Distribution of parasympathetic ganglia (red circles and dots) at base of heart

Coronary Vessels

About 5–10% of the total volume of each beat is required solely for supply to the heart muscle. The blood vessels which supply the myocardium, the coronary vessels, form a partial circuit of the systemic circulation, the "vasa privata" of the heart in contrast to the "vasa publica" of the systemic circulation.

Coronary arteries. The *right* and *left coronary arteries* (arteria coronaria dextra et sinistra) arise in the depths of the right and left pouches of the aortic valve **A1,** from a bulge in the aortic wall, the **aortic sinus** (sinus aortae). They surround the heart in the coronary groove on each side of its lower surface. After a short course the left coronary artery **A2** divides into an *anterior ventricular branch* **A3,** which runs in the anterior interventricular sulcus, and a *circumflex branch* **A6** which runs in the coronary sulcus. The right coronary artery **A4** ends in its *posterior interventricular branch* **A5** which runs in the posterior interventricular sulcus. The vessels enter the myocardium from the outside. Anastomoses between the arteries (interarterial connections) are inadequate to form a collateral circulation in case of occlusion of an artery and as a consequence "death" of part of the heart (cardiac infarct) results.

Cardiac veins. Most of the veins in the heart wall also pass to the surface of the cardiac muscle – the *great cardiac vein* **A6,** the *posterior vein of the left ventricle,* the *middle cardiac vein* and **A7** the *small cardiac vein.* They run toward the *coronary sinus* **A8** in the left atrioventricular groove at the back. The coronary sinus ends in the right atrium. Smaller veins run directly from the cardiac wall into the right atrium.

Variants of the coronary arteries: The formation and distribution of the coronary arteries varies.

Formation: In about 38% of cases additional branches arise from the aorta. In less than 1% of cases the heart is supplied by one coronary artery from the aorta. One or both coronary arteries arise from the pulmonary trunk in 1% of cases, and in the latter case life is not possible.

Distribution of coronary artery blood. Three transverse sections through both ventricles show the difference in the areas supplied by the *right (red)* and *left (white) coronary arteries:* **B** balanced type, **C** preponderance of the left, **D** preponderance of the right coronary artery.

Mechanism of blood flow in the myocardium. Blood flow through the **coronary arteries** is influenced by myocardial compression in systole and by pulsatile variations. The blood supply to the myocardium is diminished during the ventricular systole and increased during ventricular diastole. During systole the **coronary veins** are compressed. A falling oxygen tension in heart muscle plays a decisive role as a chemical factor in adaptation of the blood flow to an increased workload on the heart.

Clinical tips. If the branches of the coronary arteries are narrowed, e. g., by atherosclerosis, inadequate amounts of oxygen are supplied to the affected part of the heart muscle and it will die (cardiac infarct). In much rarer instances occlusion of the vessel may be due to a blood clot or *embolus.* For hemodynamic reasons such an embolus must originate somewhere between the pulmonary circulation and the capillary bed of the coronary vessels, usually it comes from inflamed endocardium or a damaged heart valve.

Lymph vessels carry lymph from the endocardium, myocardium and epicardium to lymph nodes at the bifurcation of the trachea (see p. 80).

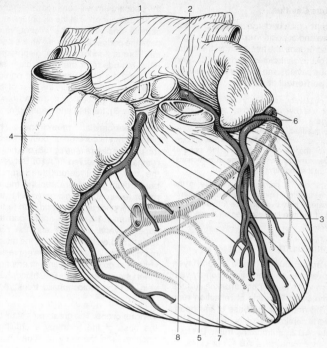

A Coronary vessels

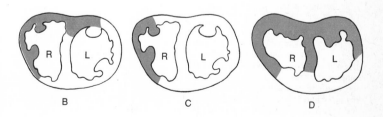

Variability of areas supplied by
coronary arteries (after Töndury)

Serous Cavities

Viscera which undergo marked changes in size and/or position in relation to neighboring organs, the heart, lungs, the major part of the gastrointestinal tract etc., lie in serous cavities: the **pericardial, pleural,** and **peritoneal cavities.**

A serous cavity is a capillary cavity closed on all sides. It is covered by a shiny, smooth serous layer, the 'serosa', and contains a small amount of serous fluid. The serous membrane covers the organs with the **visceral serosa** and the wall of the serous cavity with the **parietal serosa.** The two layers merge at reflections, e. g., at the pedical of an organ.

Clinical tips. A pathological increase in the volume of serous fluid constitutes an effusion or exudate. It may eventually be followed by fusion of the two layers of serosa if they form adhesions.

Pericardium

The **epicardium A1** is the *visceral* layer, it covers the myocardium, coronary vessels and the permanent fat on the surface of the heart. The line of reflection runs behind and over the base of the heart, several centimeters up on the wall of the aorta **AB2,** pulmonary trunk **B3** and superior vena cava **AB4,** but closer to the heart on the wall of the inferior cava **B5** and the pulmonary veins **B6.**

Embryology. The venous and arterial inflows to the heart come together only in the course of development, and accordingly the lines of serosal reflection on the two main arteries and veins are not connected with each other. Between them is a pericardial passage, the *transverse pericardial sinus,* arrow **B7.** The epicardial reflections which cover the veins form a niche, the *oblique pericardial sinus,* arrow **B8.**

The **pericardium AB9** is the *parietal* layer. Its internal structure resembles that of the epicardium, but externally it is reinforced by a layer of tough, criss-cross collagen fibers, which make it relatively inelastic, and therefore the pericardium helps to counteract any overstretching of the myocardium. However, perforating injuries of the heart will result in *cardiac tamponade,* i. e., blood which flows into

the pericardium will compress the heart. At the anterior edge of the diaphragmatic surface and around the inferior vena cava the pericardium fuses with the central fibrous tendon of the diaphragm. Fiber strands from the pericardium accompany the great vessels and participate in the stabilization of the base of the heart in the venous cross (broad black arrows in **B**), trachea and diaphragm.

Pleuropericardial membrane. The pericardium is covered on both sides by the parietal pleura (see p. 138) of which a remnant is preserved in **A10.** Between the two parietal serous membranes, pericardium and pleura, run the mainly motor (efferent) *phrenic nerve* **A11,** the *pericardiophrenic artery and vein* to the diaphragm **A12;** on the right along the superior vena cava **A4** and the right atrium, and on the left behind the lateral margin of the left ventricle. The nerve and vessels supply the pericardium amongst other structures. **A13** Brachiocephalic vein; **A14** brachiocephalic trunk; **A15** vagus nerve.

Venous cross. The great veins stabilize the base of the heart by a cruciform anchorage in the body (**B** broad black arrows, **C**). The *superior vena cava* **AB4** is anchored in the fascia of the neck, base of the skull and arms, the *inferior vena cava* **B5** in the liver and the *pulmonary veins* **B6, 16** in the elastic lungs.

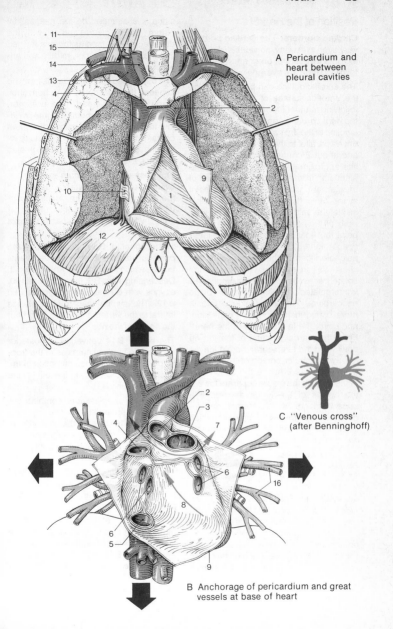

A Pericardium and heart between pleural cavities

C "Venous cross" (after Benninghoff)

B Anchorage of pericardium and great vessels at base of heart

Position of the Heart I

Cardiac borders. The cardiac borders vary even in healthy individuals, depending on age, posture, pregnancy, etc.; the following data are average values.

The cardiac outline can be projected on the anterior surface of the trunk in the following manner (see radiograph, p. 28): the *right border* extends from the sternal junction of the 3rd rib to the junction of the 6th rib parallel to the right sternal margin and about 2 cm from it. The *left border* lies approximately along a straight line extending from a point 2 cm to the left of the sternal junction of the 3rd rib to a point in the 5th left intercostal space, 2 cm inside the mid-clavicular line **A1**. The *apex of the heart* in the adult lies at this point, but that of a child lies one interspace higher, as can be shown by the palpable lifting of the apex against the chest wall in systole. The interpleural space *(recessus costomediastinalis)* **A2** is interpolated on both sides in front of the pericardium. The lungs extend into it, more during inspiration than expiration, and cover the lateral parts of the heart. From the sternal junction of the 4th rib downward the left pleural margin leaves a *lower* triangular portion of the pericardium uncovered. Above the attachment of the third rib the pleura separate toward the apices of the lungs and leave a *superior* triangular retro-sternal zone uncovered, the "thymic triangle" in which lies the *thymus* and the thymus remnants which cover the large vessels.

Right lung: **AB3** lower lobe, **AB4** middle lobe, **AB5** upper lobe. *Left lung:* **AB3** lower lobe, **AB5** upper lobe, **AB6** costodiaphragmatic recess, **A7** liver.

Mediastinum. The visceral space in the neck, which lies between the middle and deep layers of the cervical fascia, is continuous through the upper thoracic opening with the *mediastinum,* the central connective tissue space of the chest. The mediastinum extends from the back of the sternum to the anterior surface of the thoracic vertebral column. On both sides it is bounded by the mediastinal pleura (parietal pleura) and below by the diaphragm. The mediastinum is divided into the upper, posterior, middle and anterior mediastinum.

The *upper mediastinum* extends from the superior thoracic outlet to a horizontal plane above the heart. The posterior, middle and anterior mediastinums lie below this plane. The upper mediastinum contains nerve trunks entering and leaving the posterior mediastinum. It also contains a discrete organ, the thymus.

The *posterior mediastinum* extends between the thoracic vertebral spine and the posterior pericardium. Axially it contains large nerve trunks and tubular organs, which generally pass straight through it.

The esophagus **B8** passes through the *posterior mediastinum* with the vagal trunks, which lie in front of and behind it. Descending aorta **B9,** thoracic duct **B10,** azygos vein **B11** and hemiazygous vein **B12.** The sympathetic trunk **B13** lies lateral to the vertebral column, in front of the heads of the ribs.

The border **B14** between the posterior and middle mediastinum lies in the frontal plane anterior to the bifurcation of the trachea, approximately at the level of the hilus of the lung **B15.**

The *middle mediastinum* contains the heart in its pericardial sac. The mediastinal pleura **B16** lies over the pericardium, and between the two on both sides run the phrenic nerve and the pericardiophrenic artery and vein **B17.**

The *anterior mediastinum* is the cleft-shaped connective tissue space in front of the heart, between the pericardium and the chest wall.

B18 Projection of the entire width of the heart, which is covered laterally by the lungs (see *Area of relative cardiac dullness,* p. 26); **B19** projection of the chest wall of the cardiac area not covered by the lungs (see *Area of absolute cardiac dullness,* p. 26).

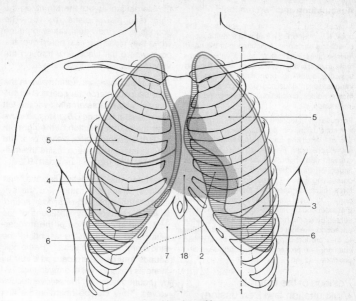

A Projection of borders of heart,
pleura and lungs on thorax

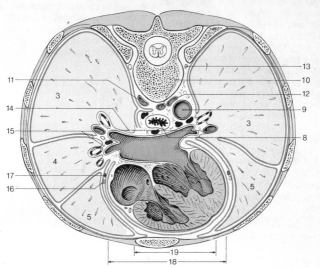

B Horizontal section through thorax
at level of 8th thoracic vertebra

Auscultation and Percussion

Auscultation is listening for sounds in the body. It is performed either with a stethoscope or with the ear alone. Auscultation of the heart mainly gives information about valvular defects. In auscultation of the lungs we distinguish vesicular breathing, bronchial breathing, and various adventitious sounds, such as rattles, friction noises, crepitations, crackling and ringing sounds, etc.

Percussion is tapping of the body surface so that conclusions can be drawn about the condition of underlying parts from variation in the sounds produced. Percussion is usually carried out with one finger, with or without putting a finger of the other hand underneath the percussing finger. The sound is judged by its force, pitch, duration and tonal nature (tympany). Dullness, i. e., a dull sound resembling that produced by percussing the thigh, is represented by a soft, high pitched sound. A resonant sound, like that found over the lungs, similar to that produced by percussing an air filled cushion, is represented by a loud, deep and prolonged sound.

Position of the Heart II (Auscultation and Percussion)

Percussion. Absolute and relative areas of cardiac dullness. In part, the heart lies immediately behind the anterior chest wall, and percussion of this region reveals the area of *'absolute'* cardiac dullness **A1**. On both sides the interpleural space extends in front of the pericardium. It is this space, *the costomediastinal recess*, into which the lung enters more during inspiration than expiration, that conceals the lateral parts of the heart. They can only be made audible as cardiac dullness by forceful percussion which penetrates the margins of the lungs: area of *"relative" cardiac dullness*, **A2**. The area of relative dullness represents the true size of the heart and corresponds to the shadow of the heart in a radiograph. The area of 'absolute' dullness merges below into the zone of dullness over the liver **A3**.

Auscultation. The heart valves lie roughly in a line which stretches from the sternal junction of the 3rd rib on the left to the sternal junction of the 6th rib on the right (see *Valve plane,* p. 16). The *sounds* which individual valves produce when they close are best heard over the point where the blood flow through the valve lies closest to the chest wall.

Sites of auscultation: Aortic valve in the right second intercostal space **B4**; *pulmonary valve* parasternally over the left 2nd intercostal space **B5**; *tricuspid valve* over the sternal junction of the right 5th rib **B6**; *bicuspid (mitral valve)* over the left 5th intercostal space inside the midclavicular line (apex of the heart) **B7**.

Heart sounds. We distinguish two heart sounds, which are rather like the syllables "lubb-dubb". The *first heart sound,* which is produced by vibrations of the walls during contraction of the myocardium, also contains the sound of the atrioventricular valves. Pathologic sounds (murmurs) are caused either by stenosis (hissing noise) or by insufficiency (pouring noise) of the atrioventricular valves. They can be heard in the first heart sound at the appropriate site of auscultation. The *second heart sound* is caused by closure of the arterial valves. Pathologic sounds (murmurs) in stenosis or insufficiency of the arterial valves can be recognized in the second sound at the corresponding auscultation sites.

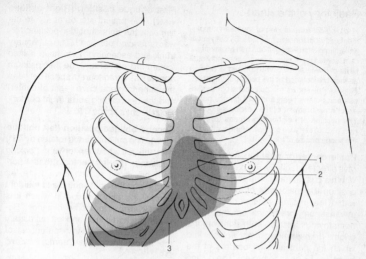

A Areas of relative and absolute
   dullness

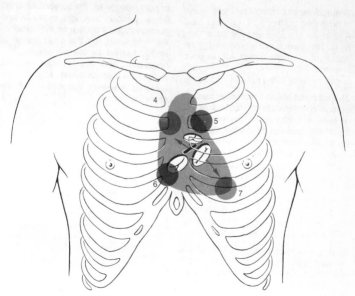

B Projection of heart valves (after Knese)
   and their auscultation sites on the anterior thorax

Radiology of the Heart

Radiological examination of the heart supplements other diagnostic methods. By putting the patient into different positions behind a fluorescent screen or a film casette, the individual parts of the heart can be examined in rotation as they are outlined at the border of the heart. *Fluoroscopy* shows the pulsations of the various parts of the heart and *radiographs* permit exact measurement of the cardiac silhouette. The position of neighboring organs (esophagus, p. 206) can also provide information about the shape or size of the heart.

Posteroanterior position. With the X-rays directed sagitally, the patient stands with his chest against the screen. In this position the major part of the heart lies within the *midline shadow* cast by the mediastinum, vertebral column, sternum, thymus, etc. On both sides are the bright (transparent) *'lung fields',* into which the right and left borders of the heart can be seen to project. The outline of the right heart border consists of two *curves,* and the left of four.

The right heart border. The *upper curve* is produced by the superior vena cava **A1.** In its shadow lies the ascending aorta. The *lower curve* is produced by the border of the right atrium **A2.** In deep inspiration the inferior vena cava may become visible at the right lower border.

The left heart border: The *upper curve* is produced by the aortic arch **A3** (aortic outline), the *second* by the pulmonary trunk **A4** (pulmonary outline) and the *third,* which is often not visible, by the left auricle **A5** (the 2nd and 3rd curves are together also called the *middle curve).* The *lower curve* is the border of the left ventricle **A6.** Where the lower curve merges into the shadow of the liver is the apex of the heart; it is often not sharply delineated because of epicardial fat deposits. **A7** Lung hilus (see p. 130).

Anteroposterior position. The patient remains in the frontal position but with his back to the screen. The findings **B** are as in **A** but reversed.

First oblique position (Right oblique view). The patient stands at 45° to the film with the right shoulder pointing forward *('fencing position').* The pulmonary trunk and ascending aorta have shifted toward the left border. The mediastinum appears as a transparent space between the superior vena cava (right of center) and the descending aorta (right border), *Holzknecht's space* **C8.**

Second oblique position (left oblique view). The patient stands rotated by 45°, with the left shoulder forward *('boxing position').* The left ventricle (basal portion) has been moved toward the left border and the right ventricle to the right border, the only position in which it is directly visible for some distance. Between the aortic arch and left pulmonary artery there appears a transparent part of the mediastinum, the *"aortic window"* **D9.**

"Laterolateral" position. The patient stands rotated by 90° with either his left or right side toward the screen. In the left lateral position the right ventricle and pulmonary trunk appear on the border of the right heart and the other border is mainly formed by the left atrium. The descending aorta lies on the far left of the view. *Holzknecht's space* **E10** is conspicuous between the heart and descending aorta.

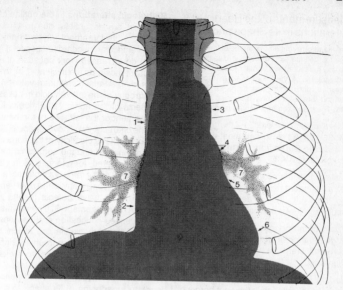

A Radiograph of heart.

Posteroanterior
position

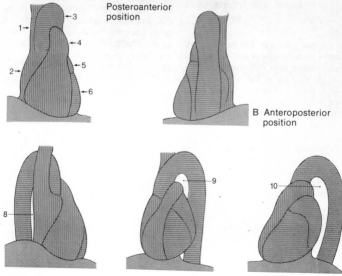

B Anteroposterior
position

C 1st oblique view

D 2nd oblique view

E Laterolateral
position

Measurements of the Heart; Alterations in its Shape and Size

An alteration in the size or shape of the heart is often a definite sign of cardiac disease. It is usually measured in a frontal radiograph. Teleradiography of the heart at 2 m focus-film distance diminishes distortion, as by this method the heart is only magnified by about 5%. The measurements usually made are:

Transverse diameter = the greatest width right **A1** + the greatest width left **A2,** as measured from the midline of the thorax (chest) **A3.**

Longitudinal diameter **A4** = the distance from the indentation in the cardiac outline on the right (the border between the upper and lower curves) to the apex.

Width of the heart at its base **A5** = the largest diameter of the heart measured perpendicular to the longitudinal diameter.

Length of the aorta **A6** = the distance from the indentation in the right border of the heart to the top of the aortic arch.

Angle of inclination = the angle between the longitudinal diameter and the horizontal line.

The normal shape of the heart varies for many reasons. *Constitutional variants* are connected with differences in physique, particularly in the shape of the chest. In asthenic types and adolescents **D** the heart can be in a more erect position and the angle of inclination may be greater than that found in subjects of more average build **C.** The heart of pyknic individuals and of elderly persons **B** is often more squat and the angle of inclination is smaller. **E** *Alterations in the shape of the heart due to respiration* depend on the position of the diaphragm. To a small extent the heart follows respiratory movements and this mainly affects the angle of inclination, which is about 45° in mid-respiration in someone of average build (after *Schinz et al.*).

Pathologic alterations in the outline of the heart are essentially due to three causes: 1. Enlargement *(hypertrophy)* of the ventricular myocardium occurs if it has to do increased work because of a defect (stenosis or insufficiency) in the succeeding arterial (semilunar) valve(s). Hypertrophy of the left ventricle in cases of defects of the aortic valve results in enlargement of the left side of the heart with the characteristic "boot" shape **F.** In this condition the elasticity of the aorta is often diminished and the aortic arch becomes elongated, for example as in old age, so the aortic arch juts out as an aortic knob. Hypertrophy of the right ventricle in mitral valve defects results in right sided enlargement **G.** The valvular defect causes congestion and backflow in the pulmonary circulation, which is reinforced by the hypertrophied right ventricle: the patient suffers from dyspnea ("cardiac asthma"). 2. *Dilatation* of the heart in myocardial insufficiency leads to broadening of its shape to approximate a round outline. 3. An *effusion* or exudate in the pericardium results in enlargement of the cardia shadow **H.** As the pericardium extends right up to the great vessels, the curves of the heart border disappear and the heart assumes a more triangular shape.

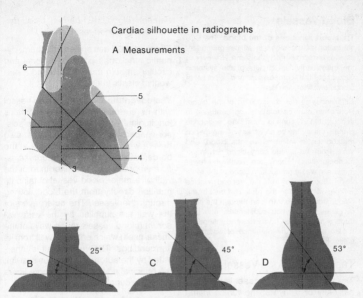

Cardiac silhouette in radiographs

A Measurements

Variability depending on physique (after Schinz)

B 25° C 45° D 53°

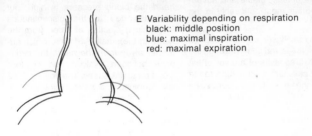

E Variability depending on respiration
black: middle position
blue: maximal inspiration
red: maximal expiration

Abnormal heart shapes

F Enlargement of
left ventricle

G Enlargement of
right ventricle

H Pericardial
effusion

Blood Vessels

Transport functions of the blood. The CO_2 saturation of the blood in all the pulmonary arteries is approximately the same. However, in the many individual circuits of the systemic circulation the composition and amount of blood flow differ, see **B.**

Differences in the composition of the blood: Hormones may be relatively concentrated in venous blood coming from the endocrine glands, there may be an excess of energy-rich metabolites in intestinal venous blood, and veins from the spleen carry blood rich in immunoglobulins. There are relatively fewer metabolic waste products in blood of the renal veins. Blood coming from the liver and muscles will be warmer than the rest. The task of the blood circulation is transport through the body of gases, energy-rich substances, end products of metabolism, hormones and antibodies. It also assists in maintenance of water and electrolyte balance.

Tasks of the Blood Vessel Walls

The **wall of a blood vessel** is made up basically of 3 layers: the *tunica intima (internal layer), tunica media* and *tunica adventitia* (middle and outer layers), simplified as the *intima, media* and *adventitia.* They differ most noticeably in the arteries.

The **intima** mainly serves the exchange of materials, fluids and gases through the vessel wall. It consists of a layer of flat *endothelial cells* arranged parallel to the long axis of the vessel **C1,** surrounded by scanty connective tissue. The intima of the arteries has in addition a fenestrated elastic membrane, the *internal elastic membrane* **C2.**

The *endothelial cells* of all blood vessels are joined to each other by *occluding contacts* (zonulae occludentes, tight junctions, see p. 40). In individual vascular sections, these contacts are variable in their numbers as well as in their tightness. In the *arteries,* the intercellular contacts are usually impermeable whilst in the *capillaries* and postcapillary *venules* of many organs they are permeable. However, in the capillaries of some organs thy form particularly impermeable barriers (*blood-brain-barrier, blood-thymus-barrier, blood-testis-barrier*).

The **media C3** has an essential hemodynamic function. It consists of spiral and circular *smooth muscle cells* and a network of elastic fibers.

The **adventitia C4** connects the vessel with its surrounding tissue. It contains longitudinally arranged cells and a scissor-shaped *fiber lattice.* Arteries also have a weaker elastic membrane at the border with the media, the *external elastic membrane* **C5.** The inner third of the wall of a major blood vessel obtains its nutrition directly from the blood flowing through the vessel. The outer layers of the wall are supplied by the *vasa vasorum* (blood vessels to the walls of the vessel itself) which enter the wall from its surroundings. The vessel wall is innervated by the autonomic nervous system; the principal nonmedullated nerve fibers lie in the adventitia.

The **hemodynamic situation** within the systemic circulation differs in the various sections, see **D.** In the *arteries near the heart* the blood pressure is high, but fluctuates due to the discontinuous (rhythmical) ejection of blood from the heart. In the smaller arteries, which are *further away from the heart,* the difference between systolic and diastolic blood pressure is less. The flow of blood is determined by a pressure gradient that extends all the way from the aorta to the venae cavae. These *differences in pressure* are due to degrees of resistance to flow in the vascular system, which depend in part on the lengths and widths of the various vessels (see Physiology). Depending on the part of the body and on posture, gravity may enhance or restrict the return to the heart of the blood in the large veins. Finally, the negative intrathoracic pressure influences blood flow through the *great veins near the heart,* see p. 144. The structure of the walls in the various parts of the vascular system reflect these hemodynamic factors. For the autonomic innervation of vessel walls, see Vol. 3.

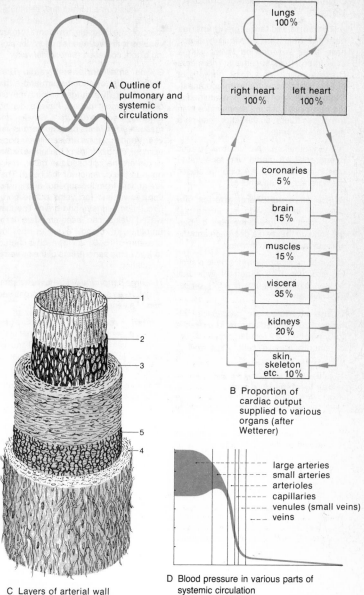

A Outline of pulmonary and systemic circulations

lungs
100%

right heart
100%

left heart
100%

coronaries
5%

brain
15%

muscles
15%

viscera
35%

kidneys
20%

skin,
skeleton
etc. 10%

B Proportion of cardiac output supplied to various organs (after Wetterer)

C Layers of arterial wall

large arteries
small arteries
arterioles
capillaries
venules (small veins)
veins

D Blood pressure in various parts of systemic circulation

Structure of the Walls of Blood Vessels

A The aorta and **large** (great) **arteries near the heart** are *elastic in type* and have *three well-defined layers* with a thick internal elastic lamina. The media consists mainly of a thick network of elastic fibers (fenestrated membranes). During fixation for histologic examination the non-elastic tissues shrink more than the elastic fibers, so the latter assume a wavy appearance.

A1 Endothelium, **A2** internal elastic membrane, **A3** media with fenestrated elastic membranes, **A4** external elastic membranes, **A5** adventitia.

B The *large* **peripheral arteries,** are *muscular in type.* With increasing distance from the heart, the elastic fiber-networks in the media decrease and the smooth muscle cells increase.

C *Small peripheral arteries* have a similar structure. The lumen of individual vessels diminishes rapidly with progressive division.

D The arterioles are *precapillary arteries* of 20–40 μm diameter. Their media is formed by 1 or 2 regular, circular layers of smooth muscle cells *("precapillary sphincter").*

From the small arteries to the end of the arterioles the mean blood pressure falls from about 80 mmHg to about 30 mmHg, and the blood flow velocity decreases from 10 cm/s to 0.2 cm/s.

E Capillaries serve the exchange of gases and substances between blood and the tissues. They originate from a further division of arterioles, form networks and, if filled with blood, their diameter varies between 5–15 μm.

The total cross-sectional area of the vessels increases in the capillary bed (2000 capillary cross sections per mm^2 have been counted in the skeletal muscle of a homothermic [warm blooded] animal). The blood pressure in the capillaries falls to 12–20 mmHg and the velocity of blood flow to 0.03 cm/s.

F The venules are *postcapillary veins.* Their walls contain irregularly distributed muscle cells which can regulate the lumen of the vessels. In some organs venules form lacunae (sinuses) for storing blood, so-called *"sinusoidal veins".*

G The *small* **peripheral veins.** Their walls consist of endothelium and a thin layer of spirally arranged smooth muscle cells, but most do not have a distinct *three-layered structure.* Small and medium sized veins have numerous valves. *Venous valves* are semilunar pockets consisting of folds of the intima, either single or one opposite the other, which open in the direction of the heart. They are absent from the superior and inferior vena cava and the veins of the portal circulation, kidneys and brain. The thin-walled veins can accommodate large amounts of blood, even at the low pressure of 15–0 mmHg which is characteristic of the venous limb of the systemic circulation.

H *Large peripheral veins* have the same structure as smaller ones. **H6** cross section of a venous valve.

I Inferior vena cava. The structure of its wall is essentially the same as that of the smaller veins. The longitudinal muscle strands in the media are arranged in small bundles.

The *structure of vein walls* varies more than that of arteries, and this is true especially of the inferior vena cava, portal and suprarenal veins and the pampiniform plexus.

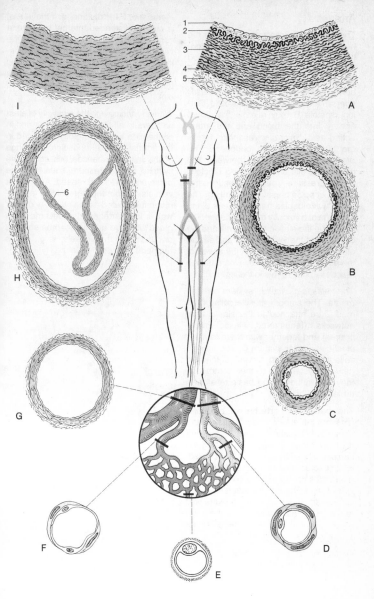

Layers in the walls of blood vessels in various parts of systemic circulation

Arteries

The **aorta** and **great arteries** (near the heart) are directly affected by the discontinuous, pulsatile ejection of blood (stroke volume) from the heart. Part of the cardiac output is at first retained in the aorta during systole **A** and the *elastic fibers* in the wall become distended. During diastole **B** they release the stored energy into the blood and the artery acting as a *"hydraulic reservoir"* forces the blood toward the periphery. The *blood pressure* fluctuates between 120 mmHg systolic and 80 mmHg diastolic blood pressure (blood pressure amplitude or pulse pressure 40 mmHg). The *pulse wave* begins here and runs along the aorta at a velocity of 4 m/s, depending mainly on the elasticity of its wall and the diameter of the vessel. With decreasing elasticity of the vessel wall the speed of the pulse wave (blood flow) increases: it reaches about 10 m/s in the leg.

The elastic-muscular system of the media. The smooth muscle cells of the media are attached to the elastic fiber networks (fenestrated elastic membranes) and together with them form an elastic-muscular system **C.** The muscle cells are controlled by the autonomic nervous system and by their contraction they can influence the basic tension in the elastic fibers.

The **peripheral arteries** of *muscular type* and the *arterioles* can influence the peripheral circulation and the blood pressure if their lumen changes (*"regulation of peripheral resistance"*). **D** Dilated, **E** contracted artery. The histological appearances do not entirely correspond to that of the living tissue because fixation causes additional contraction of the wall and thus results in waviness of the internal elastic membrane which does not shrink to the same extent.

Cutting of an artery. Depending on the position of adjacent joints (see p. 38) the arteries are more or less exposed to longitudinal tension. The *internal elastic membrane* **F1** is distended. If the vessel is cut the distended internal membrane contracts and pulls the open end of the vessel wall into the lumen, thereby closing the opening temporarily. This mechanical closure may be followed by thrombotic occlusion. i. e., by a blood clot.

With *advancing age* the elasticity of elastic fibers in all organs diminishes and in connection with the rigidity produced in the arterial walls (arteriosclerotic change of the intima enhances this effect) the systolic blood pressure rises and the *pulse pressure* is increased. As a result vessels become progressively and irreversibly dilated as the elasticity of their wall is lost. Arteries also become elongated with age and this can be seen as *undulation* of the vessels, e. g., of the temporal arteries.

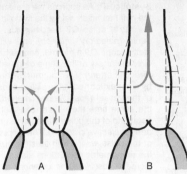

Elastic fibers of
arterial wall

Hydraulic reservoir
function

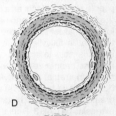

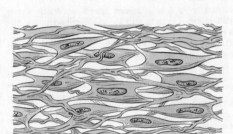

C Elastic fibers and smooth muscle cell
("tension muscles") network

Dilatation and con-
traction of arterial
wall (histologic
fixation)

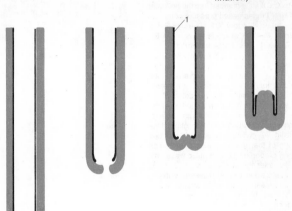

F Closure of artery after vessel has been cut
(after Staubesand)

Position of the Arteries in the Locomotor System

As a rule arteries run along the *flexor surface* of the joints. Therefore, they are not stretched over the extensor surface and are not pinched or compressed when the joint is flexed. The danger of pinching on the flexor side is minimized by their position. Arteries and their associated veins and nerves on the flexor surface of a joint are embedded in a pliable *fat pad*. It acts as a cushion when the joint is stretched and helps to lessen longitudinal tension on the vessel when the joint is fully flexed; in this way the artery is able to shorten and to retract from the danger zone. The distance between the radial epicondyle **AB1** of the humerus and the brachial artery **AB2** is greater in the flexed state **B** than in extension **A**. **AB3** Radial artery, **AB4** ulnar artery.

Estimation of the Blood Pressure (BP)

The *arterial BP* is important in the diagnosis of many diseases. It is measured during almost every medical examination by the indirect method of *Riva-Rocci*.

Method of measuring. A hollow rubber cuff **C5**, reinforced on the outside by a layer of linen, is firmly placed around the upper arm. With a rubber bulb **C6** air is pumped into the hollow of the cuff until the increased pressure compresses the brachial artery **C2**, i. e., until the pressure in the cuff exceeds the systolic BP within the artery. This is detected by disappearance of the radial pulse **C7**. A manometer **C8** indicates the pressure in the cuff. Air is now released from the cuff by a screw valve on the bulb **C6** to a point where the pressure in the cuff is just below the systolic BP. The crests of the pulse waves then briefly open the lumen of the artery with every pulse wave and a characteristic sound is heard through a stethoscope. If the pressure in the cuff is progressively reduced the lumen of the artery remains open for an increasing proportion of each pulse wave until eventually, after the pressure in the cuff has fallen below the diastolic BP in the artery, it is no longer compressed at all and a second sound is heard. During the procedure the following phenomena are observed and they serve as criteria for measuring the BP.

Systolic BP: As soon as the pressure in the cuff has fallen below the systolic BP, the radial pulse **C7** becomes just palpable *(palpatory criterion)* and with a stethoscope **C9** a murmur can be heard over the artery at the angle of the elbow (brachial artery). The murmur is caused by a vortex formed in the slitlike opening of the vessel *(auscultatory criterion)*. At the same time the patient can feel the pounding of the pulse wave in the upper arm *(subjective criterion)*. The amplitude of the pulse waves and thus the size of the manometric oscillations also start to increase *(oscillometric criterion)* at this stage; see **D** and **E**.

The *pressure levels* at which these criteria and those described below become apparent are read off the *manometer* as the systolic and diastolic BP, respectively.

Diastolic BP: When the pressure in the cuff falls below the diastolic BP the murmur in the brachial artery and the pulsation in the upper arm disappear, and the amplitude of the manometric oscillations decreases.

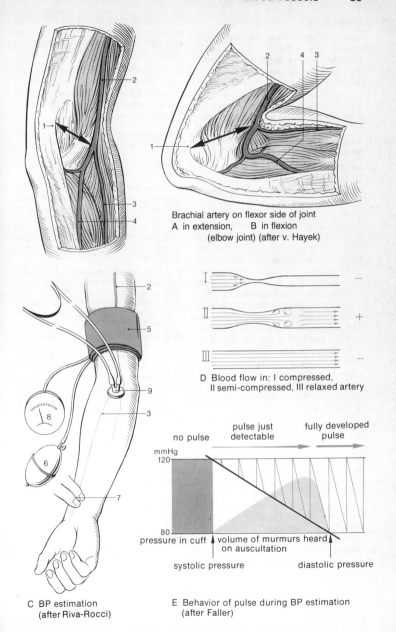

Brachial artery on flexor side of joint
A in extension,   B in flexion
(elbow joint) (after v. Hayek)

D Blood flow in: I compressed,
  II semi-compressed, III relaxed artery

C BP estimation
(after Riva-Rocci)

pulse just   fully developed
detectable   pulse

no pulse

mmHg
120

80
pressure in cuff ↑ volume of murmurs heard
              on auscultation

systolic pressure   diastolic pressure

E Behavior of pulse during BP estimation
  (after Faller)

Capillaries

The exchange of gases and substances between the blood and the tissues is facilitated by the large cross-sectional area of all the capillaries (which is about 800 times that of the aorta) and by slowing down of the blood flow (to about 0.3 mm/s as compared to 320 mm/s in the aorta). All organs have capillaries except for stratified squamous epithelia, the cornea and lens of the eye, and completely differentiated cartilage.

AB Capillaries often form three-dimensional *networks* (surface area about 6000 m^2) which are supplied by several arteries. **A** Occlusion of one artery may, therefore, be of no consequence to the organ in question. **B** If, however, a particular capillary area depends on a single artery, an *"end artery"*, without adequate cross communications *(collaterals)* with other vessels, occlusion of that artery (e. g., by *thrombosis or embolism*) results in death *(necrosis)* of the affected tissue – an *infarct*. The branches of the arteries to the liver, kidneys, spleen, brain, retina and the coronary vessels are *"end arteries"*. Capillaries are usually about 1 mm long, and their diameter, if filled with blood, is about 5–15 μm. Some organs, namely liver, bone marrow, spleen and certain endocrine glands, are characterized by having particularly wide *sinusoidal capillaries*.

C The capillary wall consists of an *endothelial cell layer* **C2** and a *basal membrane* **C1** visible under the electron microscope. In some organs, e. g., in the brain, there are also additional contractile cells attached to the outer wall, *pericytes* **C4**. The **endothelial cells** **C2**, which are 25–50 μm long, as a rule adjoin each other without gaps and form an endothelial tube. The thickness of endothelial cells varies. Transport of materials occurs from the capillaries into the surrounding tissue and vice versa, and in both cases through the endothelial cells.

The degree of intercellular transport between endothelial cells depends on organ specific differences in the tightness of occluding intercellular contacts.

C in **C** different forms of endothelium with differing degrees of transcellular transport have been drawn schematically as segments of a cross-section through a capillary.

I *Endothelia without fenestration:* the endothelial cells of many organs (e. g., muscles, brain, lungs, etc.) show only occasional "snapshots" of cytopempsis, **C5** i. e. vesicles or vesicle-like invaginations of the cell membrane.

II *Fenestrated endothelia:* in organs mainly concerned with considerable transport of substances through the capillary wall, such as the kidneys, intestinal villi and the endocrine glands etc., the endothelial walls are thinned into a *"diaphragm"* **C3** in many places. These *fenestrations* are probably an expression of extreme permeability.

III *Endothelia with intracellular gaps:* in the capillaries of the liver there are gaps **C6** between the endothelial cells. Fluid can pass directly from the capillaries into the surrounding tissue because the basal membrane also contains gaps.

D Cytopempsis: (vesicular transport) Larger molecules and particles **D7** usually move first into the cell interior **D9** by invagination of the *plasmalemma* **D8** (= cell membrane). Then they pass through the endothelial cell in a vesicular fold of the plasmalemma. Finally, in the reverse procedure, the membrane surrounding the vesicle is reincorporated into the plasmalemma and its contents are extruded through the plasmalemma on the opposite side of the cell.

Permeation. Small molecules, such as nutrients and other building materials for cells, often permeate invisibly through endothelial cells **D10**.

Prerequisites for cytopempsis and permeation are usually membrane-bound receptors, specific enzymes and, for transport through fenestrated endothelia, an appropriate surface charge on the luminal plasma membrane of the endothelial cell.

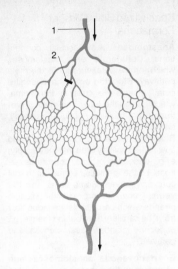

A Capillary bed supplied by
anastomosing arteries

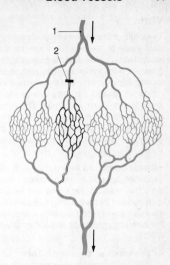

B Capillary bed supplied by
end arteries

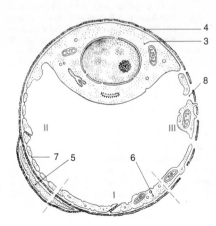

C Electron microscopic
pictures of various types
of endothelia

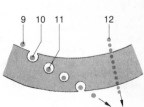

D Material penetrating through
endothelium: cytopempsis
and permeation

Veins

The walls of the veins are usually thinner than those of arteries. They can be distended even by low internal pressure, the pressure from the surrounding tissue affecting this process. The flow resistance inside veins and the pressure gradient towards the heart are small. The most important force for transporting blood in the veins is the *'muscle pump'*, aided by the *arteriovenous coupling* and *respiration*.

'Muscle pump'. As they run through the gaps and spaces in the locomotor system, the walls of the veins are readily compressed by contracting muscles and their contained blood is directed towards the heart by means of *semilunar valves* **ABC 1.**

Arteriovenous coupling. There are usually two concomitant veins that lie on either side of medium-sized and small arteries. They are anchored to the arterial wall by connective tissue **C** in such a way that the pressure of the arterial pulse narrows the lumen of the veins. The resulting movement of the blood in the veins is directed towards the heart by means of *semilunar valves* **ABC 1. AB 2** Orifice of a small vein. The arteriovenous coupling is not very important in the movement of venous blood.

Respiration. The negative pressure inside the chest (see p. 144) is transmitted to the venae cavae and so favors the return of blood to the heart.

The way in which the great veins near the heart are *built into the surrounding tissue prevents* them from *collapsing.* **D 3** In the lower region of the neck the venous walls are connected to the *cervical fascia.* **D 4** The *negative pressure* exerts suction on the walls of the veins in the thorax. **D 5** In the upper abdomen the walls of the great veins are stabilized by the surrounding *liver.*

Specialized Blood Vessel Formations

Anastomoses are vascular connections. **Collaterals** are anastomoses which form a parallel route to the main pathway. **Blocking areas** have longitudinal, muscle-like cells in the intima, which buckle the latter and make it protrude like a cushion (*'cushion arteries'*), and so block the lumen of the vessel. Blocking arteries can cut off direct blood flow to the capillary area that they supply, e. g., endocrine organs, corpora cavernosa of the genitals, uterus, umbilical cord and fingers. Some veins, **throttle veins,** have strong, circular muscle strands which can throttle the blood flow from the capillary area behind them e. g., endocrine glands, nasal mucosa and genitalia.

E Arteriovenous anastomoses form short-cuts in the precapillary part of the circulation. Two main types can be distinguished. **I** *Direct anastomoses:* an arteriovenous bridge is formed by a vessel between an artery and a vein. It has a blocking mechanism similar to those of the blocking arteries. **II** *Multiple channel anastomoses* are small organs (*glomus organs*) encapsulated in fibrous tissue, in which there is a closely packed network of one or several vessels. They may lie between an artery and a vein as **III** *glomus anastomosis,* or they may connect *precapillary vessels* **IV.** *They are found in:* the prominent ends (tips) of the limbs, the fingertips and the ball of the thumb, the tip of the coccyx, nose, the back of the tongue and the corpora cavernosa of the genitalia and of the nasal mucosa, and the cock's comb and the rabbit's ear.

A *Rete mirabile* is a capillary network on the arterial side of the main capillary area; *arterial rete mirabile,* e. g., the capillaries in a renal glomerulus: or in the venous limb of a capillary area, *venous rete mirabile,* e. g., hepatic capillaries.

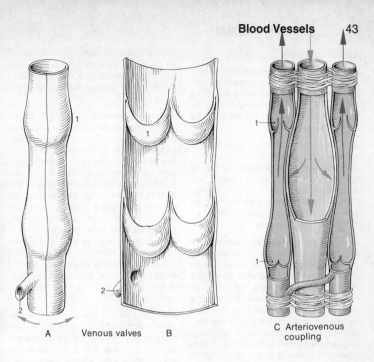

A   Venous valves   B

C   Arteriovenous coupling

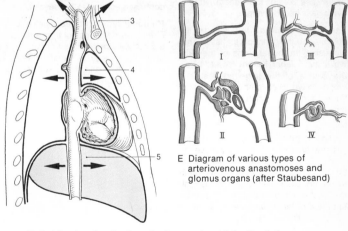

E   Diagram of various types of arteriovenous anastomoses and glomus organs (after Staubesand)

D   Stabilization of walls of great veins near heart (after Tandler)

Lymph Vessels

Fluid from the capillaries enters the tissues and organs as materials are exchanged between the blood **A1** *and the tissue* **A2**. This fluid flows mainly as *lymph* into the *lymph vessel* **A3**. The brain is an exception.

Lymph. Lymph differs from blood by its greater water content, lower protein content and lack of blood cells, except for lymphocytes. It contains fibrinogen and is capable of clotting. Lymph coming from the intestines *(chyle) after a meal rich in fat is full of fat granules and has a turbid, milky appearance.*

The **lymph vessels** are channels **A4** which run *parallel to the veins.* They conduct the lymph to the *venous angle* **A5** (confluence of the internal jugular with the subclavian vein, see p. 64). At this point, near to the heart, the lymph is *returned to the venous blood stream.* Lymph vessels provide a *drainage* system for the connective tissue regions of organs. Along the course of the lymph vessels a series of *lymph nodes* **ABC6** are interpolated which act as biological filters (see p. 98). From them lymphocytes pass into the lymph.

The *lymph vessels of the skin* and of the subcutaneous connective tissue run alongside the larger veins of the skin, while the *deeper lymph vessels* follow the course of the large arteries. The complete lymphatic pathway consists of three sections, lymphatic capillaries, conducting vessels and transport vessels.

Lymph capillaries start as blind sacs (cul de sac) **A3**, i. e., they have no permanent opening into the intercellular space, but temporary gaps do appear in the unfenestrated *endothelium.* They allow the influx of tissue fluid and the entrance of large particles, e. g., of fat globules in the villi of the intestine. Lymph capillaries lead into a *network* of interconnected **lymph vessels** with a wider

lumen. These vessels are compressed by pressure, for instance, from the surrounding muscles. The resulting lymph movement is directed centrally by *semilunar valves.*

Transport vessels. The next set of lymph vessels possesses a *muscular media* like a vein. They are composed of *valvular segments* arranged in consecutive order. Each valvular segment consists of a *semilunar valve* **B7** and a subsequent *length of lymph vessel.* The base of the semilunar pouches does not contain any muscle, is only slightly extensible and is constricted when the vessel is filled. Therefore, in radiographs, if the lymph vessels are filled with a contrast medium, they look like a string of beads **C**. The vessel wall is innervated by autonomic nerves. The lymph is transported by *contractions of consecutive valve segments* (10–12/min).

B8 Cross section through a semilunar valve: the margins of 2 pouches are cut transversely, **B9** afferent lymph vessels to a lymph node **B6, B10** hilus of the lymph node with efferent lymph and blood vessels, **C11** lymphatics filled with contrast medium, **C12** knee joint.

Regional lymph nodes receive lymph *directly* from an organ or from a region of the body. Lymph from an organ may be given up directly to several groups of lymph nodes, and the association between regional lymph nodes and a particular organ hardly ever varies.

Collecting lymph nodes receive lymph from *several regional lymph vessels.*

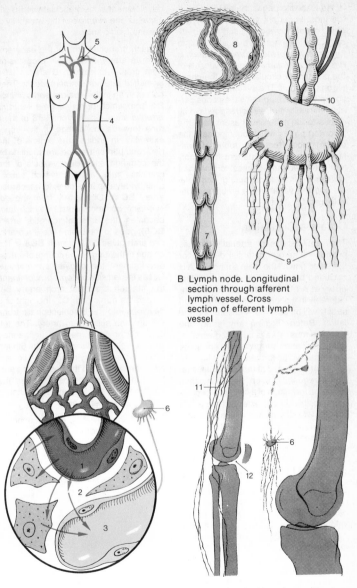

A Beginning of lymphatic capillary

B Lymph node. Longitudinal section through afferent lymph vessel. Cross section of efferent lymph vessel

C Radiograph of lymph vessels of leg which have been filled with contrast medium

In the following pages the distribution of the most important vessels of the *systemic circulation* is discussed *systematically*. Their topographic location within the *structures of the locomotor system is described in Volume 1;* for the vessels of the pulmonary circulation see the chapter on the lungs. The systemic organisation of the vascular tree varies considerably. In reality it rarely corresponds in all details to the schemata described in the following pages which only indicate the pattern most frequently found. *Deviations* of the vascular pattern from 'normal' have two main causes: formation of *collateral circulations* during postnatal life and the variability in embryonic development.

Collateral Circulation

The ramifications of arterial branches provide numerous potential detours for the blood flow. If the main channel in **A** is occluded by obliteration of the popliteal artery at **A1** the byways can widen considerably in a short time so that the blood supply will be assured = *collateral circulation.* Before ligating any artery the question must always be asked whether past experience suggests that adequate collaterals will be formed. If not, a ligature must not be placed at that particular site. Not infrequently, even small branches make the formation of a collateral circulation possible. In *end arteries* (see p. 40) no collateral circulation can develop and if blood flow through it is obstructed for any reason the organs or their affected parts die – necrosis or infarction.

Variability of the Blood vessels

Like any organ the blood vessels develop during embryonic life from primitive anlagen. Experience has shown that development is variable to a certain degree and vessel variations are therefore limited in scope and reducible to simpler stages in development. So many variants of the blood vessels are known that they cannot all be discussed here. The

development of the branchial arteries will serve as one *example* of the variability of the embryonic anlage.

In each branchial arch of the early embryo an artery develops in progression from cranial to caudal. **B** *shows all 6 branchial arteries simultaneously.* They collect in an initially paired *aortic anlage* **B2** from which *dorsal, segmental trunk arteries* arise. The seventh **B3** of them runs towards the anlage of the upper extremity. In addition the organs of the neck and part of the face are supplied by the common *ascending* vessels of the branchial arch arteries (which subsequently develop into the external carotid artery **BC4**). One artery from the *descending* common vessel, (which later becomes the internal carotid artery **BC5**)) goes to the brain. The 6th branchial arch gives off a branch **BC6** to the anlage of the lung. In the course of further development the *original symmetry is lost as the left side become predominant:* the 4th left branchial arch artery becomes the aortic arch **C7**, the right one develops into the brachiocephalic trunk **C8** and the subclavian artery. The left aorta of the trunk persists unchanged **C9,** the right one disappears; the 6th left branchial arch artery retains for a time its communication with the descending aorta as the ductus arteriosus, not so the right one. The 1st, 2nd and 5th branchial arch arteries disappear.

DE indicates the nature and frequency of variants of this 'normal' pattern.

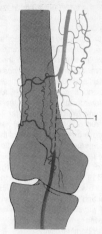

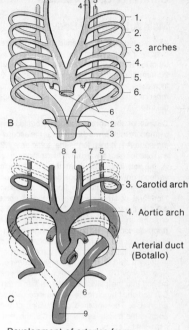

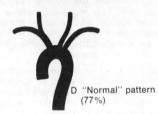

A Collateral circulation after occlusion of popliteal artery (after Loose)

B

1.	
2.	
3.	arches
4.	
5.	
6.	

3. Carotid arch

4. Aortic arch

Arterial duct (Botallo)

C

Development of arteries from branchial arch arteries (after Broman)

D "Normal" pattern (77%)

E Variants of aortic trunk (23%) (after Lippert)

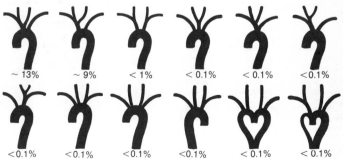

~ 13%   ~ 9%   < 1%   < 0.1%   < 0.1%   < 0.1%

<0.1%   <0.1%   <0.1%   <0.1%   < 0.1%   < 0.1%

Persistence of right-sided aorta   Duplication of aortic arch

Central or Great Vessels

Main Artery of the Body (Aorta)

All the arteries of the systemic circulation arise directly or indirectly from the aorta. The diameter of the aorta near the heart normally is about 20 mm; in old age, due to loss of elasticity, it increases to about 30 mm. The aorta is the most important "hydraulic reservoir" in the arterial tree.

Course of the aorta. It ascends behind the pulmonary trunk toward the right as the *ascending aorta* **A1**. It then forms the *aortic arch* **A2** and passes back toward the left across the hilus of the left lung **A3**. From the 4th thoracic vertebra downward the aorta descends first on the left side and then in front of the spine (vertebral column) as the *descending aorta* **AD4**. In the thoracic region the esophagus **A5** crosses in front of the descending aorta *(thoracic aorta).* After passing through the aortic hiatus of the diaphragm, the aorta is called the *abdominal aorta.* In front of the 4th lumbar vertebra it divides into the *right* and *left common iliac arteries* **C6**, each of which gives off one branch into the pelvis, the *internal iliac artery* **C7**, and one branch to the leg, the *external iliac artery* **C8**. The unpaired continuation of the aorta is the *middle sacral artery* **C9**, the remnant of a tail artery.

The **direct branches of the aorta** may be divided into 3 groups. One group is partly derived from the *branchial arch arteries,* the second from *embryonic segmental arteries* and the third from *embryonic visceral arteries.*

The *direct* **branches of the ascending aorta** are the coronary arteries, s. p. 20.

The *direct* **branches of the aortic arch** are derivatives of branchial arch arteries (p. 46), which arise on the left in the mid-line and pass to the head and neck. To the right there is the *brachio-cephalic trunk* **A10** and the *subclavian artery* **A12,** and to the left the *left common carotid artery* and the *left subclavian artery* arise directly from the aortic arch. **A13** *internal thoracic artery* to the chest wall.

The *direct* **branches of the descending aorta** are partly *trunk wall branches* and *visceral branches.*

Paired dorsal segmental arteries, trunk wall branches shown grey in **C** and **D**, divide into a *dorsal branch* to the spinal cord **D14**, the muscles of the back and skin **D15**, and a *ventral branch* **D16** to the wall of the trunk and its appendages. In the chest region the ventral branches form the *intercostal arteries* **ABD16** (**D17** lateral and **D18** anterior cutaneous branches), and they descend obliquely between the intercostal muscles to the anterior wall of the trunk. Near their origins the intercostal arteries run with the intercostal nerve and vein in a groove on the lower edge of the upper rib of each intercostal space. For this reason *puncture of the pleural space* **B** is performed from the back on top of the upper margin of the lower rib (**B19** rib, **B20** muscles of the chest wall, **B21** skin and subcutaneous tissue). The first and second intercostal arteries arise from the *highest intercostal artery* **A22,** a branch of the subclavian artery. Four *lumbar arteries* arise in the abdominal cavity.

Lateral, paired, non segmental arteries, black in **C** and **D** (the superior phrenic arteries in the chest, *trunk wall branches*) are well developed in the *abdominal cavity,* where they supply the diaphragm *(inferior phrenic arteries* **C23**, *trunk wall arteries)* and as *visceral branches,* to paired organs such as the kidneys **A24** and adrenal glands **A25** *(renal arteries* **CD26**, *suprarenal arteries* **C27**) and to the gonads *(testicular or ovarian arteries* **C28**).

Ventral, unpaired, non-segmental arteries, visceral arteries, red in **C** and **D**, supply the unpaired viscera in the chest and abdomen. *Chest:* small branches to the trachea, esophagus and mediastinum (bronchial, esophageal, pericardial, mediastinal branches). *Abdomen:* celiac trunk **C29** (with its three main branches, the splenic, left gastric and common hepatic arteries), *superior mesenteric artery* **CD30** and *inferior mesenteric artery* **C31**.

A Aorta and its branches

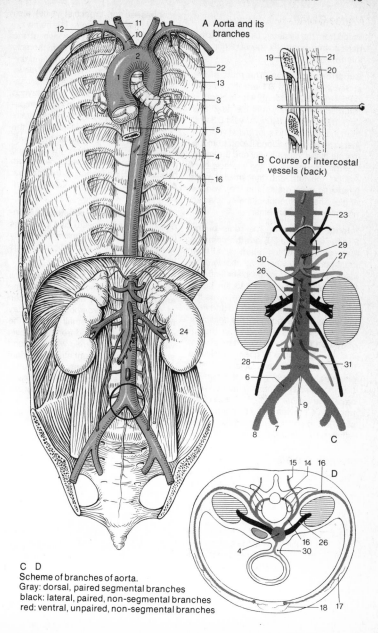

B Course of intercostal vessels (back)

C

D

C D
Scheme of branches of aorta.
Gray: dorsal, paired segmental branches
black: lateral, paired, non-segmental branches
red: ventral, unpaired, non-segmental branches

Venae Cavae

Blood from the systemic circulation flows back to the heart via the superior and inferior vena cava.

Superior vena cava. The roots of the *superior vena cava* **B1** are the *innominate veins* **B2,** through which blood is returned to the heart from the head and neck by the *internal jugular vein* **B3** and from the arms by the *subclavian vein* **B4.** It also receives additional blood from the thyroid gland, chest cavity and the neck. A large lymph vessel, the *thoracic duct* **A5** on the left, and the smaller right *lymphatic duct,* drains into the *"venous angle",* the confluence of the subclavian and internal jugular veins.

Inferior vena cava. The roots of the *inferior vena cava* **B6** are the *common iliac veins* **B7,** which return blood from the abdominal viscera via the *internal iliac vein* **B8** and from the lower limbs via the *external iliac vein* **B9** to the heart. The inferior vena cava also receives the blood from the *renal veins* **B10** and from the veins of the gonads, the *testicular and ovarian veins.* The vein from the left gonad **B11** drains into the left renal vein, and the right vein **B12** goes directly into the inferior vena cava. Just below the diaphragm, the liver gives off 3 or more *hepatic veins* **B13** into the inferior vena cava. The *lumbar veins* of the wall of the trunk **B14** provide further afferent vessels. The abdominal aorta **A15** lies to the left of the inferior vena cava.

Azygos System

Azygos and hemiazygos veins. The *segmental* veins of the dorsal chest wall, the *intercostal veins* **B16,** drain into a longitudinal anastomosis on each side of the vertebral column in the chest. In the chest region they are the *hemiazygos vein* **B17** on the left and the *azygos vein* **B18** on the right, and are upward continuations of the *ascending lumbar veins* **B19,** which receive the *lumbar veins* **B14.** Veins from the viscera of the chest (esophageal and bronchial veins) also drain into the azygos vein. The hemiazygos vein joins the azygos vein at the height of the 6th to the 10th thoracic vertebrae **B20.** The *accessory hemiazygos vein* **B21,** into which blood flows from the upper left intercostal veins, passes from above into the hemiazygos vein. Veins from the 1st and 2nd intercostal space also enter the innominate vein. The azygos vein of pencil thickness, the largest vein of the wall of the trunk, crosses the hilus of the right lung and reaches the superior vena cava from behind. **A22** Bifurcation of the trachea. **A23** Adrenal gland, **A24** kidney. Compare **B** with **A.**

Anastomoses and collaterals. Connections (anastomoses) and bypasses (collaterals) between veins are easily formed, and may include subcutaneous veins.

Clinical tips: Blood clots can form in veins after operations and in inflammations and may completely occlude the lumen of the vessel – *thrombosis.* They may remain unnoticed in small veins. If the clots become detached from the vessel wall, they can be carried with the blood stream through the right atrium and right ventricle into the pulmonary circulation to finally become trapped in the arteries or capillaries of the lungs – *pulmonary embolism.*

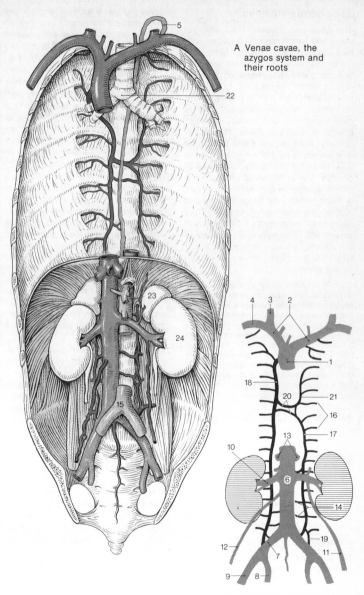

A Venae cavae, the azygos system and their roots

B Scheme of great veins

Peripheral Pathways

The *peripheral vessels,* i. e., arteries, veins and lymph vessels, together with their associated nerves form the *peripheral pathways.* In the following description the vessels are traced up to the organs which they supply. In order to protect them from mechanical injury, the peripheral vascular pathways run in cracks, niches and canals between bones, muscles and fascia. This is particularly seen in the neck and limbs where there is a lot of movement and is less marked in the chest and abdominal cavities. The pathways of an organ are joined together by connective tissue into a cable which can remain moveable against the surrounding tissues. They cross over joints on their flexor aspect (see p. 38). The peripheral pathways in the viscera frequently form a stalk about which the organs can move without pulling the vessels.

Guide structures and memory aids. The course of peripheral pathways can be recognized by observing their guide structure e. g., muscles. Such anatomic signposts are of topographic and medical importance, e. g., points where the arterial pulse is palpable, where an artery can be compressed in case of a hemorrhage (marked by a circle in the illustrations), or where a nerve can be damaged by pressure against the underlying bone or by a fractured bone, etc. The topography of the peripheral pathways in the locomotor system is discussed in Vol. I.

Arteries of the Head and Neck

The viscera of the head and neck are mainly supplied by the principal artery to the head, the *common carotid artery* **A1.** The brain and thyroid gland have an additional supply on both sides from a large branch of the *subclavian artery* **A2.** Other branches from the *subclavian artery* also go to the lateral aspect of the neck and to part of the shoulder girdle.

Arteries of the Neck I

Subclavian Artery

The *subclavian artery* leaves via the upper thoracic opening, runs across the 1st rib (the artery can be compressed against the 1st rib **A3** by pulling the arm downward and backward) and through the scalenus gap (between the scalenus anterior **A4** and medialis **A5**) into the lateral cervical region. The following arteries arise from the subclavian artery medial to the scalenus anterior muscle:

The *vertebral artery* **AB6** usually runs from the 6th cervical vertebra within the transverse foramina upward to the atlas (the uppermost cervical vertebra) where it perforates the atlanto-occipital membrane. Both vertebral arteries then unite and form the *basilar artery,* see p. 58.

The *internal thoracic artery* **B7** runs downward inside the chest wall lateral to the sternum. It gives off the *pericardiacophrenic artery* which is accompanied by the phrenic nerve, the *musculophrenic artery* to the diaphragm and the lower intercostal arteries. Eventually, as the *superior epigastric artery,* it forms an anastomosis with the *inferior epigastric artery,* see p. 72. Its branches go to the skin of the chest, the thymus, mediastinum and the root of the lung; there are communications with the upper 5–6 intercostal arteries via their *anterior intercostal branches.*

The *thyrocervical trunk* gives off the *inferior thyroid artery* **B8,** the *ascending cervical artery* **B9** that accompanies the phrenic nerve, and the *transverse cervical* **B10** and *suprascapular* **B11** arteries. The latter crosses the transverse ligament of the scapula to reach the shoulder blade muscles.

The *costocervical trunk* passes with the *deep cervical artery* **B12** to the deep muscles of the nape of the neck and with the *highest intercostal artery* **B13** to the 1st and 2nd intercostal arteries.

In 60% of cases the *transverse artery of the neck* **B14** arises *lateral* to the anterior scalenus muscle and passes through the brachial plexus to the upper medial angle of the scapula.

The *subclavian vein* **A15** lies in front of the anterior scalenus. Compare **B** and **A.**

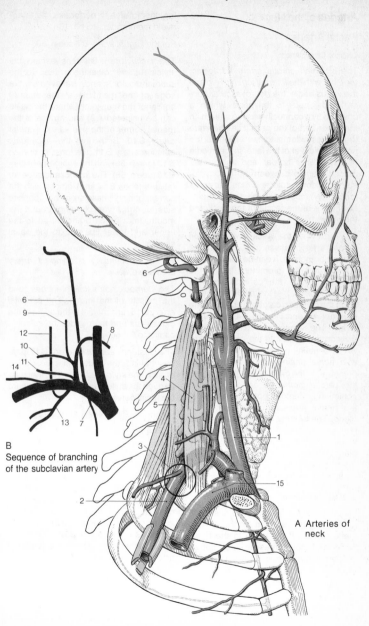

B
Sequence of branching
of the subclavian artery

**A Arteries of
neck**

Arteries of the Neck II, Facial Arteries

Common Carotid Artery

The *common carotid artery* **A1** forms, together with the *internal jugular vein* and the *vagus nerve,* the *neurovascular bundle* of the neck to the head, which is covered by a connective tissue sheath. In the lower part of the neck it lies beneath the sternocleidomastoideus. It leaves the protective cover of the latter in the middle of its anterior border and lies in the carotid triangle between the omohyoid, sternocleidomastoid and digastric (posterior belly) muscles, beneath the superficial layer of the fascia of the neck and the platysma.

The artery of the neck, the *common carotid artery,* passes through the neck without branching. It can be compressed **A2** against the prominent transverse process of the 6th cervical vertebra (carotid tubercle). At the level of the hyoid bone the common carotid divides into the *external* **A3** and *internal* **A4** *carotid arteries.* The vessel wall at the point of division is dilated into the *carotid sinus,* which contains a *pressor-receptor area* with nerve endings sensitive to blood pressure. In the dividing fork there is also the *carotid glomus,* a pea-sized organ containing chemoreceptors sensitive to the partial pressure of oxygen in the blood. The internal carotid artery enters the skull without further division to supply blood to the orbits, hypophysis (pituitary gland) and the brain. The external carotid artery divides into branches to the neck, face, skullcap and the scalp.

External Carotid Artery

The *superior thyroid artery* **B5** descends to the anterior surface of the thyroid gland giving off branches to the hyoid bone and sometimes to the sternocleidomastoid muscle, and then gives off the *superior laryngeal artery* **B6** to the larynx as, together with the superior laryngeal nerve, it perforates the thyrohyoid membrane; see p. 116.

For the *lingual artery* **B7** see p. 56.

The *facial artery* **B8** runs behind the styloideus and digastricus deep to the submandibular gland, transverses the edge of the mandible at the anterior border of the masseter (where the pulse can be palpated **A9**) and arrives at the medial corner of the eye as the *angular artery* **B10.** Branches: the *ascending palatine artery* **B11,** the tonsillar branch **B12** (see p. 56) and the *submental artery* **B13** (see p. 56). The *inferior* and *superior labial arteries* **B14** anastomose with the corresponding vessels on the opposite side to form a vascular ring around the mouth, from which branches go to the upper and lower lips and to the nasal septum.

For the *ascending pharyngeal artery* **B15** see p. 56.

The *sternocleidomastoid branches* go to the muscle of that name. The *occipital artery* **B16** runs medial to the origin of the digastricus and then under the splenius capitis to the occiput.

The *posterior auricular artery* **B17** passes behind the ear, and gives off a branch, the *stylomastoid artery,* that runs through the stylomastoid foramen toward the facial nerve.

For the *maxillary artery* **B18,** see p. 56.

The *superficial temporal artery* **B19,** a superficial terminal vessel, divides in the temporal area into *frontal* and *parietal branches.* The pulse wave in it is visible and palpable **A20.** In old age the artery becomes quite prominent and coiled due to loss of elasticity and the deposition of calcium in its wall. Two branches, the *zygomaticorbital artery* **B21** and the *transverse facial artery* **B22** accompany the zygomatic arch. The medial temporal artery **B23** enters the temporal muscle. Compare **B** and **A.**

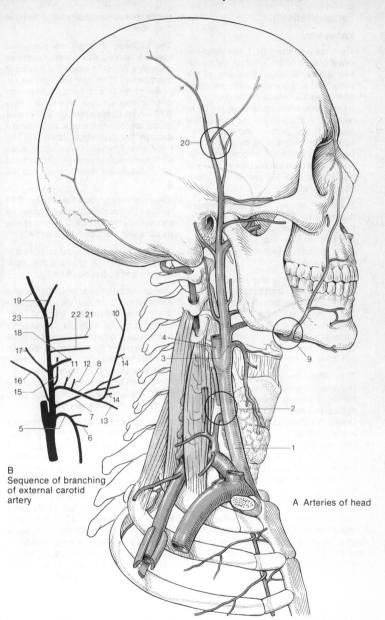

20

19
23
18
17
16
15
5

22 21
11 12 8
7
13
14
14
10

6

B
Sequence of branching
of external carotid
artery

4
3
2
1

9

A Arteries of head

Facial Arteries II

Maxillary Artery

The *maxillary artery* **BC1**, the *deep terminal branch* of the *external carotid artery* **ABC2**, is larger than any of the superficial end branches. The maxillary artery passes behind the temporomandibular joint, obliquely through the deep region of the face to the pterygopalatine fossa. Three sections can be distinguished in the route of the artery, the section behind the temporomandibular joint, the section through the muscles of mastication and the part in the pterygopalatine fossa.

Pathway around the temporomandibular joint. Branches: small branch to the temporomandibular joint, external auditory meatus and middle ear (the deep auricular artery and anterior tympanic artery).

The *inferior alveolar artery* **BC3** gives off the *mylohyoid branch* before passing together with the inferior alveolar nerve into the mandibular canal. It supplies the teeth, the bone and the soft parts of the lower jaw. Finally, as the *mental artery* **B4,** it traverses the mental foramen to run to the skin of the chin.

The *middle meningeal artery* **ABC5** is the most important artery of the dura mater. It passes through the foramen spinosum into the middle cranial fossa. During its extradural course it divides into an anterior and posterior branch. The middle meningeal artery is usually accompanied by 2 branches of a cutaneous nerve, the *auriculotemporal nerve* **A6.** Small branches go to the soft palate, the trigeminal ganglion, the tympanic cavity and the facial canal.

The pathway through the muscles of mastication. Branches to the masticatory muscles: *masseteric artery* **B7**; *deep anterior* **B8** and *posterior* **B9** *temporal arteries* and *pterygoid branches*. *Buccal artery* **B10** to the cheek and buccal mucus membrane (inner lining of the cheek).

The pathway through the pterygoid fossa. Branches: The *superior posterior alveolar artery* **B11** enters the maxilla at the maxillary tuberosity and supplies the posterior teeth and gums. It forms an anastomosis with the *infraorbital artery* **B12** which passes through the inferior orbital fissure into the orbit and leaves it through the infraorbital canal and the infraorbital foramen. On its way it gives off the *anterior superior alveolar arteries* which supply the anterior teeth and the gums.

The *descending palatine artery* **B13** reaches the palate in the pterygopalatine canal. Note the *greater* **C14** and *lesser palatine arteries*. Compare **B** and **A**.

(The *artery of the pterygoid canal* runs in the pterygoid canal to reach the upper pharynx and Eustachian tube.)

The *sphenopalatine artery* **BC15** enters the nasal cavity through the sphenopalatine foramen and supplies its lateral, posterior and medial walls via the *posterior lateral nasal artery* and *septal arteries* **C16**. The roof of the nasal cavity receives the *nasal branches* **C17,18** of the *anterior ethmoidal artery* **C19** (from the ophthalmic branch of the internal carotid artery **C20**). The latter also gives off the small anterior meningeal artery **C21.**

The *lingual artery* **C22** gives off branches **C23** to the sides of the tongue (the *dorsal lingual branches),* and to its body and tip *(deep lingual artery)*. The *sublingual artery* **C24** goes to the sublingual gland and the gums.

The *posterior auricular artery* is shown as **A25,** the *submental artery* as **C26** and the *tonsillar branch* **C27** of the *facial artery* **C28.** The *ascending pharyngeal artery* **C29** supplies the lateral wall of the pharynx and the tympanic cavity. Its terminal branch is the *posterior meningeal artery* which passes through the jugular foramen.

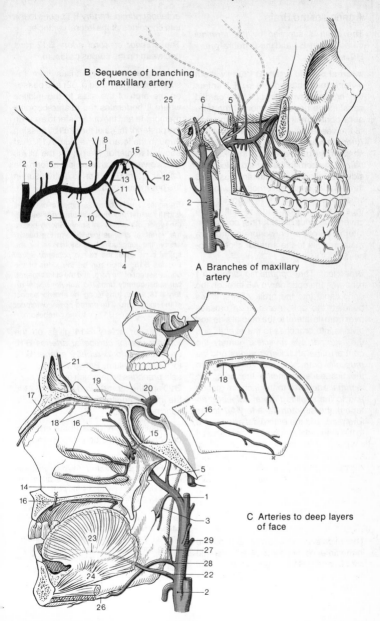

B Sequence of branching
of maxillary artery

A Branches of maxillary
artery

C Arteries to deep layers
of face

Arteries of the Brain

The brain is supplied by the *terminal carotid artery* **B1** and the *vertebral artery* **B2,** see p. 52.

Internal carotid artery B1 runs without dividing to the base of the skull. It enters the cranial cavity through the carotid canal in the petrous portion of the temporal bone (small branches from this part of the vessel go to the tympanic cavity). After leaving the canal it runs in an S-shape in the middle fossa of the skull (the *carotid syphon)* in the carotid sulcus beside the sella turcica (pituitary fossa) **A3** through the sinus cavernosus giving off small branches to the hypophysis, the trigeminal ganglion and the wall of the cavernous sinus. In its further course the internal carotid artery perforates the dura mater medial to the anterior clinoid process.

Branches: The *ophthalmic artery* **A4** runs with the optic nerve **A5** through the optic canal into the orbit. At first it lies beneath the optic nerve, then crosses over from its lateral to the medial side and divides into branches to the *eyeball* and the *external eye muscles,* namely the *central retinal and ciliary arteries,* and the *muscular branches,* (see Vol. III), and to the tissues surrounding the orbit: the *lacrimal artery* **A6** to the lacrimal glands and to the eyelids *(lateral palpebral arteries),* the *supraorbital artery* **A7** to the forehead, and the *ethmoidal arteries* **A8** to the ethmoidal air sinuses and the nasal cavity. From the anterior ethmoidal artery arises the *anterior meningeal artery* **A9** which runs to the dura mater. The *medial palpebral arteries,* the *supratrochlear artery* and the *dorsal nasal artery* go to the skin of the forehead, the medial corner of the eye and the bridge of the nose, respectively.

The *middle cerebral artery* **B10** is the immediate continuation of the internal carotid artery **B1** and runs in the lateral cerebral sulcus (of Sylvius). For its distribution in the brain see Vol. 3. The *anterior choroidal artery* **B11** goes to the tela choroidea of the lateral ventricles.

The *anterior cerebral artery* **B12** runs backward on the corpus callosum.

The **vertebral artery** arises from the subclavian artery (see p. 52) and passes on the arch of the atlas to the midline where it perforates the atlanto-occipital membrane and the dura mater to arrive in the posterior fossa of the skull by passing through the foramen magnum **A13.** Just above its anterior margin on the clivus the two *vertebral arteries* join to form the *basilar artery* **B14.** For its distribution in the brain see Vol. 3.

Branches: The horizontal segmental branches and the *spinal* and *muscular branches* leave in the region of the cervical spine. Shortly before the vertebral arteries join to form the basilar artery, the longitudinal anastomoses for the spinal cord arise; the *paired posterior spinal arteries* **B15,** which run on the sides of the dorsal aspect of the cord, and the initially paired but subsequently unpaired *anterior spinal artery* **B16,** which runs along the anterior aspect of the spinal cord. The paired *posterior inferior cerebellar arteries* **B17** go to the cerebellum.

The **basilar artery B14** gives off the *anterior inferior cerebellar arteries* **B18** and the *superior cerebellar arteries* **B19** to the cerebellum. Between these vessels *branches* arise which go *to the pons* **B20** and the *artery to the labyrinth* **B21** to the internal auditory meatus. **A22** Abducens nerve.

The basilar artery ends in the *posterior cerebral artery* **B23** above the cerebellar tentorium.

Cerebral arterial circle *(of Willis).* The *posterior* **B23** and *middle* **B10** *cerebral arteries* on each side are connected by the *posterior communicating artery* **B24.** Together with the *anterior communicating artery* **B25,** between the two anterior cerebral arteries, they form an arterial circle around the sella turcica at the base of the brain. Numerous short arteries arising from it pass into the neighboring basal part of the diencephalon (midbrain). **A26** Crista galli, **A27** oculomotor nerve, **A28** middle meningeal artery. Compare **B** and **A.**

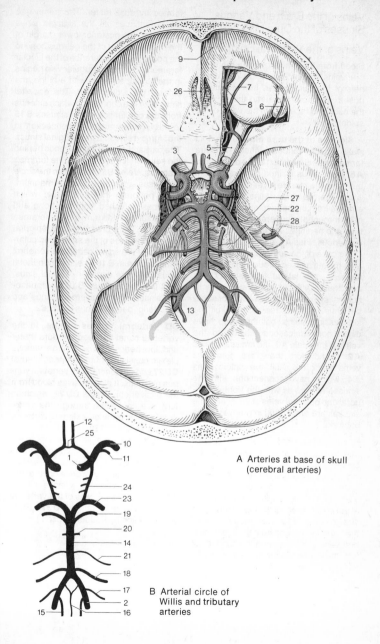

A Arteries at base of skull
(cerebral arteries)

B Arterial circle of
Willis and tributary
arteries

Veins of the Brain and Venous Sinuses of the Dura Mater I

Veins of the Vertebral Column

Blood from the brain flows via the cerebral veins first into the rigid blood channels of the dura mater, the venous sinuses, and from there into the veins of the neck and face, or into the veins of the spinal canal.

A Sinuses of the dura mater. The rigid walls of the venous sinuses **A1** are formed by reduplication of the dura mater **A5**. Each sinus is lined by endothelium but lacks the muscle layer of the venous wall. The cerebral veins **A8,** which emerge from the surface of the brain **A9,** pass through the subarachnoid space of the leptomeninges **A6** to enter the nearest venous sinus **A1,** either directly or, as in the case of the superior sagittal sinus, after having passed through lakelike confluences, the lateral lacunae **A7.** An arachnoid villus (granulatio arachnoidealis) extends into the superior sagittal sinus. At the base of the skull the sinuses join veins, but they also have direct communications leading to the veins of the scalp **A3,** the emissary veins **A2.** In addition, there are numerous veins inside the skull, the diploic veins **A4,** which have connections with the sinuses as well as with the frontal, temporal and occipital veins of the scalp. The sinuses are embedded in grooves in the bony skull.

B The superior sagittal sinus **A1, B10** runs along the attachment of the falx cerebri **B11** to the cranium. The falx is formed by the dura mater and juts out between the two hemispheres of the brain. The inferior sagittal sinus **B12** lies at the lower border of the falx cerebri. The straight sinus **B13** takes the blood from the former sinus and from the great cerebral vein (see Vol. III), to the confluence with the superior sagittal sinus, the confluence of the sinuses **B15.** There the transverse sinus arises, at the origin of the tentorium cerebelli **B14,** which is also

formed by dura mater. The transverse sinus continues as the sigmoid sinus **B23** along the posterior lower margin of the pyramidal part of the petrous bone to the posterior circumference of the jugular foramen. The internal jugular vein begins in the jugular foramen with a bulge, the bulb of the jugular vein. The occipital sinus **B24** is a connection which extends from the confluence of the sinuses **B15** to the foramen magnum (connection to the sigmoid sinus via the marginal sinus). The basilar plexus **B21** descends behind the sella turcica **B20** toward the foramen magnum. Veins **B22** arise from the circle of sinuses around the foramen magnum. They perforate the dura mater which becomes detached from the bone and periosteum at the rim of the foramen magnum. They drain into the epidural venous plexuses of the spine, particularly those of the spinal canal, the vertebral venous plexus. **B16** The sphenoparietal sinus, **B17** the superior petrosal sinus, **B18** the frontal sinus, **B19** the sphenoidal sinus. For the cavernous sinus and the inferior petrosal sinus see p. 62.

CD Vertebral venous plexus. In the vertebral canal between the dura mater and the periosteum is a large venous plexus consisting of the anterior internal **CD27** and posterior **CD26** vertebral venous plexuses. It also receives blood from the basivertebral veins **CD29,** which, in turn, are connected through the vertebrae to the anterior external vertebral venous plexus **CD30** in front of the vertebral column. This plexus is connected to the azygos and hemiazygos veins **C28.** In between the vertebral arches veins run to the poorly developed external posterior vertebral venous plexus **CD25** along the process of the vertebral column. Intervertebral disk **D31.** (From an illustration by Rauber-Kopsch.)

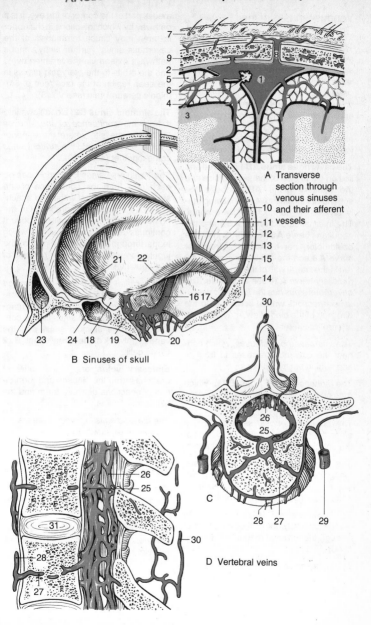

A Transverse section through venous sinuses and their afferent vessels

B Sinuses of skull

C

D Vertebral veins

Venous Sinuses of the Dura Mater II

The venous sinuses at the base of the skull have two main confluences on each side: in the *cavernous sinuses* **AB1** in the middle fossa of the skull, which receives numerous afferent vessels and drains into several efferent channels; and the *sigmoid sinus* **B2** in the posterior fossa of the skull. It is the largest sinus and carries blood through the jugular foramen to the internal jugular vein.

The **cavernous sinus AB1** is paired and lies on either side of the sella turcica and the pituitary gland **AB3,** above the sphenoidal air sinus **A4.** It is incompletely divided by connective tissue septa. The internal carotid artery **AB5** and the abducens nerve **A6** pass through it. The oculomotor nerve **A7,** the trochlear nerve **A8** and the ophthalmic nerve **A9** run in the lateral wall of the sinus, and the maxillary nerve **A10** lies at its base. The *cavernous sinus has the following vascular connections* which, according to the direction of the blood flow, may either be afferent or efferent.

The *intercavernous sinuses,* which connect the cavernous sinuses in front of and behind the hypophysis.

The *sphenoparietal sinus* **B11,** which runs on both sides along the crest of the lesser wings of the sphenoid bone.

The *superior petrosal sinus* **B12** runs along the superior margin of each side of the petrous bone to the sigmoid sinus.

The *inferior petrosal sinus* **B13** passes along the posterior aspect of the inferior margin of each petrous bone to reach the anterior part of the jugular foramen.

The *basilar plexus* **B14** is connected with the venous network of the spinal canal through the foramen magnum **B15** (see p. 60).

The *superior ophthalmic vein* **B16** connects the cavernous sinuses with the facial veins via the *angular vein* in the medial part of the corner of the eye. It is a pathway by which infections in the upper lip area may cause thrombosis of the cavernous sinus. Further away venous pathways extend via the foramen ovale and the orbits to the *pterygoid plexus* in the deep region of the face (see p. 64), middle cerebral vein (see Vol. 3).

The **sigmoid sinus B2:** blood flows into it from the *confluence of the sinuses* **B17.** It is the site where the *superior sagittal sinus* **B18** and the *straight sinus* come together (see p. 60).

The *transverse sinus* **B19** originates on each side from the confluence of the sinuses. It runs at first within the attachment of the tentorium cerebelli. On the posterior surface of the petrous bone it continues as the *sigmoid sinus,* which leads through the posterior part of the jugular foramen into the *superior bulb of the jugular vein.*

The *occipital sinus* **B20** is an unpaired venous sinus which divides at the posterior circumference of the foramen magnum. As the *marginal sinus* **B21** it connects with the sigmoid sinus and the venous network of the spinal canal. **B22** Crista galli.

Emissary veins are further efferent branches from the sinuses and form direct connections between them and the extracranial veins.

The *mastoid emissary vein* connects the transverse sinus with the occipital vein.

The *parietal emissary vein* connects the superior sagittal sinus with the veins of the scalp.

The *condyloid emissary vein* connects the sigmoid (transverse) sinus with the external vertebral venous plexus.

The *occipital emissary vein* connects the confluence of the sinuses with the occipital vein.

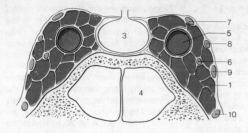

**A Transverse section
through cavernous sinus**

**B
Sinuses at base of skull
showing efferent
connections**

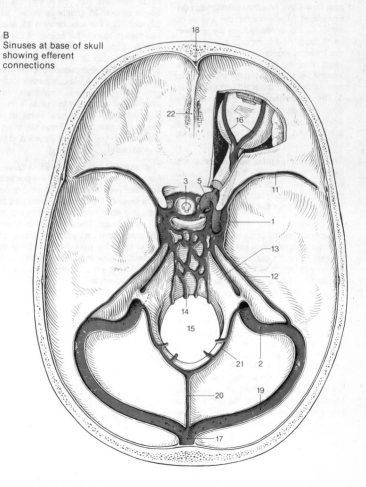

Veins of the Face and Neck

Blood coming from the facial part of the skull, from the upper cervical viscera and most of the blood from the cranium, as well as from the distribution areas of the internal and external carotid arteries, is directed via the great vein of the neck, the *internal jugular vein* 1, and several smaller veins on the upper thoracic aperture. There the internal jugular vein with the *subclavian vein* 2 from the shoulder and arm forms the *brachiocephalic vein* (innominate). The innominate veins on either side combine to form the superior vena cava. At the *venous angle,* where the internal jugular and the subclavian veins combine, large lymph vessels flow into the veins: on the left the *thoracic duct* and on the right the *right lymphatic duct* 3. The subclavian vein lies in front of the anterior scalenus 4 and the subclavian artery 5 passes between the scalenus anterior and medialis 6 (in the scalenus gap).

The **internal jugular vein** starts at the jugular foramen at the base of the skull where it has a dilatation, the *superior bulb of the jugular vein.* Here it receives its *afferents from the sinuses of the dura mater:*

7 Superior sagittal sinus, 8 inferior sagittal sinus, 9 confluence of the sinuses, 10 straight sinus, 11 great cerebral vein, 12 transverse sinus, 13 superior petrosal sinus, 14 sigmoid sinus, 15 inferior petrosal sinus, 16 cavernous sinus, 17 superior ophthalmic vein.

The *internal jugular vein* lies with the *vagus nerve 18* and the *internal carotid artery* in a common connective tissue sheath, the *vagina carotica,* which is connected to the pretracheal cervical fascia. For this reason the lumen of the internal vein is kept open by the pull of the lower hyoid muscles and the return of the blood to the heart is facilitated. In perforating injuries to the vein (e. g., knife wound) air can be aspirated into the vein leading to air embolism.

Blood also flows into the internal jugular vein from the *face* and *neck:* from the *pterygoid plexus* 19, lying in the deep areas of the face between the masticatory muscles, which is connected with the *cavernous sinus* 16, and with numerous veins, among them veins from the orbits and the face and with the first part of the retromandibular vein via the maxillary veins. The pterygoid plexus and the *superficial temporal veins* 20, as well as the veins from the ear, drain into the *retromandibular* 21 *(posterior facial) vein* 22, which, as the *angular vein* 23, anastomoses in the medial corner of the eye with the *superior ophthalmic vein* 17, into the internal jugular vein. The *superior thyroid vein* 24 also passes into the internal jugular vein. A venous network lying on the pharyngeal muscles, the *pharyngeal plexus,* often ends with its own veins, the *pharyngeal veins,* in the internal jugular vein. These veins may also carry blood from the larynx and the thyroid gland.

The *external jugular vein* 25 originates from the occipital veins and the posterior auricular vein. It runs between the superficial layer of the cervical fascia and the platysma and often flows directly into the subclavian vein. The *anterior jugular vein* 26 comes from the hyoid region and also lies outside the main fascia. The *vertebral* and *internal thoracic* 27 veins usually empty into the brachiocephalic vein near the venous angle.

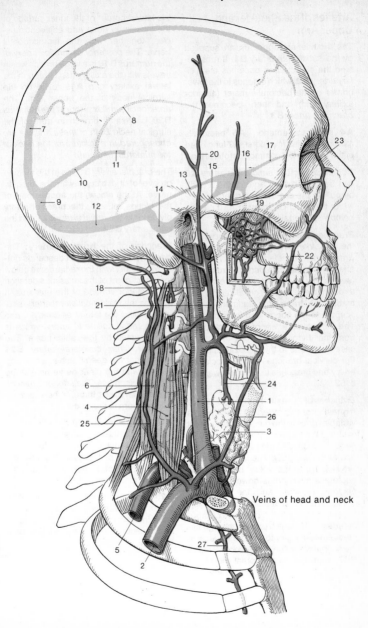

Veins of head and neck

Arteries of the Shoulder and Upper Arm

The shoulder and upper arm are supplied by the *subclavian artery* **B1**. It runs behind the clavicle and becomes the *axillary artery* **B2**, which extends to the lower border of the pectoralis major (anterior axillary fold) and there becomes the *brachialis artery* **B3**.

A4 scalenus anterior, **A5** pectoralis minor, **A6** latissimus dorsi, **A7** long head of the triceps, **B8** common carotid artery.

Compression: The subclavian artery **A9** can be compressed against the first rib by pulling the arm backward and downward; the *brachial artery* at **A10** can be *compressed* against the humerus.

Branches of the **subclavian artery B1** to the *shoulder girdle:* the *suprascapular (transverse scapular) artery* **B11** which arises from the *thyrocervical trunk* **B12** (see p. 52), runs parallel to the clavicle through the lateral cervical region and across the transverse ligament of the scapula to the supra- and infraspinatus. The *transverse artery of the neck* arises from the thyrocervical trunk in 40% of cases and runs to the trapezius muscle. For cervical branches of the subclavian artery and branches to the chest wall see p. 52.

Branches of the **axillary artery B12:** the *highest thoracic artery* **B13** (which is variable) goes to the muscles of the chest wall and the *thoracoacromial artery* **B14** sends branches to the pectoral muscles and to the shoulder. The *lateral thoracic artery* **B15**, at the lateral border of the pectoralis minor, runs downward to the pectoral muscles and the mammary gland. The *subscapular artery,* a short vessel, divides into the *circumflex scapular artery* **B16**, which passes through the medial axillary gap to the dorsal face of the scapula, and the *thoracodorsal artery* **B17** (with the thoracodorsal muscle to the lateral margin of the inner surface of the latissimus dorsi). The latter reaches

the lateral border of the inner surface of the latissimus dorsi and adjacent muscles, together with the thoracodorsal nerve. The *posterior circumflex artery of the humerus* **B18** is a large branch which travels with the axillary nerve through the lateral axillary gap **A19** to lie on the surgical neck of the upper arm. The *anterior circumflex artery of the humerus* **B20** is a small branch in front of the surgical neck. *Both of these arteries are endangered in a fracture of the neck of the humerus (unusual).*

The *brachial artery* **B3** lies in the median groove of the biceps with the median nerve. At the elbow, the aponeurosis of the biceps muscle, the brachial artery divides into the *radial artery* **B21** and the *ulnar artery* **B22.**

Branches of the **brachial artery:** The *deep brachial artery* **B23** descends outward together with the radial nerve below the attachment of the latissimus dorsi on the posterior surface of the humerus. It ends as the *middle collateral artery,* situated behind the lateral epicondyle, and as the *radial collateral artery,* which is accompanied by the radial nerve. The *superior ulnar collateral artery* **B24** arises after the deep brachial artery and passes with the ulnar nerve behind the medial epicondyle. The *inferior ulnar collateral artery* runs through the intermuscular septum to the elbow.

Arterial network around the elbow: The collaterals of the brachial arteries and the recurrent branches from the antebrachial arteries (see p. 68) form together the arterial anastomotic network around the olecranon, the *rete articulare cubiti.*

B25 recurrent interosseous artery, **B26** common interosseous artery, **B27** anterior interosseous artery, **B28** posterior interosseous artery, see p. 68. Compare **B** and **A.**

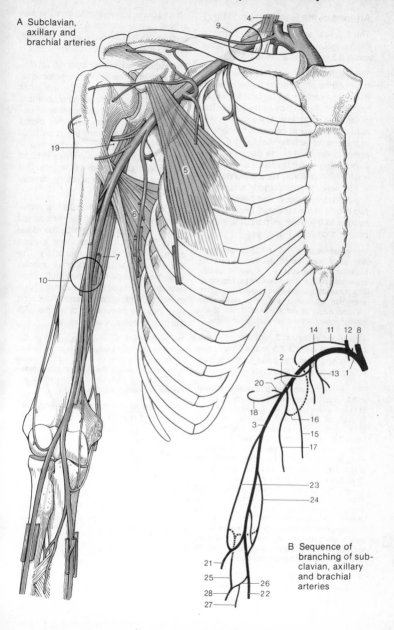

A Subclavian, axillary and brachial arteries

B Sequence of branching of subclavian, axillary and brachial arteries

Arteries of the Forearm and Hand

The *brachial artery* **B1** divides between the aponeurosis of the biceps into the *radial* **B2** and *ulnar arteries* **B3**.

The **radial artery** runs with the superficial branch of the radial nerve beneath the brachioradialis **A4** (a guideline structure) to the thumb side of the hand. The *pulse* can be palpated at **A5** in the lower third of the forearm medial to the tendon of the brachioradialis muscle. The radial artery at the level of the carpus passes through the radial foveola ('anatomist's snuff box') (see Vol. 1) onto the dorsum of the hand. It runs dorsally around the 1st metacarpal bone and between the two heads of the 1st dorsal interosseous muscle as part of the *deep palmar arch* **AB6**. Branches:

The *radial recurrent artery* **B7** to the rete cubiti, and small branches to the carpus and the superficial palmar arch – the *palmar carpal, superficial palmar and dorsal carpal branches*. The *dorsal metacarpal artery* goes to the dorsal aspect of the thumb and index finger. The *principal artery of the thumb* **BC8** runs to the palmar aspect of the thumb and index finger.

The **ulnar artery B3** runs beneath the flexor carpi ulnaris **A9** to the little finger side of the wrist and ends as the superficial palmar arch. Branches:

The *recurrent ulnar artery* **B10** goes to the *rete cubiti*, and the *common interosseous artery* **B11** leaves the ulnar artery in the cubital fossa. It gives off the *recurrent interosseous artery* **B12** which divides immediately into the *posterior* and *anterior interosseous arteries*. The posterior vessel comes to lie between the oblique cord and the interosseous membrane on its dorsal aspect. From there it runs with the deep branch of the radial nerve between the superficial and deep extensor muscles toward the hand. The *anterior interosseous artery* runs along the palmar aspect of the interosseous membrane toward the hand. There are also smaller branches of the recurrent interosseous artery which go to the carpus and to the deep palmar arch: the *dorsal and palmar carpal branches* and the *deep palmar branch*.

The arteries of the hand lie mainly in the muscles of the palm where the metacarpal and finger arteries arise from the *two arterial palmar arches*. A weaker dorsal carpal branch of the radial artery gives off arteries to the back of the hand.

The superficial palmar arch AB13, distally near the heads of the metacarpal bones, is supplied mainly by the *ulnar artery* **B3** the *superficial palmar branch*, of the radial artery is an inconstant vessel). The *three common palmar digital arteries* **B14** arise from the superficial palmar arch and each divides into *2 common palmar digital arteries* **B15** to the adjacent sides of pairs of fingers. The little finger receives its artery directly from the palmar arch.

The deep palmar arch AB6 lies proximally near the bases of the metacarpal bones. It is supplied mainly by the *radial artery* **B2,** but it also receives a *deep palmar branch* from the ulnar artery. The *3rd and 4th palmar metacarpal arteries* arise from the deep palmar arch. They form anastomoses with the common palmar digital arteries of the superficial arch and through *perforating branches* with the dorsal metacarpal arteries.

The *dorsal carpal branch* **BC16,** a small branch of the radial artery, passes to the dorsal aspect of the carpus. It gives off the *dorsal metacarpal arteries* from which the small *digital arteries* arise. The carpal branches form networks: *the palmar and dorsal carpal rete*.

D Course of the dorsal and palmar arteries and nerves of the fingers. Compare **B** with **A, C** and **D.**

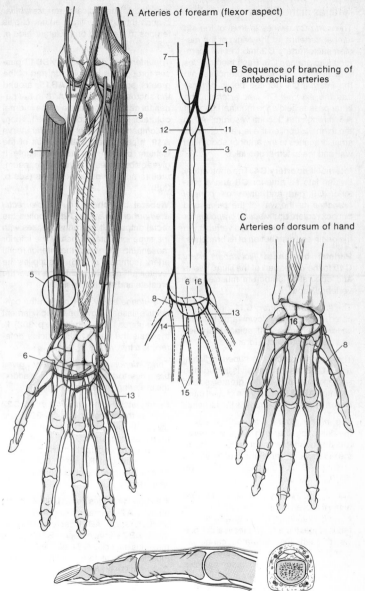

A Arteries of forearm (flexor aspect)

B Sequence of branching of antebrachial arteries

C Arteries of dorsum of hand

D Arteries and nerves of finger

Arteries of the Pelvis

The *aorta* **C1** divides in front of the 4th lumbar vertebra (at the origin of the *median sacral artery* **C2** into the *two common iliac arteries* **C3**. Each then divides without giving off further branches into the *internal* **C4** (for the pelvic organs) and the *external* **C5** *iliac arteries*. The latter passes below the inguinal ligament **A6**, through the "lacuna vasorum" **A7** to the thigh. It supplies the leg and gives off small branches to the anterior abdominal wall and the external genitalia.

Internal iliac artery C4. The short artery divides into an anterior **C8** and a posterior **C9** part and gives off *parietal branches* for the wall of the pelvis and buttock region, and *visceral branches* for the viscera. There is marked variability in the origin and distribution of its branches.

Parietal branches: *Iliolumbar artery* **C10** to the upper part of the ilium, psoas and quadratus lumborum muscles and vertebral canal.

The *lateral sacral artery* **C11**, often duplicated, runs laterally downward along the sacrum and gives off spinal branches into the anterior sacral foramina.

The largest branch, the *superior gluteal artery* **C12**, leaves the pelvis through the suprapiriform part of the greater sciatic foramen **AB13** to supply the gluteal muscles; a deep branch supplies the hip joint.

The *inferior gluteal artery* **BC14** leaves the pelvis together with the sciatic nerve and the internal pudendal artery through the infrapiriform part of the greater sciatic foramen **AB15**. It supplies the glutei and adductor muscles.

The *obturator artery* **C16** passes forward along the lateral wall of the pelvis and has a *pubic branch* that runs toward the inferior epigastric artery. It leaves the pelvis through the obturator canal. Its branches outside the pelvis are: an *anterior branch* to the adductor muscles and the skin of the external genitalia, a *posterior branch* to the deep external

muscles of the hip, and an *acetabular branch* through the ligamentum capitis femoris to the head of the femur (see p. 72).

The *internal pudendal artery* **BC17** passes through the infrapiriform part of the greater sciatic foramen **AB15**, around the sacrospinal ligament **A18** in the pudendal canal (in the fascia of the internal obturator muscle, Alcock's canal). Then, accompanied by the pudendal nerve **A19**, it passes along the ramus of the ischium, beneath the subpubic angle. It gives off branches to the anus, perineum, urethra and the penis or clitoris (see p. 308).

Visceral branches: The *middle rectal (hemorrhoidal) artery* **C20** supplies the rectal ampulla. It has anastomoses with the superior rectal branch of the inferior mesenteric artery and the inferior rectal artery (compare p. 246). It supplies the levator ani muscle and in the male the prostate and seminal vesicles.

The *uterine artery* **C21** runs in the connective tissue below the broad ligament of the uterus to the cervix (see p. 302). In the male the artery to the ductus deferens is a thin vessel.

The *inferior vesicular artery* gives branches to the fundus of the bladder, the ureter and prostate or vagina.

During fetal life, the *umbilical artery* **C22** runs to the placenta (see p. 324) and gives off the *superior vesical artery* and branches to the ureter. After birth the rest of it becomes obliterated and forms the medial umbilical ligament. **C23** is the common stem of the obturator and umbilical arteries.

A24 Lacuna of muscles, **AB25** piriformis muscle, **A26** sacral plexus, **A27** lesser sciatic foramen, **AB28** sacrotuberal ligament, **AB29** sciatic nerve, **B30** gluteus maximus muscle, **B31** gluteus medius muscle, **B32** gluteus minimus muscle. Compare **C** with **A** and **B**.

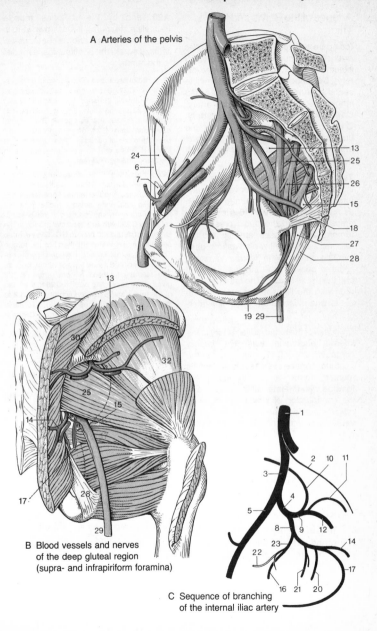

A Arteries of the pelvis

B Blood vessels and nerves
of the deep gluteal region
(supra- and infrapiriform foramina)

C Sequence of branching
of the internal iliac artery

Arteries of the Pelvis and the Thigh

(compare with p. 70.) The *internal iliac artery* C1: C2 superior gluteal artery, C3 iliolumbar artery, C4 common iliac artery, C5 abdominal aorta, C6 middle sacral artery, C7 lateral sacral artery, C8 inferior gluteal artery, C9 uterine artery, C10 inferior rectal (hemorrhoidal) artery, C11 obturator artery after exit of the umbilical artery, C12 internal pudendal artery, C13 piriformis, and C14 sacrospinal ligament.

The **external iliac artery** C15 reaches the terminal line from above and runs through the lacuna vasorum to the thigh as the *femoral artery* (A16 site of compression against the pubic crest). The *femoral artery* passes laterally and the *femoral vein* and *lymph vessels* pass medially through the *lacuna varosum,* situated between the inguinal ligament and the psoas fascia. (The femoral nerve runs with the iliopsoas muscle through the lacuna musculorum which lies laterally.)

Branches of the external iliac artery to the internal *abdominal wall:* The *inferior epigastric artery* C17, which runs to the posterior surface of the rectus abdominis muscle, forms a longitudinal anastomosis with the internal thoracic artery, the continuation of which, the *superior epigastric artery,* runs across the transversus abdominis muscle into the rectus sheath. The inferior epigastric artery crosses the posterior wall of the inguinal canal and sends a *pubic branch* to the obturator artery. The *deep circumflex iliac artery* C18 runs in an arch along the iliac crest and gives off branches to the abdominal wall before anastomosing with the iliolumbar artery.

The **femoral artery** C19, the continuation of the *external iliac artery,* passes along the groove between the vastus medialis muscle and the adductor muscles. It is covered by the fascia lata and the aponeurosis of the adductor muscle

A20, and by the sartorius muscle (guideline structure). It runs downward in the *adductor canal* and as the *popliteal artery* it enters the popliteal fossa medial to the femur.

Superficial branches: The *superficial epigastric artery* C21 goes to the outer surface of the abdominal wall, the *superficial circumflex iliac artery* C22 runs parallel to the inguinal ligament to the anterior superior iliac spine. The *external pudendal artery* C23, often duplicated, supplies the external genitalia. Other branches go to the extensor and adductor muscles and to the wall of the inguinal region.

Deep branches: The largest branch is the *deep femoral artery* BC24 which supplies the thigh muscles and the femur. Other branches that also commonly arise directly from the femoral artery are the *medial circumflex femoral artery* C25, with its *superficial branch* to the superficial, and *deep branch* BC26 to the deep adductor muscles and the head of the femur. There are anastomoses between it, the obturator artery and a variable acetabular branch. The *lateral circumflex femoral artery* BC27 gives off an *ascending branch* C28 to the head of the femur and a *descending branch* C29 to the quadriceps femoris muscle. Terminal branches of the deep femoral artery C30 are represented by 3(–5) *perforating arteries* to the adductors and the dorsal muscles of the thigh.

The *head* B31 and *neck of the femur* and the capsule of the hip joint are thus supplied by both circumflex femoral arteries; the head receives an additional supply from the obturator artery by its acetabular branch in the ligamentum capitis femoris B32.

The *articular rete of the knee* in front of the knee joint has the following afferent vessels: the *descending genicular artery* C33 arises from the femoral artery and penetrates the aponeurosis of the adductor muscle A20 on which it descends. The lateral C34 and medial C35 superior genicular arteries, the lateral C36 and medial C37 inferior genicular arteries, are all branches of the *popliteal artery.* The anterior C38 and posterior recurrent tibial arteries arise from the tibial arteries. The anterior tibial artery is C39. The *medial genicular artery* C40 penetrates the *knee joint.* Compare C with A and B.

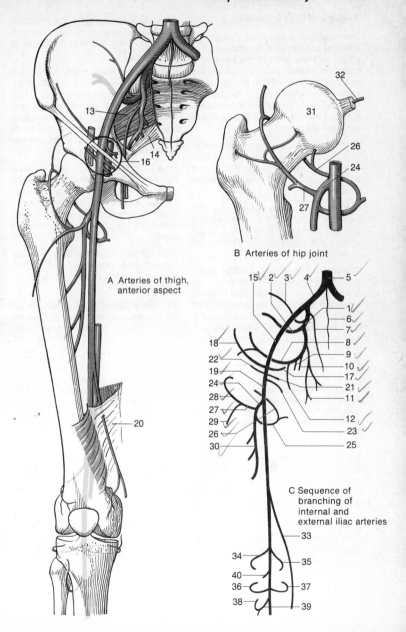

A Arteries of thigh,
anterior aspect

B Arteries of hip joint

C Sequence of
branching of
internal and
external iliac arteries

Arteries of the Leg and Foot

The **popliteal artery** runs through the adductor canal to enter the popliteal fossa at the flexor side of the knee. It is called the popliteal artery until its division into the *anterior* **ABC1** and *posterior* **AB2** *tibial arteries.* Lying in the popliteal fossa **A3** from dorsolateral to ventromedial, are the tibial nerve, popliteal vein and popliteal artery.

Branches to the *genicular rete* (see p. 72): Lateral **AC4** and medial **AC5** superior genicular arteries, and lateral **AC6** and medial **AC7** inferior genicular arteries.

The *medial genicular artery* **A8** perforates the capsule of the knee joint and supplies the connective tissue space within it. The sural arteries supply the calf muscles.

The **anterior tibial artery ABC1** perforates the interosseous membrane at the lower border of the popliteus muscle, where the anterior **B9** and posterior **B10** recurrent tibial arteries join the genicular rete. The anterior tibial artery runs with the deep peroneal nerve, embedded in the extensor muscles to the dorsum of the foot, where it continues as the *dorsal artery of the foot* **B11.**

The lateral and medial malleolar rete above the ankles are supplied by branches from the anterior lateral and anterior medial malleolar arteries: the posterior lateral and medial malleolar branches.

The *dorsal artery of the foot* **B11** lies with the deep peroneal nerve, lateral to the tendon of the extensor hallucis longus muscle, where the pulse can be palpated **C12**. It gives off the *lateral tarsal artery* **B13**, and together with these arteries it forms a vascular arcade, the *arcuate artery* **B14**, across the heads of the metatarsal bones. The *dorsal metatarsal arteries* **C15** are branches of the arcade and from them arise the *dorsal digital arteries* **C16** to the toes, and also the *perforating branches* and *deep plantar*

branch between the 1st and 2nd metatarsal bones which go to the sole of the foot. Superficial peroneal nerve **C17,** fibula **C18,** tibia **C19.**

The **posterior tibial artery** is the larger of the two tibial arteries. It passes beneath the tendinous arch of the soleus muscle between the superficial and deep flexor muscles, runs behind the medial malleolus together with the tibial nerve and finally, covered by the abductor hallucis, it reaches the sole of the foot.

Branches: The *peroneal artery* **AB20,** that runs along the fibula to the lateral malleolus, supplies the soleus muscle and the peroneus muscles and feeds the lateral malleolar and calcaneal retia (which are linked to the posterior tibial artery).

The *medial plantar artery* **BD21** gives off a superficial branch to the great toe (hallux) and a deep plantar artery to the *plantar arch* **D22**.

The *lateral plantar artery* **BD23** crosses the quadratus plantae muscle, and decends deep to the lateral border of the foot.

The **plantar arch D22** anastomoses with the deep plantar branch from the dorsal artery of the foot. The plantar arch gives off the *plantar metatarsal arteries* **D24,** from which arise the *plantar digital arteries* **D25** to the toes, and *perforating branches* to the dorsal metatarsal arteries. Compare **B** with **A, C** and **D**.

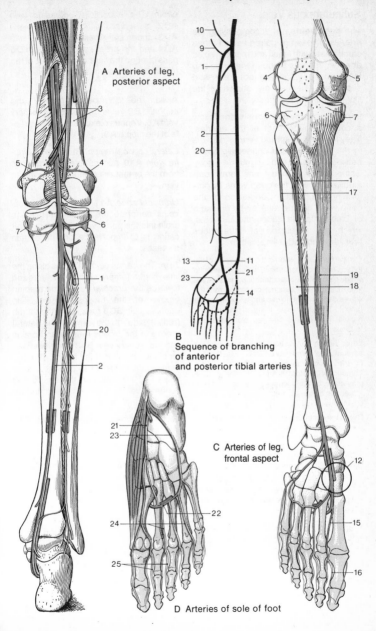

A Arteries of leg, posterior aspect

B Sequence of branching of anterior and posterior tibial arteries

C Arteries of leg, frontal aspect

D Arteries of sole of foot

Subcutaneous Veins

The **deep veins** run accompanying the arteries. These are paired veins (venae comitantes) with small and medium sized arteries, such as those in the arms and legs. These veins are not discussed separately but they are shown in the illustrations of the arteries.

In contrast, **subcutaneous veins** follow a course independent of the arteries. They form a widespread *venous network* in the subcutaneous connective tissue between the skin and the muscle fascia. Some larger named veins form communications with the deep veins by perforating the fascia. Congestion of the deep veins can be relieved by drainage of blood into the subcutaneous veins, which aid regulation of intravenous pressure and body temperature.

Clinical tips: Congestion in the deep veins of the legs (deep varicose veins, *varices*) leads to congestion in the subcutaneous veins of the legs. The lumens become dilated so that the valves can no longer close and become incompetent. The excessive filling of these veins makes their course become elongated and tortuous.

When there is congestion in the portal vein, the paraumbilical veins may direct blood to the skin of the abdominal wall. Compare p. 248. Tumours of the thoracic and abdominal cavities may act as space-occupying masses and lead to venous congestion extending into the subcutaneous veins.

Due to the enlargement of the mammary glands in pregnancy, the veins of the breasts become more numerous and engorged.

The blood conveys body heat which is generated by chemical reactions mainly in the muscles and liver. Infrared photographs of the body surface therefore show the subcutaneous venous network most spectacularly.

Trunk: At seven points on the trunk the subcutaneous veins converge like a star in order to pierce the fascia and open into the deep veins.

Groin: The *superficial epigastric vein* **AD2,** *superficial iliac circumflex vein* **AD3,** *superficial external pudendal vein* **AD4** and the *great saphenous vein* **D5** pass through the saphenous hiatus of the fascia lata to drain into the *femoral vein* **D1.**

Axilla: The *thoracoepigastric* **A7** and *cephalic* **B8** veins drain into the *axillary vein* **A6.** Venous plexus around the areola of the nipple **A9.**

Lateral cervical region: The *anterior jugular vein* **A10** and cutaneous branches from the breast drain into the subclavian vein.

Umbilical zone **A11:** The *paraumbilical veins* perforate the abdominal wall to drain into the portal vein. This communication is of importance only in certain diseases.

Arm: The *basilic vein* **B12** runs along the little finger (ulnar) side of the forearm towards the *brachial vein* in the median groove of the biceps muscle. The *cephalic vein* **BC8** runs along the thumb (radial) side of the arm in the lateral groove of the biceps muscle and drains into the *axillary vein* **B13** (the vein used for catheterisation of the heart). The *median cubital veins* **B14** lie in the cubital fossa and run between the basilic and cephalic veins; they are used for intravenous injections.

Leg: The *small saphenous vein* **E15** runs behind the lateral malleolus along the lateral side of the calf and drains into the *popliteal vein* **E16.** The *great saphenous vein* **D5** runs anterior to the medial malleolus, along the medial side of the leg and drains into the *femoral vein* **D1.**

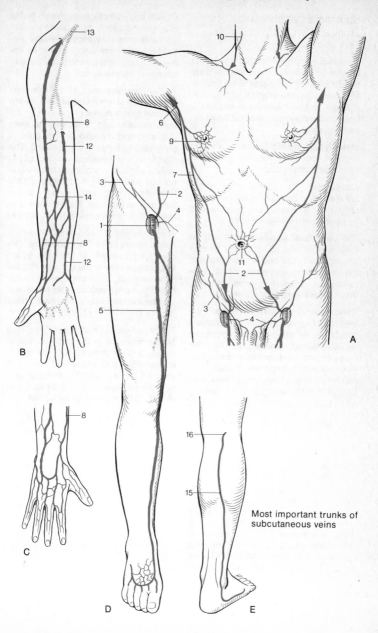

Most important trunks of
subcutaneous veins

Superficial Lymph Vessels of the Trunk and Lymph Nodes of the Arm and Leg

Regional lymph nodes are the *first* to receive lymph from a region of the body or from an organ. A region of the body or an organ may drain lymph into several different regional lymph nodes, and a lymph node may receive blood from several organs.

Mass lymph nodes lie proximal to the regional nodes and *receive lymph* from several of these nodes.

Clinical tips: The regional lymph nodes are the first to be affected by the spread of an infection along the lymphatics and by seeding of a malignant tumour.

The *superficial lymph vessels* vary in their number and course. They pass to the regional lymph nodes in *parallel* with the *subcutaneous veins.* Most subcutaneous lymphatics arise from intracutaneous lymphatic capillaries.

Clinical tips: Extensive movements of the skin may tear intracutaneous lymphatic capillaries. In areas of high lymphatic capillary density (palms and soles) this may produce lymphorrhagia "water blisters" beneath the epithelium. Lymphatic capillary damage is also implicated in the formation of blisters after burns, freezing and chemical damage. In *infections* of the skin, the subcutaneous lymphatics may be seen as *red streaks,* and regional lymph glands become painful due to increased tension of the capsule.

The *regional subcutaneous lymph nodes* of the trunk and the extremities lie on the *flexor side* of the joints **A.** If there is no local inflammation or generalised, immunobiological reaction, and no malignant tumours have invaded them (seedlings, metastases from carcinomas in their drainage area), the subcutaneous lymph nodes are only millimeters in size and are difficult to palpate and are mobile over their beds.

Clinical tips: Pathologically, lymph nodes may grow to one or more centimeters in diameter, they may become visibly protuberant and may fuse to form an immobile mass. Lymph node palpation may often lead to the diagnosis.

Axillary lymph nodes AB 1 and **superficial inguinal lymph nodes A 2** are the regional lymph nodes of the skin, of the trunk (anterior and posterior), the arms and legs, and are also the last point of entry of lymph into the lymphatics. The margin between the area which drains to axillary and that which drains to the inguinal lymph nodes lies in a girdle about the level of the navel. Lymph from the skin of the external genitalia, the perineum and anus drains to the inguinal lymph nodes.

The superficial **B 3** and deep **C 4** *cubital lymph nodes* lie at the *elbow joint.* The deep lymph vessels of the arms, which follow the course of the arteries, receive subcutaneous lymphatics even in the hand. Some lymphatics from the thumb and the extensor side of the hand **D** pass directly to the axillary lymph nodes.

The superficial **G 5** and deep **H 6** *popliteal lymph nodes* lie in the *popliteal fossa.* Some deep lymph nodes lie in front of the interosseous membrane and below the patella **F 7.** Lymphatics from the great toe and the extensor side of the foot **E** run directly to the inguinal lymph nodes.

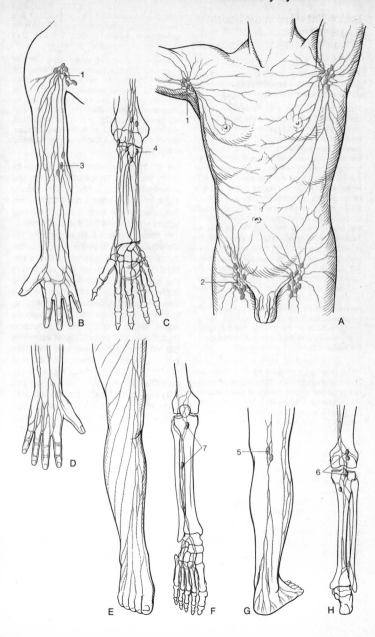

Lymph Vessels (Lymph Nodes) of the Head and Neck, and the Deep Lymph Vessels (Lymph Nodes) of the Trunk

The lymphatic drainage of the various organs and the regional lymph nodes of the viscera are discussed under the appropriate organs.

The lymph nodes for the forehead and upper eyelid lie in front of the ear, *the parotid lymph nodes* **A1**. The lymph nodes of the middle and lower facial regions including the teeth, of the maxilla and mandible, and those of the tongue are situated at the lower border of the mandible, *submandibular lymph nodes* **A2**. Lymph from the lower lip drains into the *submental lymph nodes* **A3**. The lymph nodes of the posterior nasal cavities lie in the retropharyngeal space in front of the vertebral column, *retropharyngeal lymph nodes*. Lymph from the posterior half of the scalp flows to the *occipital* **A4** and *retroauricular* **A5** *lymph nodes*. Lymph from the superficial tissues of the neck and the parotid gland flows into the *retroauricular lymph nodes* **A5**. Lymph from the superficial tissues of the neck and the parotid gland runs into the *superficial cervical lymph nodes* **A6**. More proximal lymph nodes for this catchment area, the *deep cervical lymph nodes* **A7**, surround the neurovascular trunk deep in the neck. Lymph from the base of the tongue and the tonsils flows directly into the upper part of this group, the superior cervical lymph nodes.

The *axillary lymph nodes* from several groups which receive lymph from the arm, chest wall and mammary glands. They are situated beneath the pectorales major and minor muscles **A8**, above it **A9**, deep to the clavipectoral fascia **A10**, at the lower border of the pectoralis major **A11**, and deep in the axilla **A12**. Other lymph nodes draining the mammary gland are the *parasternal lymph nodes* **A13**, which also receive lymph from the liver, diaphragm, pericardium and intercostal spaces. Some lymph from the breast passes to nodes near the venous angle **A14**. Lymph from the shoulder drains into the subscapular lymph nodes **A15**. The deep lymphatic vessels of the wall of the trunk lie segmentally together with the intercostal blood vessels; the nodes lie in the intercostal spaces on either side of the vertebral column (paravertebral group) and the sternum (parasternal glands) **A13**.

The *bronchopulmonary lymph nodes* **B16** are collection nodes for lymph drainage from the lungs. *Tracheal lymph nodes* **B18** lie under and

over the bifurcation of the trachea **B17** and along it. Other lymph nodes are situated in the anterior and posterior mediastinum **B19**. **B20** pulmonary lymph nodes.

The superficial inguinal lymph nodes continue in groups of lymph nodes situated along the external and internal iliac artery **B21** and the aorta **B22**. Afferent vessels to this group come from the pelvic organs and the intra- and retroperitoneal viscera **B23**. The *celiac (pre-aortic) lymph nodes* **B24** around the celiac trunk are important. Lymph from the liver, gall bladder, stomach, spleen, duodenum and pancreas drains into them.

Lymph from the entire abdominal cavity flows via large **lymphatic trunks,** the *intestinal* **C25** and *lumbar* **C26** *lymphatic trunk,* into the **cisterna chyli C27,** which lies between the aorta and the vertebral column at the level of the aortic hiatus. The **thoracic duct C28** arises from the cisterna chyli. It ascends in front of the thoracic vertebral column, crosses behind the aorta to the left side and enters the left venous angle from above and behind (see p. 50). On its way it receives intercostal lymph vessels. The following lymph vessels drain into the left venous angle, either direct or via the thoracic duct: the *jugular trunk* **C29,** which drains the left side of the head and neck, the *bronchomediastinal trunk* **C30** draining the left half of the posterior mediastinum and the left lung, the *subclavian trunk* **C31** from the left axilla and the left arm. The lymph from the right side of the head and neck, from the right axilla, the right arm and from the right half of the anterior and posterior mediastinum, including the right lung, drains via the **right lymphatic duct C32** into the right venous angle.

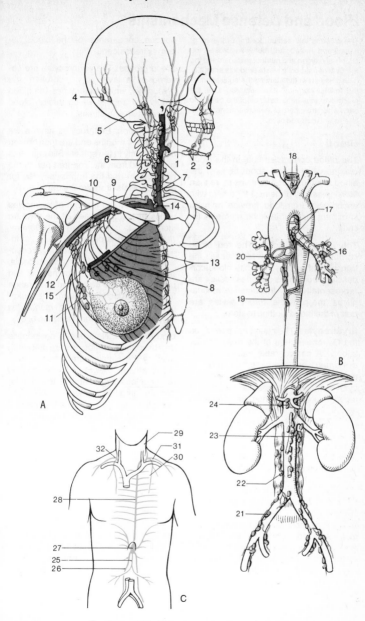

Blood and Defense Mechanisms

The *blood* is the medium for the exchange of gases and metabolites between cells – the supply of oxygen and nutrient materials and the removal of metabolic waste products and CO_2. It also assists in maintaining the equilibrium of the "milieu interieur" (homeostasis), i. e., the electrolyte and water balance of the body and serves for transport of heat, hormones, antibodies and defense cells.

Blood

The *blood corpuscles* float in the *blood plasma* which forms about 56% of the blood volume. Blood contains more than 90% plasma proteins, together with amino acids from food, provide the most important nutrient building materials for cells.

The blood corpuscles: The red blood corpuscles, *erythrocytes,* serve the transport of gases; the white blood corpuscles, *leukocytes* (colorless cells), are representatives of the defense mechanisms; the platelets, *thrombocytes,* are part of the blood clotting system.

Erythrocytes. Their number depends on the O_2 requirement of the body and O_2 supply. A considerable increase *(polycythemia)* or decrease *(anemia)* is pathologic. The mature human erythrocyte does not possess a nucleus, is biconcave, can change its shape, is homogenous in appearance and measures 7.7 μm in diameter. More than 90% of its dry substance consists of *hemoglobin,* the iron-containing blood pigment. This gives blood its red appearance – *oxygenated hemoglobin* in arterial blood is bright red, and *deoxygenated hemoglobin* in venous blood is dark red. About 5–15% of the erythrocytes are *reticulocytes,* immature cells, see p. 86. The number of reticulocytes increases after loss of blood. Irregularly shaped and abnormally large erythrocytes circulating in the blood after birth are pathologic. Erythrocytes shrink in hypertonic solutions ("thornapple form") and they burst in hypotonic solutions, when the normal opaque appearance of the blood becomes translucent.

Blood groups. The erythrocytes are covered by a layer of glycoprotein, the *glycocalyx,* which determines the blood group and against which foreign blood may contain agglutinins.

Disintegration. Erythrocytes have a life span of 3–4 months and are then broken down mainly in the spleen. Bile pigments are formed from the iron-free part of the hemoglobin and the iron is recycled for erythropoiesis (iron circulation).

Leukocytes. The number of leukocytes varies at different times of the day. An increase to more than 10 000/mm³ *(leukocytosis)* is pathologic, e. g., in inflammations or in tumor growth; a reduction to below 2000/mn³ *(leukopenia, agranulocytosis)* is also pathologic and may, for example be caused by damage to the erythropoietic system. The leukocytes include *granulocytes, monocytes* (see p. 86) and *lymphocytes* (see p. 86).

Granulocytes. Neutrophils, eosinophils and basophil granulocytes are differentiated by the staining properties of their cell granules. *Neutrophil granulocytes* (cell nucleus segmented 3 to 4 times) contain small granules – *azurophilic granules* with lysosomal enzymes and peroxidase, and *specific granules* with lysozymes and bactericidal materials. *Eosinophilic granulocytes:* the strongly acidophilic large granules, which stain with eosin, also contain proteolytic enzymes. *Basophilic granulocytes* also have large granules which contain heparin and histamine.

Thrombocytes. The *thrombocytes* (platelets) disintegrate easily and liberate *thrombokinase,* a group of enzymes closely concerned with blood clotting. Thrombocytes are 1–3 μm in size. They do not contain a nucleus and so are really cell fragments.

Blood cells
(from Faller)

A Derivatives from red bone marrow

Red blood corpuscles
(erythrocytes)

Blood platelets
(thrombocytes)

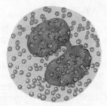

Neutrophil
granulocytes

Eosinophil
granulocytes

Basophil
granulocytes

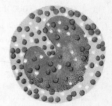

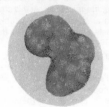

Monocyte

B Derivatives from lymphatic organs

Small lymphocyte

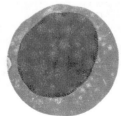

Large lymphocyte

Numbers of Blood Cells

One cubic millimeter (mm^3) of blood contains:

4.5–5 million erythrocytes
4000–8000 leukocytes of which:

granulocytes:	neutrophils	55 –68%
	eosinophils	2.5– 3%
	basophils	0.5– 1%
	"stab cells"	
	(with horseshoe shaped nucleus) 2	– 3%
monocytes:		4 – 5%
lymphocytes:		36 –20%
200,000–300,000 thrombocytes		

Hemopoiesis after birth. After birth hemopoiesis normally takes place only in the *bone marrow*. **A** With the termination of bone growth in length, blood formation is restricted to the marrow in the *epiphyses* of the long bones and the *short, flat bones*. In a case of chronic blood loss or of interference with medullary (bone marrow), hemopoiesis after birth, blood formation can be resumed in the marrow of the diaphyses of the long bones, in the connective tissue of the liver and in the other sites of fetal hemopoiesis.

The total weight of *bone marrow* is about 2600 g, or about 4.6% of the body weight. The bone marrow exceeds in mass all other organs except the blood, the muscles and the skeleton. It contains up to 10% of the total blood volume. The red, hemopoietic bone marrow contains mainly the cells responsible for *granulopoiesis* (formation of the granule-containing, colorless or white blood corpuscles, leukocytes) and *erythropoiesis* (formation of red blood cells) **B1**, as well as the giant cells of the bone marrow (megakaryocytes) **B2** (which are responsible for *thrombopoiesis,* the production of platelets), and also fat cells **B3** and trabeculae of bone **B4**. The survey **B** shows a colorful, complex picture.

The common pluripotential *stem cell* of all the blood cells is the *hemocytoblast*. It is functionally but not morphologically characteristic and resembles a middle-sized lymphocyte. Hemocytoblasts appear during early embryonic develop-

ment; in the adult they are found in red bone marrow and also singly in the blood.

Hemopoiesis during fetal life. The following periods of hemopoiesis can be distinguished **C:**

Megaloblastic period: The very earliest erythropoiesis starts in the *extraembryonic connective tissue of the yolk sac* and abdominal ridge of the germinal tissue, about 2 weeks after fertilization. The blood corpuscles are large nucleated erythrocytes, *megaloblasts*. At this stage leukocytes are absent. The megaloblastic period lasts until the end of the third month.

Hepatolienal period: Toward the end of the second month hemopoiesis starts in the *liver* and *spleen* and to a less extent in the *lymph nodes*. Leukocytes now appear and the erythrocytes attain the normal adult size. The hepatolienal period ends before birth.

Medullary (Bone marrow) period: During the 5th fetal month hemopoiesis commence in the *bone marrow* in all bones. This is the final site of hemopoiesis throughout life, in the red bone marrow.

Lymphocytes are formed in the 4th month first in the liver and then in the bone marrow. From there they migrate partly into the thymus and thence into the lymphatic organs in which they multiply, and partly they leave the bone marrow to pass directly into the peripheral lymphatic organs. Even after birth, lymphocytes migrate from the bone marrow into lymphatic organs.

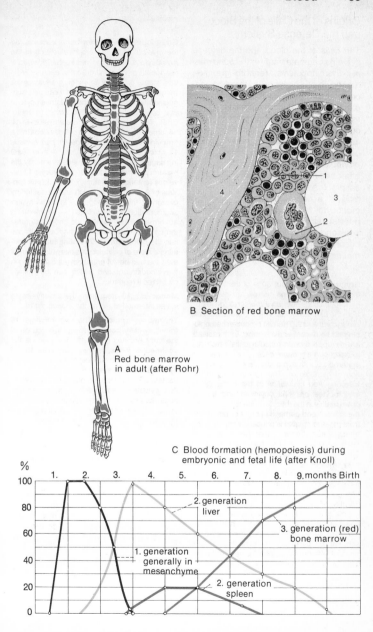

A
Red bone marrow
in adult (after Rohr)

B Section of red bone marrow

C Blood formation (hemopoiesis) during
embryonic and fetal life (after Knoll)

%

1.   2.   3.   4.   5.   6.   7.   8.   9. months Birth

2. generation
liver

3. generation (red)
bone marrow

1. generation
generally in
mesenchyme

2. generation
spleen

Origins of the Cells of the Blood and the Defense Systems

The cells of the blood and the defense systems originate partly in the *bone marrow* (hematopoiesis); (erythro- granulo-, monocyto-, lympho-, thrombocyto-poiesis) and partly in the *lymphatic organs* (cells of the immunological system). The common stem cell of all blood cells is the *hemocytoblast* 1. Two cells are produced during its mitosis. They behave differently. One remains a *pluripotential hemocytoblast,* whilst the other becomes an *irreversibly unipotential precursor cell* which lies at the beginning of a cell *line* (erythro-, granulo-, monocyto-, lympho-, thrombocyto-poiesis). Under the influence of active substances, *poetins,* the precursor cells become *"blast"* cells and then via intermediate stages they become mature blood cells.

Erythropoiesis. About 30% of the immature blood cells of the bone marrow are involved in erythropoiesis. The *erythroblast* 3 develops from the *hemocytoblast* 1 via the precursor cell, *proerythroblast* 2. This cell produces hemoglobin and so becomes acidophilic. Erythroblasts are grouped around reticular cells which are looked upon as *"nurse cells".* They supply the erythroblast with iron necessary for the formation of hemoglobin. The iron enters into the following *iron circulation* in the body: senile erythrocytes are phagocytosed and broken down mainly in the reticular connective tissue of the spleen (and perhaps in other organs). Iron from the hemoglobin is principally stored in the phagocytes of the reticular connective tissue as *hemosiderin* (Berlin blue reaction). In cases of massive blood destruction there is macroscopically visible iron storage, *hemosiderosis.* Hemosiderin in turn is broken down to *ferritin, a* molecular combination of 6 atoms of iron that can be followed in the electron microscope. Ferritin, bound to a protein, is carried through the bloodstream to the bone marrow. Here it is taken up by the reticular cells which eventually release it to the neighboring erythroblasts. Subsequent cell divisions result in nests of *normoblasts* 4. After their nuclei have been extruded they become *erythrocytes* 5. Erythrocytes which are not yet quite mature still contain a network-like residue of basophilic ribosomes *(substantia granulofilamentosa), reticulocytes* 6.

Granulopoiesis: Development proceeds via the *myeloblast* 7, and *pro-myelocyte* 8 to the *myelocyte* 9, when the neutrophil, eosinophil (acidophil) or basophil granules appear. From this stage on three lines of granulopoiesis (neutrophil, eosinophil, basophil) can be distinguished, which lead via the metamyelocyte to the polymorphonuclear leukocyte (granulocyte). The nucleus of the *metamyelocyte* 10 becomes elongated and the cell develops into a granulocyte with a *horsehoe-shaped nucleus* 11 which is not yet quite mature. With increasing segmentation of the nucleus eventually the mature, *polymorphonuclear granulocyte* 12 develops (criterion of maturity). The nucleus consists of 2–4 segments which are connected by extremely thin (nucleogenic) threads; hypersegmentation of the nucleus (> 4 segments) is abnormal. The granulocyte now migrates through the walls of the bone marrow capillaries into the blood stream. The bone marrow contains *a store of granulocytes* many times larger than the number of granulocytes circulating in the bloodstream and which can be quickly mobilized if required.

Monocytes 13 stem from the pro-myelocyte via an intermediate form, the *monoblast* 14.

Thrombocytopoiesis. The *megakaryocyte* 15 (giant cell of the bone marrow), the precursor of the thrombocytes (platelets) develops via the intermediate forms, the *megakaryoblast* 16 and the *immature megakaryocyte* 17. Megakaryocytes have a diameter of more than 50 μm and contain a large lobulated nucleus. Granules in their cytoplasm move into peripheral, pseudopodialike protrusions and ultimately, *thrombocytes* 18 are separated off. After repeated thrombocyte formation the megakaryocyte dies.

Lymphopoiesis: The precursor cell which is still immunologically incompetent, becomes the *differentiated* (T- or B-) *lymphocyte* 19. After primary contact, they produce a T- or B-*immunoblast* 20, which by T differentiation produces *immunocytes* 21, or by B differentiation produces *plasma cells* 22, or *memory cells* 23 either of T- or B-cell type (see p. 90ff).

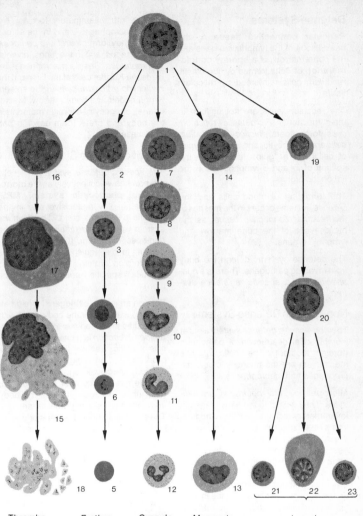

Thrombo-
poiesis
(blood
clotting)

Erythro-
poiesis
(gas
transport)

Granulo-
poiesis
(non-
specific
defense)

Monocyto-
poiesis
(specific
defense)

Lympho-
poiesis
(specific
defense)

Formation of blood cells and cells
of defense systems at sites of
hemopoiesis

Defense Systems

Reticular connective tissue A, the basic tissue of the lymphatic organs and the bone marrow, is a **spongy conglomeration of cells** formed by fibroblastic *reticulum cells* reinforced by reticulum fibers.

The reticulum cells are not uniform in their functions. *Fibroblastic reticulum cells* form fibers, *histiocyte reticulum cells* are phagocytic, and two other types of cell have a "guide function" in the settling of T- and B-lymphocytes in the lymphatic organs.

The formation of blood cells and the defense systems occurs in the meshes of the *reticular connective tissue,* as the basic tissue of the bone marrow and lymphoid organs.

The defense systems defend the body from invading pathogens. There is a *non-specific* and a *specific resistance system.*

Non-Specific Defense System

The *non-specific defense system* aims at instant, local destruction of pathogens (foreign bodies). The cells of this mechanism are the phagocytic cell, *microphages* and *macrophages.*

Microphages are *neutrophil granulocytes.* They collect in the early stages in inflammatory foci attacked by pathogens and substances released by cell breakdown. The neutrophil granulocytes digest the phagocytosed material **B1** (streptococci in this instance) with the help of their lysosomes, the neutrophil granules. At the same time they produce proteolytic enzymes to soften the inflammatory infiltrate (abscess formation). This kills the granulocytes, thereby producing *pus cells.* **B2** Cell nucleus, **C3** phagolysosomes.

Macrophages are derived from *monocytes* and are mobile or (temporarily) sessile. They are found in all tissues as "tissue macrophages", *histiocytes.* They

migrate into inflammatory foci as "exudate macrophages", and in the serous cavities they form *pleural* and *peritoneal macrophages,* and in the lungs, *alveolar macrophages.* Sessile macrophages include the *Kupffer cells* of the liver and the *histiocytic reticulum cells* in the spleen, lymph nodes and bone marrow. Macrophages phagocytose foreign materials in the body and break them down in part enzymatically.

Specific Defense System

The specific defense system. The cells are now known collectively as the **mononuclear phagocytic system, MPS.** They also play a decisive role in the specific defense system. The *immune system* works *quite differently* from the non-specific system. It enables the organism to distinguish its *own proteins* from foreign proteins, *antigens,* and to produce specific *antibodies* against them.

Foreign antigens (pathogens, foreign albumins) which enter the body, react with antibodies produced there. The *antigen-antibody reaction* is a chemical process. It is specific. Each antigen evokes the production of an (antigen homologous) antibody specifically against itself, which may be reproduced after a long period, even up to decades. It gives the body *immunity* against that particular antigen. The specific defense system is principally situated in the lymphatic organs.

The cells of the *specific defense system* are the *immunologically competent T-* and *B-lymphocytes.* They perform their functions in association with the *accessory cells of the specific defense system* (macrophages, specialised reticulum cells of lymphatic organs). The T- and B-lymphocytes represent the two arms of the immune system, the *cellular* and *humoral immune responses.* Both types of lymphocyte arise from the stem cells of the bone marrow and develop their defence characteristics (immunocompetence) stepwise through precursor cells.

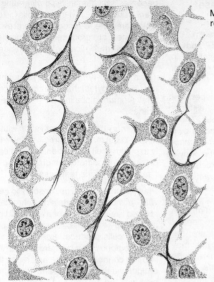

Microscopic section through reticular connective tissue

A

Non-specific defense system: granulocyte system

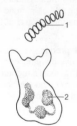   

B Phagocytosis of chain of streptococci by neutrophil granulocyte

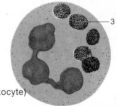

C Neutrophil granulocyte (leukocyte) with phagosome

Cells of the Specific Defense System

The cells of the specific system, the immunologically competent T- and B-lymphocytes, **A**, fulfill their functions in *co-operation* with the *accessory cells* of the specific defense system.

The **T-lymphocytes** (*t*hymus dependent lymphocytes) are prepared for their functions in the cortex of the thymus. There they 'learn' to tolerate the body's own (autologous) tissues (*general T-lymphocyte tolerance* of self, i. e. of the body's own tissues) and to develop resistance only against non-self material, *antigens*. They leave the thymus either as *regulatory cells* (*T-helper cells* or *T-suppressor cells*) or as *cytotoxic T-cells* and migrate in the blood stream into the T-regions of the lymphatic organs. From there, the immunocompetent T-lymphocytes may again enter the blood stream *through* the lymph vessels and so *recirculate*. T-lymphocytes are always available throughout the body (T-lymphocyte reservoir).

The *T-helper-cells* stimulate B-lymphocytes to produce and secrete antibodies, either directly by cell *membrane-bound* factors or indirectly by soluble helper-factors. The B-cell-response to most antigens is dependent on this T-cell help. This function of the T-helper cells presupposes that they are able to 'recognize' the antigen. It is 'presented' to them by the accessory cells with which they come into close contact. *T-suppressor-cells* may under certain conditions, suppress the immune response of the B-cells, the T-helper cells and the cytotoxic T-cells.

The *cytotoxic T-cells* may destroy antigenic cells by direct contact ("cellular immune response") without themselves being damaged. They are stimulated by T-helper cells.

The specificity of each of these functions is developed at the first antigen contact, the *primary contact*, when the T-cell first becomes activated into a proliferating *T-immunoblast* **B**. "Memory cells", too, which are able to recognize the provoking antigen, are produced during proliferation (cell multiplication) of T-immunoblasts.

The mature **B-lymphocytes** (*b*one marrow lymphocytes) have antigen receptors (immunoglobulin) on their membranes.

The *primary contact* (by activation with T-helper-cells) may lead directly or indirectly to plasma cell production. *Plasma cells* produce and secrete humoral antibodies ("humoral immune response").

Plasma cells are large, basophilic cells with a nucleus containing chromatin shaped like the spokes of a wheel (**C** light microscopy, **D** electron microscopy). Plasma cells produce large amounts of *immunoglobulins*, which are discharged into the connective tissue and reach the antigen via the blood stream. They combine with the antigen and destroy it. The humoral immune response proceeds rapidly in a matter of minutes or hours. **D1** ergastoplasm = the site of immunoglobulin formation. **C** magnification about × 1000.

In *direct* plasma cell production, proliferative *B-immunoblasts* develop from B-cells and from these identical daughter cells are formed (clone formation) which then differentiate into plasma cells.

Indirect production of plasma cells occurs in *lymph follicles* by the formation of germinal centers (secondary follicles). The B-lymphocyte develops (through intermediate stages of centroblasts and centrocytes) into the "memory cell", forming receptors for the antigen concerned, and then (even years later) after renewed contact with the same antigen (secondary contact) it reacts rapidly by differentiation into plasma cells.

E Macrophage, see p. 92.

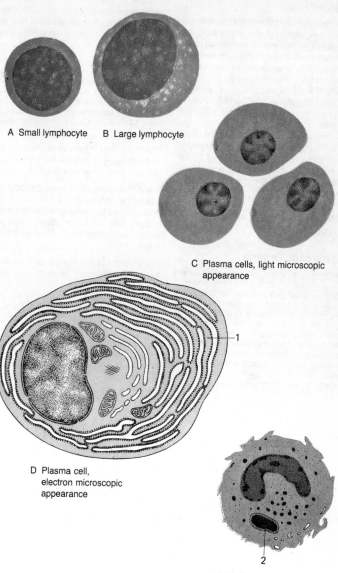

A Small lymphocyte B Large lymphocyte

C Plasma cells, light microscopic
appearance

1

D Plasma cell,
electron microscopic
appearance

2

E Histiocyte with phagosome

Accessory Cells of the Immune System

The **accessory cells** of the immune system are required for processing immunogenic information as well as in the immune response. Phagocytic and nonphagocytic accessory cells may be distinguished. The *phagocytic accessory cells,* macrophages (p. 90) (tissue macrophages, Kupffer cells of the liver, alveolar macrophages) may amongst other activities "present" antigens to immunocompetent lymphocytes. The *nonphagocytosing accessory cells* (specific reticulum cells) permit settling of T-lymphocytes (in the T-region) and B-lymphocytes (in the B-region) in the lymphoid organs.

Origin of Humoral and Cellular Immunity

Lymphatic Organs

Different roles of the immunobiologic system are allocated to different lymphatic organs, – thymus, lymph nodes, spleen, tonsils, lamina propria of the intestinal mucosa. It is therefore not possible for one organ to completely take over the function of another. The *thymus* is irreplaceable as the primary immunologic organ for the development of cellular immunity. During childhood active immunologic maturation processes are accompanied by exaggerated, often dramatic growth of the lymphatic organs. By about the 6th year they have already attained their adult size, during the 10–12th year about double that size is attained and then the organs shrink to return to their final size by the age of 20. A further regression occurs in old age.

Thymus

The **thymus** consists of two oval lobes joined together. It is situated in the mediastinum above the pericardium, in front of the left innominate vein and the superior vena cava, between the two mediastinal layers of the pleura and behind the sternum. During childhood it is well developed and extends downward to the sternal attachment of the 4th rib,

and may produce broadening of the shadow at the base of the heart in a radiograph. The lobes reach to the lower border of the thyroid gland. They lie under the middle layer of the fascia of the neck.

During childhood, the superficial surface of the thymus is arranged in lobules, which hang on a common, central vascular cord. These are, in turn, composed of smaller lobules in which microscopically a cortical and a medullary zone can be distinguished, see p. 94. **A1** Innominate veins, **A2** heart, red triangle = triangular outline of thymus.

The *thymus* is fully developed in the *child* and *juvenile,* but in the adult it becomes largely transformed into adipose tissue.

T-lymphocytes (*t*hymus lymphocytes) reach the cortex of the thymus as T-precursor cells, *thymocytes* where they mature into *regulator cells* (T-helper cells, T-suppressor cells) and *cytotoxic T-cells.*

In the *cortex of the thymus,* before leaving it, T-lymphocytes 'learn' to differentiate autologous (self) and foreign antigens. The many lymphocytes which are unable to perform this function die, and the remainder are tolerant of autologous antigens. This is the basis for the general *T-lymphcyte tolerance* of the body. T-lymphocytes respond in the thymus and in subsequent immune responses by comparing *the major histocompatibility complex, MHC* presented to them by thymic epithelial cells with possible antigens. The *maturation* of the T-lymphcyte subgroups, the T-helper cells, T-suppressor cells and cytotoxic T-cells can be evidenced by demonstration of their characteristic *membrane-bound antigen patterns.* Descendants of T-lymphocytes are the carriers of the *cellular immune response* as regulator-T cells or cytotoxic T-cells.

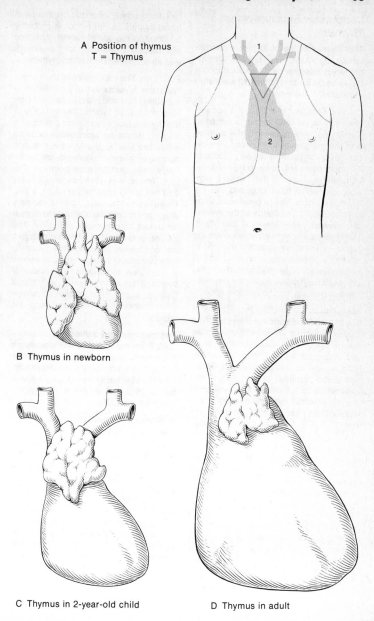

A Position of thymus
T = Thymus

B Thymus in newborn

C Thymus in 2-year-old child

D Thymus in adult

Microscopic Structure of the Thymus

The thymus of the child has a characteristic structure. The lobules are attached to vascular cords; in each lobule the *medulla* **B1** and *cortex* **B2** are distinguishable.

Cortex. The stroma **B3** has the appearance of a loose reticulum. It is arranged like an epithelium towards the surface and the margins of the lobules. The reticulum of the tissue is densely filled with T-lymphocytes **B4**. In the marginal zone, immediately beneath the capsule, lymphocytes which have migrated into the thymus multiply. The population of small lymphocytes in the cortex of the thymus is renewed every three to four days. Lymphocytes from the thymus enter the blood stream continously, though in smaller numbers with increasing age. A large number of lymphocytes die in the cortex of the thymus during the development of general lymphocyte tolerance.

Medulla. In the medulla the network of the stroma **B3** is narrower than that of the cortex and there are fewer lymphocytes (thymocytes) **B4**. Occasionally, there are concentrically arranged cells in the stroma. *Hassal's corpuscles* **B5**. There are about 1 million of the latter in the newborn and 1.5 million in juveniles: in the 20 year old they still number about 700 000 but after the 40th year there are only 250 000. A Hassal's corpuscle can consist of a few cells, or it can develop into a cyst 0.1–0.5 mm in diameter which contains cellular debris. The *corpuscles* are round lamelleted structures formed from a few or many concentrically arranged cells of the basic tissue. Hassal's corpuscles may develop rapidly and then disappear or may form cysts. The function of these corpuscles is unknown, but they appear to be formed in connection with immunologic processes. The stroma of the thymus is separated from its surface and from blood vessels by a basement membrane. In addition, thymus differs from other lymphatic organs by a blood-thymus-barrier (which is probably confined to the cortical area). Antigens are trapped in the walls of the vessels and do not penetrate into the thymus.

Function. The importance of the thymus lies in the development of the different T-cell subgroups and in the formation of the humoral thymic factors. The *removal* of the thymus in newborn mice produces a wasting syndrome which leads to death within two weeks. The lymphatic organs contain very few lymphocytes, are abnormally small and have no germinal centers. The spleen does not form follicles and the blood lymphocyte count is low. There is an *immunobiologic insufficiency* which affects cellular immunity to a greater extent than humoral immunity. The longer removal of the thymus is delayed, the less severe the syndrome. Comparable pathologic states may be observed *in man* due to *aplasia, hypoplasia* or *hypofunction* of the thymus. A humoral thymus factor thymopoetin stimulates differentiation of the peripheral lymphatic organs.

Postpubertal involution (age involution). From the 18th year onwards thymic tissue undergoes a process of involution to become the *thymic remnant* (still functional) and is partly replaced by adipose tissue **C6** *(adipose body of the thymus, lipomatous atrophy).*

Accidental involution which may be due to infection, malnutrition, radiation etc., goes hand-in-hand with emigration of lymphocytes from the cortex and shrinkage of the entire organ. In infections, the number of Hassal's corpuscles increases considerably – *infection type,* and in malnutrition they decrease – *hunger type* of accidental involution. In cases of sudden death in children an abnormal enlargement of thymus is occasionally found (status thymolymphaticus, "thymic death") due to misdirected immunologic processes.

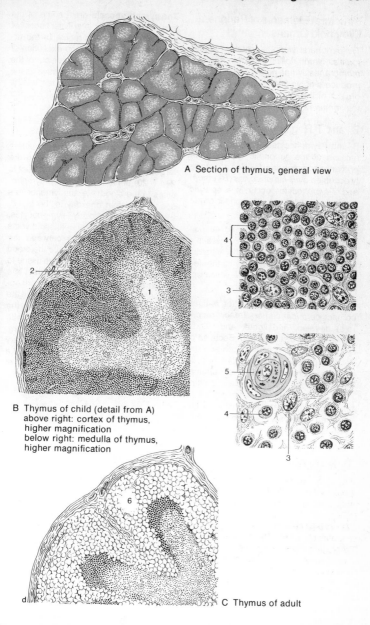

A Section of thymus, general view

B Thymus of child (detail from A)
above right: cortex of thymus,
higher magnification
below right: medulla of thymus,
higher magnification

C Thymus of adult

Structural Elements of Peripheral Lymphoid Organs

The peripheral lymphoid organs, *thymus, tonsils, lymph follicles of the mucous membranes, lymph nodes* and *spleen* as producers of lymphocytes are important organs of the specific defense system, the *immune system*.

B- and T-Regions

B- and T-lymphocytes occure in different regions of the lymphatic tissues. The primary and secondary follicles are mostly occupied by *B-lymphocytes (B-region)* and *T-lymphocytes* occur in a specific region of each individual lymphatic organ *(T-region)*.

In the *primary follicle,* prior to antigen contact (e. g. in the newborn), there is a homogenous mass of differentiated, small, lymphocytes, evenly distributed even the entire lymph follicle which appears darkly stained throughout.

In the *secondary follicle,* after antigen contact *germinal centers* (reaction centers) develop, and lighter and darker zones can be distinguished. Because of antigen contact germinal centers develop in phases. The two types of cells which are characteristic of a germinal center, appear, *centroblasts* and *centrocytes,* intermingted with *lymphocytes.* In the germinal centers, in addition to the activated cell types, there are B-lymphocytes and T-Lymphocytes (as T-helper cells).

Types of Reticulum Cells

Four types of reticulum cells, histiocytes, fibroblastic, dendritic and interdigitating reticulum cells can be distinguished in the lymphatic organs. *Histiocytes* and *fibroblastic* cells are found in all parts of lymphatic tissues. The *dendritic* and *interdigitating* cells associated with the B- and T-regions as *accessory cells of the immune system,* control the further differentiation of B- and T-lymphocytes, respectively.

The lymph vessels drain some of the tissue fluid from the "interstitial paces" of the connective tissue region of the organs and tissues (with the exception of brain tissue). The fluid passes out of the blood vessels into the connective tissue during transport of metabolites and is returned to the venous blood via the lymph vessels.

Recirculation of Lymphocytes

The mass of lymphocytes which enters the blood stream in 24 h. through the largest lymph trunk, the thoracic duct is 2–4 (–20) times the number of lymphocytes in the blood; i. e. the lymphocytes spend less than one day in the blood. Lymphocytes, especially T-lymphocytes, leave the post-capillary epithelioid veins of the lymphatic organs between the endothelial cells, and so leave the blood to enter the reticular connective tissue to be returned later via the lymph tract into the blood – *lymphocyte recirculation.* About 98 % of all lymphocytes are held in the interstitial connective tissue and in the reticular tissue of lymphatic organs, with only 2 % 'under way' in the blood stream. *"Epitheloid venules"* – post capillary veins whose endothelial cells look like cuboid epithelial cells are the site of lymphocyte recirculation.

Two arms of the Immune System (after *Müller-Hermelink, Würzburg*)

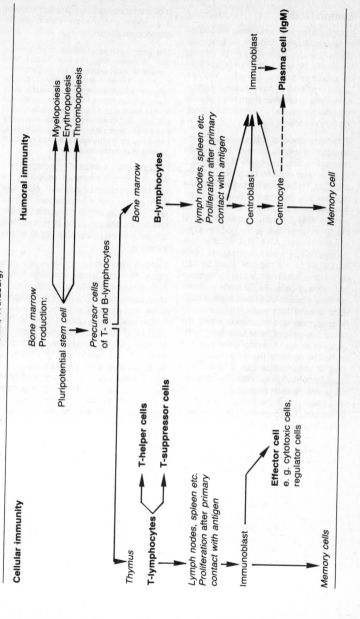

Cellular immunity

Thymus

T-lymphocytes → **T-helper cells**
→ **T-suppressor cells**

Lymph nodes, spleen etc.
Proliferation after primary
contact with antigen

Immunoblast → **Effector cell**
e. g. cytotoxic cells,
regulator cells

Immunoblast → *Memory cells*

Humoral immunity

Bone marrow
Production:

Pluripotential stem cell → Myelopoiesis
→ Erythropoiesis
→ Thrombopoiesis

Precursor cells
of T- and B-lymphocytes

Bone marrow

B-lymphocytes

lymph nodes, spleen etc.
Proliferation after primary
contact with antigen

Centroblast → Centrocyte

Immunoblast → **Plasma cell (IgM)**

Centrocyte → *Memory cell*

Lymph Nodes

Lymph nodes are found along the course of lymph vessels and act as biological filters. Lymph nodes near to organs which are the first to receive lymph from an organ or from a limited region are called *regional lymph nodes*. Lymph nodes which receive lymph from several regional lymph nodes are called *collecting lymph nodes*.

A **lymph node** or gland (*nodus lymphaticus*) is usually covered by a fibrous connective tissue capsule **B1** so that it forms a bean-shaped body about 1 mm in length. Trabeculae **B2** extend from the capsule into the interior. Several lymph vessels carry lymph to the node on its convex side (*afferent vessels AB3*). At the hilum lymph leaves the node in a few *efferent vessels AB4*. The reticular connective tissue forms a *marginal sinus* **BCE5** below the capsule, lined by flattened cells, which is transversed by solitary reticular cells. Radial sinuses (*intermediate sinuses* **BC6**) lead into the central sinuses of the medulla. There are both lymphocytes and macrophages in the sinus. The cells of the sinus belong to the phagocytic *reticulo-endothelial* system of cells. The *cortex* of the lymph node consists of lymph follicles **BC7**, and **D**, the *medulla* of medullary cords. The lymphocytes are densely packed in the cortex and usually form the secondary follicles (*B-region*). In the *medulla* they are less densely arranged in the cords of the reticular connective tissue (medullary cords). Between the cortical follicles and the medullary cords there is a paracortical zone which mainly contains T-lymphocytes (*T-region*). With increasing age the number of lymphocytes decreases, the reticular connective tissue is reduced and both are replaced by fat and connective tissue. At the hilum blood vessels enter the parenchyma, subdivide within the trabeculae and finally pass to the cortical follicles surrounded by lymphatic tissue. The cortical follicles are highly vascularised by circular vessels. In **C** an oval encloses the B-lymphocyte zone of a segment and a circle shows the T-lymphocyte zone.

The *lymph* is in contact with lymph node tissue over a wide area. Foreign bodies and pathogenic microorganisms are phagocytosed by macrophages (*"biological filter"*). Cancer cells may establish secondary deposits (metastases) in lymph nodes. Inflammatory processes lead to swelling and pain through distension of the capsule. Lymphocytes enter the blood stream with lymph from the vas efferens **AB4**. Lymph nodes are the most important *source of lymphocytes* during postfetal life.

"Epitheloid venules", postcapillary veins, whose endothelial cells take on the form of cuboidal epithelial cells, are the region in which *lymphocytes recirculate*. Epithelial venules are the characteristic structures of the lymph nodes, tonsils and the lymphatic tissue of the intestine.

Lymph. According to investigations done on cats, lymph distal to lymph nodes can be virtually devoid of cells or it may contain from 200 to 2000 lymphocytes per mm^3. When lymph leaves a lymph node it contains 1500–150 000 cells mm^3. Whilst the efferent lymph of non-stimulated lymph nodes contain only about 1% of newly formed lymphocytes, 60 hours after antigenic stimulation of a lymph node almost 100% of lymphocytes are newly formed. Thus' as a consequence of the immune response provoked by the antigen, the efferent lymph carries specifically sensitised lymphocytes and plasma cell precursors into distant, nonregional lymph nodes.

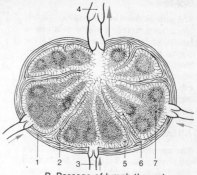

B Passage of lymph through
lymph gland (schematic)

A Lymph node

D Lymph follicle (magnification of
upper square in C)

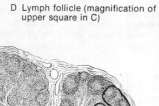

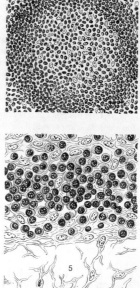

C Section through lymph gland

E Cortical sinus (magnification of
lower square in C)

Spleen

The **spleen** *(lien)* is interpolated into the *blood stream*. It is the size of a fist, is bluish-red in color, soft and shaped like a coffee bean. The spleen is 10–12 cm long, 6–8 cm wide, 3–4 cm thick and weighs 150–200 g. The organ lies at the back in the left upper abdomen below the diaphragm, at the level of the 9th–11th rib, with its longitudinal axis parallel to the 10th rib **A1. A2** Lower border of the lung, **A3** lower border of the pleura.

The upper *diaphragmatic surface, facies diaphragmatica,* **B** is convex, the *faceted lower surface, facies visceralis,* is concave and directed toward the viscera. The anterior superior margin of the spleen is narrow and has indentations: the posterior, inferior border is broad and blunt. The posterior (upper) pole of the spleen reaches to within 2 cm of the transverse of the 10th thoracic vertebra, the anterior (lower) pole to the medial axillary line. It is difficult to palpate. The spleen is mainly held in position by the *phrenicocolic ligament* which extends from the left colic (splenic) flexure to the lateral wall of the trunk. It forms the floor of the space in which the spleen lies.

The *hilum* is the point of entry and exit of the vessels and nerves; it is situated at the visceral surface **C**. It is narrow and elongated and extends along the *hilar groove.* The visceral surface of the spleen **D4** is situated behind the hilum and touches the left kidney **D5**, the surface in front of the hilum touches the stomach **D6**, the tail of the pancreas **D7** and the left colic flexure. **D8** Liver. The spleen is covered by the peritoneum and lies intraperitoneally. The *gastrosplenic ligament* **CD9** stretches from the hilum of the spleen to the greater curvature of the stomach **D6**. The *phrenicolienal ligament* **CD10** goes to the posterior wall of the trunk. The splenic recess of the *bursa omentalis* (arrow) extends to this ligament (see p. 242). The short gastric artery and vein and the left gastroepiploic artery run in the gastrosplenic ligament. The phrenicolienal ligament is shorter than the gastrosplenic ligament and contains the splenic artery and vein **C11**. It extends to the diaphragm. The spleen is displaced during breathing. If the body lies on its right side the spleen can follow the traction of the abdominal viscera and move a little lower down and forward.

Accessory spleens are derived from scattered splenic anlagen. They are either solitary or multiple and are the size of a pea to a chicken egg. They lie mainly beside the main spleen or along branches of the splenic artery, or they may appear at the greater curvature of the stomach, in the great omentum or elsewhere.

Blood vessels and nerves. The splenic artery (see p. 240) runs behind the upper border of the pancreas **D7** to the left (phrenicolienal ligament) and to the hilum of the spleen. It has a thick muscular wall and is often very tortuous. The first branches arise in the phrenicolienal ligament so that the artery enters the organ with 6 or more branches. The *splenic vein* is composed of several veins from the spleen and is one of the 3 great root veins of the portal vein, see p. 248. The *lymph vessels* come partly from the subserous connective tissue. *Lymph vessels* around the proximal parts of the central arteries and in the trabeculae show fluid flow in the opposite direction to the blood stream. *The lymph nodes* are: the pancreaticosplenic and celiac lymph nodes, see p. 80. The *nerves* contain viscerosensory and viscero- or vasomotor nerve fibers from the celiac plexus. The myofibroblasts of the trabeculae and the trabecular arteries (up to the divisions in the lymph tracts and the follicles) are supplied by adrenergic nerve fibers which contribute to the contractility of the trabecular-capsular system.

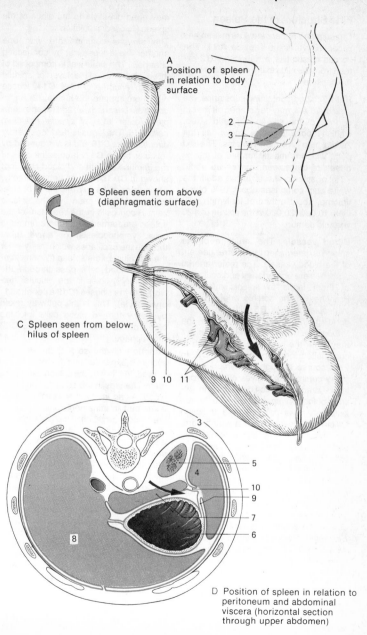

A Position of spleen in relation to body surface

B Spleen seen from above (diaphragmatic surface)

C Spleen seen from below: hilus of spleen

D Position of spleen in relation to peritoneum and abdominal viscera (horizontal section through upper abdomen)

Fine Structure of the Spleen

Capsule and trabeculae. The spleen has a connective tissue capsule **AB1**. The fibrous elastic tissue trabeculae **B2** extend from the hilus to the opposite area of the capsule.

Pulp. A blood-containing substance, the *"red pulp"* **A3**, may be scraped off the cut surface of the spleen. Pinhead sized, whitish spots, together known as the *"white pulp"* **A4** then remains. The red pulp, which is mostly formed of the sinuses of the spleen, makes up about 77% of splenic tissue. In the adult, the white pulp comprises about 19% of the volume, about a third of its lymph follicles, 10,000–20,000 in the spleen of a 20 year old person.

Blood vessels. The structure of the spleen is determined by the arrangement of the blood vessels. The major arterial branches are end arteries and occlusion of them leads to splenic infarction. Branches of the splenic artery run as *trabecular arteries* **B5** with the corresponding *trabecular veins* **B6** in the trabeculae **B2**. As central arteries **B7** they enter the lymphoid cords which at some sites form lymph follicles **B9**. In their course through the lymph cords, every central artery gives off numerous lateral *branches* which open into the network of the *marginal zone* or into the *red pulp*. At those points where a lymph follicle has developed in a lymph cord, the central artery is displaced laterally and so crosses the follicle eccentrically **B8**. The central artery ends in about 50 terminal pulp arterioles (penicilliary arteries) **B10**. The latter penetrate the neighboring red pulp and subdivide into capillaries. For a short stretch the capillaries are enveloped by a fusiform coat of densely packed macrophages, the *Schweiger-Seidel sleeve* **B11** – sleeve capillaries. The capillaries mostly drain through a *reticular cord* of the reticular connective tissue **B12** surrounding the sinus into the broad *splenic sinus* **B13** ("open circulation"). A few capillaries

may open directly into the sinus of the spleen ("closed circulation").

The *sinuses* communicate with one another like a loose net by connecting channels. *The sinus wall* is composed of three elements: (1) a layer of spindle shaped, *endothelial cells* **C14**, whose nuclei project into the lumen; (2) an *incomplete basal lamina*, and (3) an incomplete, outer layer of *special reticulum cells* **C15**. The sinus is held together by *circular fibers* **C16** and is surrounded by reticular tissue **C17**. It contains cells of the immune system and blood cells (not shown in **C**). The sinuses open directly, or via *veins of the pulp* into the *trabecular veins* and these drain into the splenic vein. Blood cells in the reticulum of the spleen are formed either there (immune cells, lymphopoiesis) or arrive there through the capillaries which freely enter the reticulum of the spleen ("open circulation"). Blood cells pass through the (temporary?) holes in the reticular network into the sinuses (**C18** a penetrating erythrocyte). This is the pathway taken by newly formed blood cells into the blood of the sinus. **C19** Mitosis, **C20** macrophage.

Function. In the spleen, the vascular system is expanded and the blood flow is slowed. In this way the spleen can cause many changes in the composition of the blood. As an *immune organ*, the spleen takes part in immunobiological processes. *Blood formation:* If the bone marrow becomes inadequate and in other disease situations, granulopoiesis and erythropoiesis restart in the spleen as they did temporarily during fetal development. *Storage.* The spleen can store discarded metabolic products and thereby become much increased in size (storage diseases). *Blood cell destruction:* An excess of hemoglobin iron from blood cell breakdown may be stored in the spleen and may be apparent microscopically *(hemosiderin)*, and in extreme cases macroscopically as brown staining of the organ *(hemosiderosis)*. In man the spleen does not function as a *blood storage organ*.

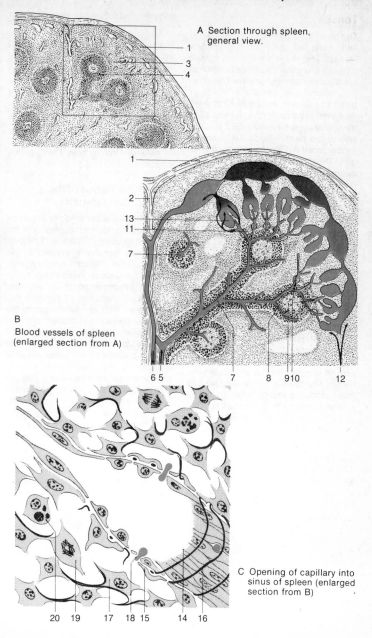

A Section through spleen, general view.

B
Blood vessels of spleen (enlarged section from A)

C Opening of capillary into sinus of spleen (enlarged section from B)

Tonsils

The tonsils surround the exits of the oral and nasal spaces into the pharynx (throat). They form a circle and have a protective function by early activation of specific defense mechanisms. They form the "pharyngeal circle" of Waldeyer which consists of the *unpaired pharyngeal tonsil* in the roof of the pharynx **AC1**, the paired *palatine tonsils* **AB2** on each side of the palatine arch **AB3** and the *lingual tonsil* **A4** in the retrolingual region. There is some additional lymphatic tissue in the lateral wall of the pharynx *(lateral cord)* which, at the entrance into the Eustachian tube, is condensed into the *tonsilla tubaria* **A5**. **A6** Entrance into the larynx.

The lymphoid tissue in the *tonsils* consists of *aggregated lymph follicles* **D7** lying directly beneath the covering epithelium, the surface of which is deeply fissured by *crypts* **D8** (invaginations). The lymphocytes immigrate into the intercellular gaps between the epithelial cells. In the crypts there are plugs of detritus from shed epithelial cells and leukocytes. The marginal wall of lymphocytes in the secondary follicles, which is thickened toward the epithelial surface, sits like a semilunar cap on the germinal center. The tonsil is demarcated from its surroundings (connective tissue, glands and muscles) by a taut fibrous capsule which allows it to be surgically enucleated. Its lymph vessels lead to regional lymph nodes.

The **pharyngeal tonsil AC1** protrudes from beneath the roof of the pharynx. It lies behind the choanae (see p. 114) and in early school age can block them during the period of most marked development of the immune system, which occurs at this age. The consequences of such blockage are interference with sleep, lack of concentration at school, breathing through the continuously open mouth, and malformation of the facial bones. The tonsil is covered by stratified ciliated epithelium. **C9** Sella turcica, **C10** soft palate.

The **palatine tonsils** *(tonsilla palatina)* **AB2** have 10–20 crypts and are covered by a stratified non-keratinized squamous epithelium.

The **lingual tonsil** *(tonsilla lingualis)* **A4** is flat, has shallow crypts and is covered by a stratified, non-keratinized squamous epithelium. Into the crypts empty mucous glands, the *lingual glands,* which lie between the muscle fibers of the tongue.

Lymphatic Tissue of the Mucous Membranes

Lymph follicles occur singly or in groups in mucous membranes, especially in the *intestinal mucous membrane.* **The gut-associated lymphatic system** in the stomach, small intestine and appendix mucosa **E** founds an independent organ complex which responds to ingested antigens. It consists of lymphocytes which are *diffusely distributed in the mucosal connective tissue,* and *intraepithelial lymphocytes* particularly from the *folliculi lymphatici aggregati,* Peyer's patches of the small intestine and appendix mucosae **E11**. Where the latter occur there are no villi or crypts. *M-cells* (membranous cells) particularly phagocytose antigens from the intestinal lumen and present them to the underlying lymphocytes and macrophages. B-lymphocytes which have been stimulated in this way migrate into other parts of the intestine and so spread defensive responses. **E12** Lumen of gut, **E13** circular muscle layer, **E14** longitudinal muscle layer.

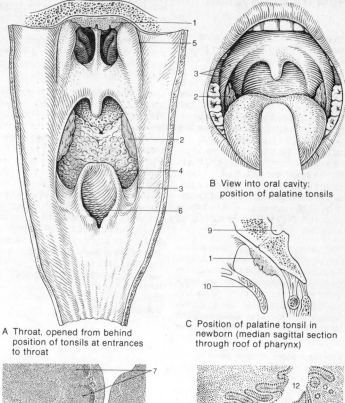

A Throat, opened from behind
position of tonsils at entrances
to throat

B View into oral cavity:
position of palatine tonsils

C Position of palatine tonsil in
newborn (median sagittal section
through roof of pharynx)

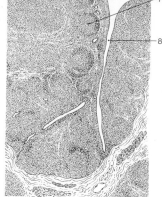

D Section through palatine tonsil

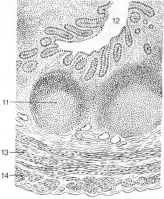

E Section through appendix

Respiratory System

Under *respiration* we understand the *transport of gases* to and from cells – accessible to anatomical examination – as well as the *biological processes of oxidation* which take place within cells with the help of oxygen, and which are studied by biochemical methods (physiological chemistry). In animals which breathe through lungs the gases are carried via the *respiratory tract,* the trachea and the bronchi, into the lungs. Here oxygen from the inhaled air diffuses into the blood and CO_2 from the blood into the exhaled air. The *inhaled air* contains about 20.93% O_2, 0.03% CO_2 and 79.04% N_2 and *exhaled air* contains about 16% O_2, 4% CO_2 and 80% N_2, as well as the noble gases.

The transport of gases between the lungs and the cells of organs and tissues is achieved via the *blood stream.* During the process of intracellular biological oxidation large molecular nutrients are broken down into smaller, energy-poorer metabolites and energy is released: The combustion of fats and carbohydrates into H_2O and CO_2.

The **respiratory organs** serve for the transport of gases. We distinguish respiratory organs through which air flows, i. e., the *nasal cavities, pharynx, larynx* and *trachea,* from the *lungs* which serve the immediate exchange of gases between air and blood. A large part of the respiratory tract (air passages), the *bronchi,* is within the lungs. The larynx serves also for the production of sound. The olfactory organ (the nose) controls the inhaled air, aids orientation in the environment and, together with the sensory nerves of the nasal mucosa, helps to protect the individual (see Sense Organs, Vol 3). One part of the air passages is situated in the head; the *upper respiratory tract:* the nose with the nasal cavities, the nasal sinuses and the pharynx (throat). From a clinical point of view the oral cavity, too, is often included with these structures. The other part is situated in the neck and in the trunk, the *lower respiratory tract:* the larynx, trachea, bronchi and the lungs. In all parts of the respiratory tract the inhaled air is prepared, i. e., it is cleaned in various ways, moistened and warmed.

Unlike other mammals, the larynx and trachea of adult humans do not immediately follow the nasal cavity, since they have a longer pharynx and the upright position has produced a right-angled bend with respect to the long axis of the trachea. The larynx and nasal cavity are separated by the middle (oral) part of the pharynx *(mesopharynx).* Since the mesopharynx provides a passage for the *respiratory tract* and the *alimentary canal,* which cross here in man, to permit the ingestion of food **1** the respiratory tract **2** must be closed for short periods for swallowing (see p. 204). **3** Trachea. In the infant, whose pharynx is not yet stretched, breathing and ingestion of food can proceed simultaneously. **4** Hilus of lung: *red* = pulmonary veins, *blue* = pulmonary arteries.

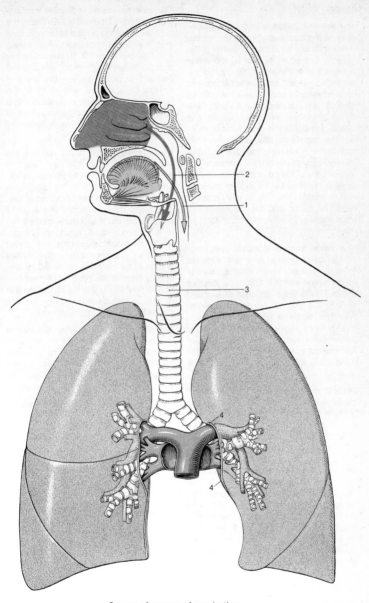

Survey of organs of respiration

Nose

The **external nose** is thrown into relief against the lips and the cheeks by the *nasolabial fold* **A1**. The framework of the nose is formed at its *root* by the *nasal bone* **A2** and toward its tip by hyaline cartilages, which are moveable against each other and which keep the nostrils open. The *nasal cartilages* **A3** stiffen the tip of the nose with their middle and lateral crura on each side. A facial muscle, the nasalis muscle, arises from the root of the nose, and together with the levator labii superioris, it controls the movements of the nose (see Vol. 1). The nose is covered by the skin of the face and it contains sebaceous glands in the region of the tip and the nasolabial fold (comedones, blackheads, may form in them).

The **nasal septum** consists of a bony and a cartilaginous part and extends from the nasal cavity into the external nostril where it ends in fibrous connective tissue. Its cartilage consists of a *septal lamella* **B4** and paired *dorsal lamellae* **A5**. The cartilage is interpolated posteriorly and above between the bony parts of the septum: the *perpendicular plate* (lamina perpendicularis) of the ethmoid bone **B6** and the *vomer* **B7**. **B8** Skin.

On both sides of the border between the cartilaginous and the bony part of the septum the vomer forms a ledge, the septal crest *(crista septi)*. Foreign bodies which are introduced into the nose can slide along this crest upward into the back of the nasal cavity (N. B. risk of perforation of the ethmoid bone in children!) The septal crest has a highly vascularized thickening of the mucosa, *Kiesselbach's point,* which bleeds easily in nasal injuries. Anteriorly in the septal mucosa on both sides of the septum there is a *cavernous body* which can narrow the atrium (vestibule) at the level of the middle meatus. The septum of the nose is frequently deflected to one side or the other – *septal deviation.*

The **external nares (nostrils)** lead into the *vestibule* of the nose. This lies in the mobile part of the nose and is separated from the nasal cavity by a curved elevation, the *limen nasi* **B9**. The vestibule has a circle of hairs, *vibrissae,* which curve outward to guard against the entry of foreign bodies. The epithelium of the facial skin merges in the vestibule into a stratified, squamous epithelium. Further up, at the limen nasi, it is replaced by columnar epithelium.

Nasal cavity. The *nasal cavities* are separated from each other by the *septum* and from the oral cavity by the *palate.* The lateral wall starts at the floor of the nasal cavity 10–15 mm from the septum. The nasal cavity narrows like a gable towards the top. Each nasal cavity ends in a posterior nasal aperture, *choana,* which communicates with the throat (pharynx). The floor of the nasal cavities is formed by the *hard palate* **BC10** (anteriorly the palatine process of the maxilla and posteriorly the horizontal lamella of the os palatinum) and by the *soft palate* **BC11** *(velvum palatinum).*

The *roof of the nasal cavity* is a groove which slopes forward under the bridge of the nose towards the nostrils, and posteriorly along the anterior wall of the sphenoid bone **BC12** in the *sphenoethmoidal recess* toward the choanae.

The *lateral wall* is enlarged by 3 elevations, the *conchae* **C13–15** (see p. 110), **BC16** frontal sinus, **BC12** sphenoidal sinus, **C17** pharyngeal orifice of the Eustachian tube.

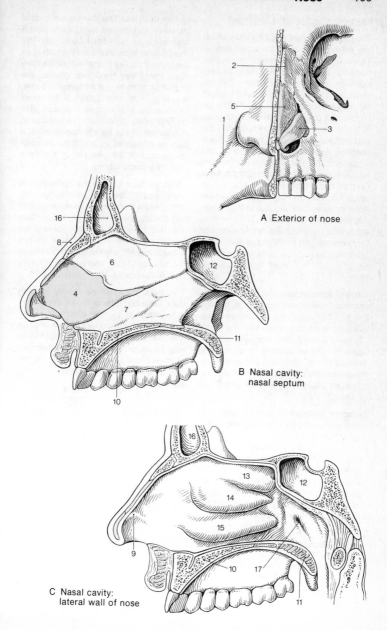

A Exterior of nose

B Nasal cavity:
nasal septum

C Nasal cavity:
lateral wall of nose

Nasal Conchae and Meatus I

The **nasal conchae** are thin bones which are covered by a mucous membrane. There is a nasal passage or meatus beneath each of the *three conchae, superior, middle and inferior.* The inferior, the longest concha **A1,** consists of an independent bone **B2.** Into the *inferior meatus* **A3** opens the *nasolacrimal duct,* which collects the lacrimal fluid from the conjunctival sac of the eye and via two ductlets, *lacrimal canals,* collects and empties off the fluid through the *lacrimal sac.* The bone of the middle concha **A4** is part of the ethmoid bone. The *frontal sinus* **A6,** the *maxillary sinus* **A7** and the *anterior ethmoidal sinuses* **A8** all open into the *middle meatus* **A5.** The *posterior ethmoidal sinuses* **A9** drain through 1 or 2 openings into the *superior nasal meatus.* The superior concha, the smallest of the 3, is formed by the ethmoid bone.

Clinical tips: Because the opening of the maxillary sinus into the middle meatus lies just below its roof, discharge of secretions from the sinus into the nasal cavity is greatly restricted. An artificial passage can be created by perforating the lateral wall of the nasal cavity in the inferior meatus **A10. A11** Access to the maxillary sinus by perforation of the root of a tooth.

Mucous membrane of the nose. We distinguish 2 sections: 1) The *respiratory region* covers the inferior and the middle concha, the corresponding area of the septum and the floor of the nasal cavity. It serves to condition inhaled air. 2) *The olfactory region* covers the superior concha, the part of the septum opposite it and the roof of the nasal cavity **A12.** It contains the olfactory organ.

Respiratory region. The mucosa consists of a *ciliated columnar epithelium* **B13, C14.** Its cilia beat toward the pharynx and distribute the mucus, which is produced by the goblet cells **C15** and by small *nasal glands* **B16** over the surface of the mucosa. The mucus traps dust particles and carries them away and it also helps to humidify inhaled air. The mucosa contains veins **B17** which form *cavernous bodies* in the wall of the conchae and, if they become engorged, the mucosa can swell to a thickness of 5 mm, resulting in an obstruction of the nasal cavity and a 'stuffy' nose. The veins give off heat to the inhaled air. The undulating movement of the cilia is coordinated. It can be observed in the living mucosa (e. g., roof of the pharynx of the frog) under the microscope.

Olfactory region. Here the epithelium is composed of sensory and supporting cells. Below the mucosa lie serous *olfactory glands,* the olfactory glands of Bowman and the non-myelinated fibers of the *olfactory nerves,* bundles of which run through openings in the lamina cribosa of the ethmoid bone toward the *olfactory bulb of the brain* **A18.**

The *air stream* passes mainly through the middle and inferior meati. On its way it is sampled by the olfactory organ. Sniffing produces currents which whirl air into the olfactory region and prolong its stay inside the nasal cavity. Deformities of the walls of the nasal cavity influence the air-current conditions. **A19** Orbit, **A20** nasal septum.

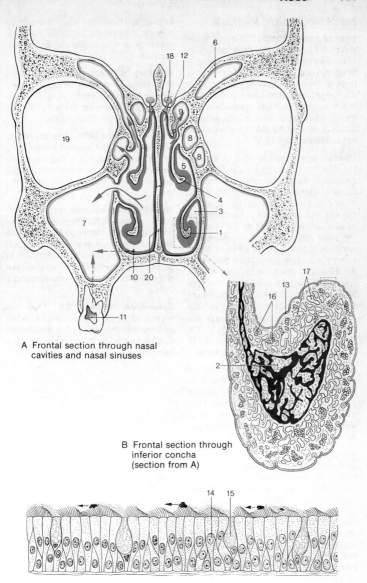

A Frontal section through nasal
cavities and nasal sinuses

B Frontal section through
inferior concha
(section from A)

C Epithelium of respiratory part of nasal mucosa (section from B)

Nasal Sinuses and Meatus II

Inferior nasal meatus. The *nasolacrimal canal* **A1** opens into the anterior part of the inferior nasal meatus. If the head is bent forward lacrimal fluid runs out through the external nares (nostrils), and if it is bent backward it runs out through the choanae. The pharyngeal opening of the auditory tube *(ostium pharyngeum tubae auditivae)* **A2** lies in the extension of the inferior concha.

Middle nasal meatus. After removal of the middle concha in **A** a curved gap, the *semilunar hiatus,* becomes visible. Its upper border is formed by the *ethmoidal bulla* **A3**, a rounded elevation open on the top *(the opening of the anterior ethmoidal sinuses).* A fold of the mucosa, the *plica semilunaris,* in which lies the uncinate process of the ethmoid bone, forms the lower border. The *frontal sinus* **AB5** leads into the anterior upper end **A4** and the *maxillary sinus* **A6** into the posterior lower end of the hiatus. About 1 cm behind the middle concha in the sphenoethmoidal recess beneath the mucous membrane of the sphenopalatine foramen, there is an opening into the pterygopalatine fossa.

Superior nasal meatus. It bears 1–2 openings of the *posterior ethmoidal sinuses.* The *sphenoethmoidal recess* runs between the posterior border of the superior concha and the anterior wall of the sphenoid bone. The *sphenoidal sinus* **AB8** enters it from behind **A7**.

Nasal sinuses. The sinuses of the nose *(paranasal sinuses)* mainly develop after birth. In the newborn child the frontal and maxillary sinuses are present as recesses which only start to grow slightly towards the end of the first year. They reach about half their final size at about 10 years, and become fully developed with the lengthening of the face between the 15th and 20th year. The nasal sinuses are lined by the nasal mucosa, which is invaginated where they open into the nose. The air sinuses warm the inhaled air. Infections which are difficult to get at may persist in the sinuses. The following sinuses should be distinguished.

Maxillary sinuses (Highmor's antra) **B9**. They extend below the orbit and their floor is separated from the roots of the molars and premolar teeth by a bony plate only a few millimeters thick.

Frontal sinuses AB5. The degree to which they extend into the frontal bone varies greatly and usually they are asymmetrical. In their neighborhood lie the anterior cranial fossa and the roof of the orbit.

The **ethmoidal sinuses** consist of *ethmoidal air cells* **B10**. They develop at an early age. They border on the orbits, the anterior cranial fossa and the nasal cavities and are separated from them in places by paper-thin bony lamellae. We distinguish anterior, middle and posterior cells.

Sphenoidal sinuses B8. They are separated by a sagittal septum, often incomplete. The roof of the sphenoidal sinus is formed by the sella turcica at the base of the skull.

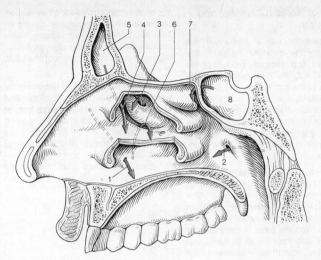

A Openings of nasal sinuses
and nasolacrimal canal

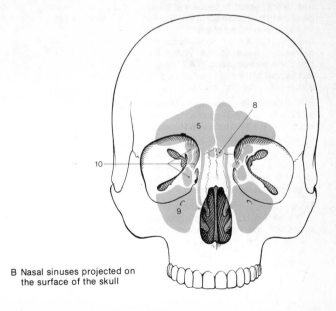

B Nasal sinuses projected on
the surface of the skull

Posterior Nares (Choanae) and Soft Palate

The **posterior nares** *(choanae)* are framed by the base of the skull, the vomer **A1,** the pterygoid process of the sphenoid **A2,** the hard palate and the muscles of the soft palate. The orifice of the *Eustachian tube* **A3** points toward the choanae, hence the possibility of spread of infected mucus from the nose through the Eustachian tube into the middle ear if the nose is blown under pressure.

Soft palate *(velum palatinum).* This consists of a sheet of muscle and tendon covered by mucous membrane. The muscles are the *levator veli palatinum* **A4** and the *tensor veli palatini* **A5,** which originate from the base of the skull and the latter also from the tip of the Eustachian tube. Both these muscles radiate into the aponeurosis of the palate. The levator elevates and the tensor stretches the soft palate in its course around the pterygoid hamulus **A2.** The ostium of the tube is dilated by contraction of the muscles, which permits adjustment of the air pressure in the middle ear to the external air pressure by the act of swallowing. The *uvular muscle* **A6** runs sagittally into the connective tissue of the uvula. The *palatopharyngeus muscle* **A7** passes into the wall of the pharynx (see Deglutition, act of swallowing, p. 204).

In the infant the choanae can be blocked by an enlarged *pharyngeal tonsil* **AB8,** see p. 104. **B9** Sella turcica, **B10** dens of the axis, **B11** soft palate.

Rhinoscopy. By use of a laryngeal mirror the anterior part of the nasal cavity can be examined through the choanae *(anterior and posterior rhinoscopy).*

Blood vessels and nerves. The arteries: 1) The terminal branches of the *maxillary artery,* viz. the posterior lateral nasal arteries and the arteries of the septum enter through the foramen sphenopalatinum into the nasal cavity; 2)

the anterior and posterior ethmoidal arteries stem from the *internal carotid artery,* see p. 58. The *veins* communicate with the orbital veins, the pterygoid plexus and the venous sinuses of the skull, see p. 64.

The *lymphatics* drain into lymph nodes which are situated in front of the 2nd cervical vertebra, on the greater horn of the hyoid bone and near the parotid gland. *Nerves:* Sensory fibers from the 1st and 2nd branch of the trigeminal nerve. Secretomotor postganglionic parasympathetic fibers from the carotid plexus. The olfactory nerves are also involved.

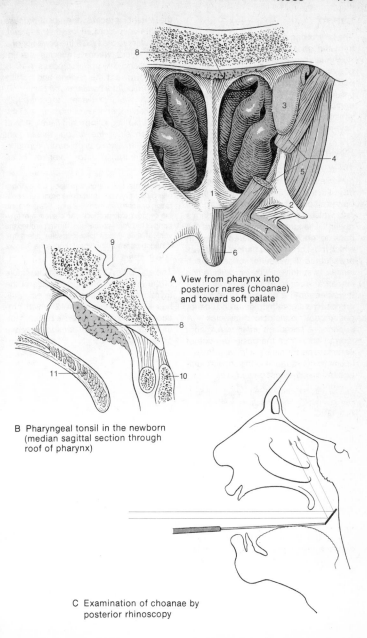

A View from pharynx into
   posterior nares (choanae)
   and toward soft palate

B Pharyngeal tonsil in the newborn
   (median sagittal section through
   roof of pharynx)

C Examination of choanae by
   posterior rhinoscopy

Larynx

The **larynx** can *shut off* the lower airways from the pharynx (coughing, vomiting). All the laryngeal muscles except for the posterior cricoarytenoideus muscle take part in the closure. In addition, the larynx produces *tones*. The larynx consists of a *cartilaginous skeleton* with *ligaments*, which carries the *muscles* and is mostly covered by *mucous membrane*.

Skeleton of the Larynx

The *thyroid cartilage* **A1** consists of two quadrilateral laminae which are fused in front like the bow of a ship. At the tip of the bow is a notch, the *superior thyroid notch* **A2,** which can be felt as the "Adam's apple". The laminae diverge toward the back to enclose a sheltered space in which most of the other parts of the larynx are situated. The anterior surface of each lamina is subdivided by an *oblique line* into an anterior facet for the origin of the thyroideus and a posterior facet for the origin of the sternothyroideus and inferior pharyngeal constrictor muscles. A *superior* **A3** and an *inferior* **A4** *horn* (cornu) arise from the posterior edge of each lamina forming a joint with the articular (outer) surface of the cricoid cartilage (arrow), the *cricothyroid joint*.

Movements: The thyroid and cricoid cartilages can tilt around a transverse axis: *see arrows in* **A**, p. 119.

The *cricoid cartilage* **A5** is shaped like a signet ring. Its *lamina* **A6** is 2–2.5 cm high and lies at the back. The *upper edge of the lamina* has *two articular surfaces* for the arytenoid cartilages **A7**, the lateral surface on each side has an *articular surface* for the inferior horn (cornu) of the thyroid cartilage.

Both *arytenoid cartilages* **A7** sit on the upper edge of the lamina of the cricoid cartilage. Each has the shape of a triangular pyramid. There are 3 lateral surfaces, medial, dorsal and lateral, a basal articular surface and 3 processes. From the anterior process, the *vocal process* **B8,** the vocal cord arises; on the lateral, *muscular process* **AC9** the posterior and lateral cricoarytenoid muscles are inserted; and the upper process **A10** is inclined toward the midline and carries the *corniculate cartilage (of Santorini)*. The arytenoid cartilages, together with the rolled upper edge of the lamina of the cricoid cartilage, form a cylindrical joint on each side, *the cricoarytenoid joint* (arrow). Its axis runs obliquely from dorso-medio-cranial to ventro-lateral-caudal.

Movements: Lateral gliding along the cylinder axis by the arytenoid cartilages produces separation or approximation by 2 mm. Tilting around the cylinder axis produces an elevation or lowering of the vocal processes; turning around the vertical axis (loose joint capsule) alters the distance between the vocal processes (arrows in **C**, p. 119).

The *epiglottis* **A11** lies against the middle of the internal (posterior) surface of the thyroid cartilage (arrow). The epiglottis has a stem *(petiolus)* **A12** which produces the *epiglottic tubercle* beneath the mucosa, and an oval *lamina,* concave posteriorly.

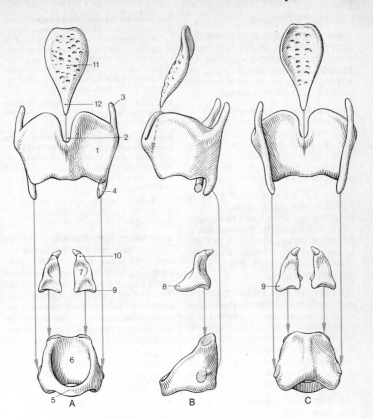

Laryngeal cartilages separated

Laryngeal Ligaments

These may be divided into *inner* and *outer laryngeal ligaments*. The inner ones join the different parts of the laryngeal skeleton together, and the outer ligaments fix the larynx between the hyoid bone and the trachea.

Inner laryngeal ligaments. The *fibroelastic membrane of the larynx* is the membrane composed of a dense elastic fiber net which lies below the mucous membrane of the larynx and corresponds to the tela submucosa. It varies in thickness in the three areas of the larynx.

In the region of the *cavum infraglotticum* it is called the *conus elasticus* (see p. 122). This is a short tube which begins with a round lumen on the inner side of the cricoid cartilage and ends in a slit-shaped, sagitally orientated fissure beneath the mucous membranes of the left and right vocal cords.

The *vocal cords* are the two, thickened, upper ends of the conus elasticus. They are attached posteriorly to the vocal processes of both arytenoid cartilages and anteriorly to the inner surface of the angle of the thyroid cartilage. The conus elasticus and the two vocal cords form the elastic wall of the tone producing mechanism („lip pipe") in the larynx.

The *cricothyroid ligament* **A1** is a firm fiber band in the conus elasticus which runs anteriorly in the midline between the cricoid cartilage and the lower border of the thyroid cartilage.

Clinical tips: As the cricothyroid ligament lies below the glottis, in life-threatening obstruction of the glottis, e. g., swelling of the mucous membrane, or glottal oedema, the air passage may be opened artificially by a cut or puncture through the ligament, *coniotomy*.

The *quadrangular membrane* is the poorly formed part of the fibroelastic membrane of the larynx, which lies beneath the mucous membrane of vestibulum of the larynx.

The *vestibular ligament ("false vocal cord"*, see p. 121) is the lower, free margin of the quadrangular membrane. The band-like marginal zone of the membrane lies in the vestibular fold and is attached on both sides to the anterior surface of the arytenoid cartilage and the inner surface of the bow of the thyroid cartilage above the insertion of the vocal cord.

The elastic *posterior cricoarytenoid ligament* medially strengthens the lax joint capsule of the arytenoid-cricoid joint.

The *cricopharyngeal ligament* extends from the corniculate cartilage to the posterior surface of the lamina of the cricoid cartilage and more fiber tracts attach it to the connective tissue of the mucous membrane of the oesophagus, which lies behind it.

The *thyroepiglottic ligament* **C2** attaches the petiolus of the epiglottis to the inner surface of the bow of the thyroid cartilage.

External laryngeal ligaments. The *thyrohyoid membrane* **ABC3** stretches between the upper margin of the thyroid cartilage and the hyoid bone. Thickened fiber tracts of the membrane are called the *median thyrohyoid ligament* (between the superior thyroid notch and the body of the hyoid bone **AB4**) and the *thyrohyoid ligament* **BC5** (between the upper horn of the thyroid cartilage and the posterior end of the greater horn of the hyoid bone). A small cartilage, the cartilage triticea, **BC6** lies in the thyrohyoid ligament.

ABC7 Epiglottic cartilage, **ABC8** lesser horn of the hyoid bone, **ABC9** greater horn of the hyoid bone.

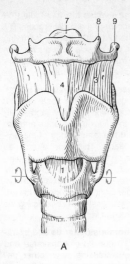

A

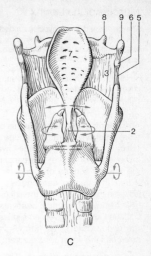

C

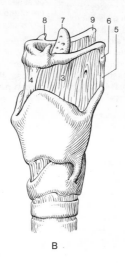

B

Laryngeal cartilages joined by ligaments

Muscles of the Larynx

The *upper* and *lower hyoideus muscles* (see Vol. 1) elevate, lower and tilt the larynx and fix it in position. Quite distinct from them are the *laryngeal muscles* which move various parts of the larynx against each other. According to position and origin they are divided into *external* cricothyroid muscles in front of the cricoid cartilage and *internal* laryngeal muscles inside the thyroid cartilage.

The **cricothyroideus A1** arises on each side anteriorly from the bridge of the cricoid cartilage and extends toward the lower edge of the thyroid cartilage and the interior surface of the lower cornu. When the thyroid cartilage is fixed in position this muscle tilts the cricoid cartilage backward around a transverse axis, because of the movements possible in the cricothyroid joints (arrows in **A**). *It stretches the vocal cord,* diagram in **A**.

The **cricoarytenoideus posterior BCD2** (called 'posticus' clinically) arises on both sides from the dorsal surface of the cricoid plate and runs laterally and upwards to the muscular process of the arytenoid cartilage **B3**. It pulls the muscular process backward and thus displaces the vocal cords sideways. The movement is mainly around the longitudinal axis with a slight tilt toward the back. *It widens the rima glottidis* and is in fact its only dilator: diagram in **B**. **B4** Vocal cord = upper margin of the elastic cone.

The **cricoarytenoideus lateralis BD5** (called 'lateralis') arises from the upper margin and outer surface of the side of the cricoid cartilage. It, too, runs upward and backward to the muscular process of the arytenoid cartilage. It is the antagonist of the cricoarytenoideus posterior because it moves the muscular process forward and so pulls the vocal cords toward the middle. *It narrows the rima glottidis:* movement mainly around the longitudinal axis and slight backward tilting: schema in **B**.

The **vocalis B6** is paired. It arises from the posterior surface of the thyroid carti-

lage and runs within the vocal fold (the true vocal cord) to the vocal process **B7** of the arytenoid cartilage. By isometric contraction the muscle *controls the tension of the true vocal cord,* thereby affecting the pitch at which it vibrates.

The **thyroarytenoideus CD8** is paired. It arises together with the vocal muscle, which it continues laterally, and is attached by thin fiber tracts to the anterior surface and the lateral edge of the arytenoid cartilage. This muscle acts as *constrictor of the rima glottidis* (movement mainly around the longitudinal axis). The *thyroepiglottic muscle* **D9** consists of fiber bundles extending toward the epiglottis; *it assists in constricting the entrance into the larynx.*

The **arytenoideus obliquus** and **transversus C10** consist of transverse and dorsal fibers crossing each other between the posterior surfaces of the arytenoid cartilages. They approximate the arytenoid cartilages toward each other by a sliding movement along the upper rim of the cricoid plate (diagram **C**) and so aid *closure of the rima glottidis.*

The **aryepiglotticus D11** continues the course of the arytenoideus obliquus toward the epiglottis. It runs in a fold of the mucosa, the aryepiglottic fold, and assists *in narrowing the entrance into the larynx;* diagram in **D**.

The *aryepiglottic fold* envelops the *corniculate cartilage* **CD12** (of *Santorini*) and the inconstant *cuneiform cartilage* **CD13** (of *Wrisberg*). Both are visible through the mucosa as a yellowish sheen.

Laryngeal muscles
and their functions

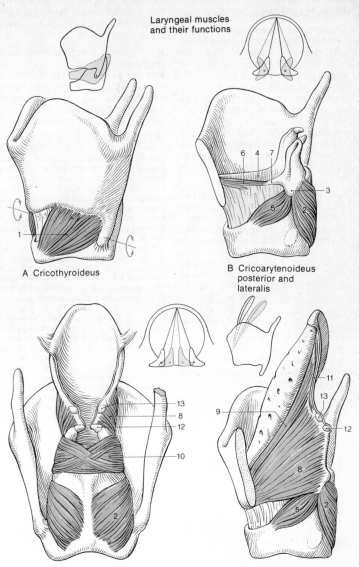

A Cricothyroideus

B Cricoarytenoideus
posterior and
lateralis

C Laryngeal muscles from back.
Action of arytenoideus transversus and obliquus

D Laryngeal muscles seen from side
(lamina of thyroid cartilage
removed). Action of thyro-
epiglotticus and aryepiglotticus

Mucous Membrane of the Larynx

The framework, ligaments and muscles of the larynx are largely covered by *mucous membrane*. It extends from the root of the tongue to the upper anterior side of the epiglottis and forms *3 folds* and *2 grooves* on its way between them. The *plica aryepiglottica* (the aryepiglottic fold) runs from the lateral margin of the epiglottis on both sides to the tips of the arytenoid cartilages. It embraces the oval entrance into the vestibule of the larynx. A reserve fold in between the arytenoid cartilages permits their lateral movement.

Lateral to the *aryepiglottic fold* and medial to the *thyroid cartilage* **AC1** on both sides is a groove in the mucous membrane, the **piriform recess AC2,** which diverts food away from the entrance into the larynx and into the esophagus, arrow in **AC**. Another mucosal fold, the *plica laryngea* runs across the piriform recess. It is caused by the underlying superior laryngeal blood vessels and superior laryngeal nerve, which pierce the thyrohyoid membrane and run to the internal laryngeal muscles.

The space below the entrance to the larynx is subdivided into *sections by 2 pairs of sagittal folds* **AC3, 4** lying above each other.

Vestibule of the larynx. The vestibule of the larynx extends to the upper pair of the folds, the *ventricular folds* **AC3** (false vocal cords). The 4–5 cm long anterior wall of the *vestibule* is formed by the epiglottis **C5,** whose stem protrudes as a tubercle. In the submucosa of the vestibular folds are some muscle fibrils and the weak, elastic, quadrangular membrane.

Laryngeal ventricle. The space between the *vestibular folds* and the lower pair of folds, the **vocal folds** *(true vocal folds)* **AC4** widens into the 1 cm high *laryngeal ventricle* **AB6.** The laryngeal saccule, an extension of it, may reach as far up as the hyoid bone.

The widened space below the vocal cords is called the **infraglottic cavity A7.**

A8 Thyrohyoid muscle, **A9** thyroepiglottic muscle, **A10** sternothyroid muscle, **A11** vocalis muscle, **A12** cricothyroid muscle, **AC13** cricoid cartilage, **AC14** first tracheal cartilage, **AC15** hyoid bone, **C16** root of the tongue, **C17** thyrohyoid ligament, **C18** cricothyroid ligament.

Fine structure of the larynx. The framework of the larynx consists of hyaline cartilage with the following exceptions: the epiglottic, corniculate, and cuneiform cartilages and the vocal process of the arytenoid cartilage, all of which consist of elastic cartilage. From the second decade of life the hyaline cartilages of the larynx may ossify to a varying degree. The mucosa on the upper surface of the epiglottis is covered by stratified squamous epithelium, and on the lower surface stratified, ciliated, columnar epithelium on which seromucous glands open. A two-row (stratified) *ciliated columnar epithelium* **B19** starts in the vestibule of the larynx and extends into the bronchioles. The *true vocal cords* are covered by stratified, partly cornified *squamous epithelium* (extent shown by arrow in **D**) **B20.** They are subject to severe mechanical stress during phonation. The cords stand out at *laryngoscopy* by their whitish-gray appearance against the reddish color of the rest of the mucosa. The mucosa covering them is immovably fixed to the connective tissue by its strong papillae. In contrast, the mucosa above the vocal cords and everywhere else in the larynx is more loosely attached to the surrounding tissue: hence the danger of edema of the glottis, swelling of the mucosa inside the vestibule of the larynx, which can result in death by asphyxiation. The vestibular folds and the ventricle of the larynx contain numerous *glands* **B21,** their secretions drip onto the almost glandless vocal folds. **B22** Vocalis muscle. Elastic fibers in black.

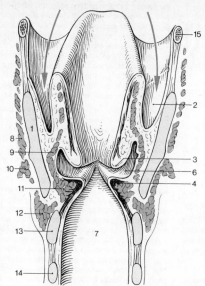

A Frontal section through larynx

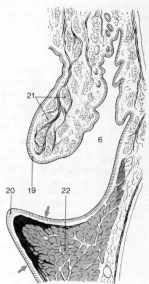

B Frontal section through vestibular and vocal folds; latter is true vocal cord. Section from A

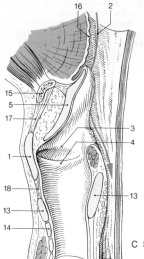

C Sagittal section through larynx

Glottis, Phonation

The structures concerned with phonation, i. e., all those which adjoin the rima glottidis, are collectively termed the *"glottis"*. The *vocal fold, plica vocalis,* often called the vocal cord, is the free margin of the vocal lip, a mucous tissue fold.

The anterior ⅔ of the **rima glottidis,** the *intermembranous part,* is bordered by the vocal folds **A1,** its posterior ⅓, the *intercartilaginous part,* by the vocal processes of the arytenoid cartilages (**A2** apex of the vocal process). The rima glottidis in the male is 2.0–2.4 cm long, its width in quiet respiration is 0.5 cm and in forced respiration up to 1.4 cm. These dimensions are smaller in children and in females. **E** When viewed by the *laryngeal mirror* the intercartilaginous part appears at the bottom, and the intermembranous part and the epiglottis **A3** appear at the top. **A4** Ventricular fold, **A5** cuneiform tubercle, **A6** corniculate tubercle.

The *shape of the rima glottidis* changes. **D** in *quiet respiration* or whispering the vocal processes usually lie adjoining each other. The intermembranous part is closed, the intercartilaginous part is opened and forms a triangle due to the tension exerted by the lateral cricoarytenoid muscle. **A** During *medium respiration* both the intercartilaginous and the intermembranous parts are opened (pulled by the cricoarytenoideus posterior) and the rima glottidis becomes a rhomboid, **B** *forced respiration.* The movements are *symmetrical.* In unilateral paresis of the recurrent nerve the affected vocal fold is retarded in its movement.

C Closure of the glottis. For *phonation* and during *coughing* the rima glottidis is first closed – *closure of the glottis,* the *position for phonation.* During *coughing* the closed glottis is burst open by thrusts of expiration. During *the act of swallowing* the larynx is closed by the root of the tongue, which presses the epiglottis against the entrance to the larynx *(tongue-epiglottis mechanism),* and by contraction of the thyrohyoid muscles. An adipose tissue body, lying beside the thyroid cartilage, is pushed between the thyroid cartilage and the epiglottis, which in turn is pressed against the ventricular folds *(adipose body-epiglottis mechanism).* The food slides across the epiglottis and through the piriform recesses into the esophagus. Closure of the larynx is assisted by the action of the aryepiglotticus, which lowers the epiglottis, and by forward tilting of the arytenoid cartilages. If the epiglottis is destroyed it can effectively be replaced by the action of the root of the tongue.

Phonation. The voice is produced when the closed, taut vocal folds in the position of phonation are opened and made to vibrate by an air stream expelled from the lungs, causing sound waves to be produced. The *volume of sound* depends on the force of the air stream, and the *pitch* is determined by the frequency of vibration. The latter depends on the length (sex and age differences), tension and thickness of the vocal cords, as in the strings of a musical instrument. The gross adjustment is controlled by the cricothyroideus and the other muscles inserted into the muscular process, and the fine adjustment by the vocalis muscle. The air passages above the larynx i. e., pharynx, mouth and nasal cavities, act as *sound resonators.* The air column vibrating within the passage gives the voice its *quality* (timbre) chest-head voice. The timbre will change if different parts of the passage are blocked, e. g., obstruction of the nose by the soft palate or a head cold.

Speech. Because of the low position of the larynx in man the expiratory air stream reaches the palate and mouth where it can be articulated into speech. *Vowels* are formed by alterations within the passage and *consonants* by sound formation in the mouth. In case of loss of the vocal cords, the ventricular folds **A4** can also produce sound. A sound can also be produced in another way, by articulation in the mouth. Thus, after removal of the larynx a "belch language" (esophageal speech) can be learned, the patient articulates sounds produced by eructation of swallowed air.

Laryngoscopic view of
glottis (after Pernkopf)

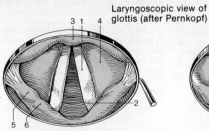

A Mid-position in normal respiration

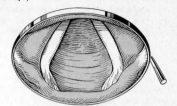

B Forced respiration

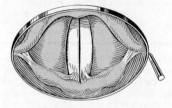

C Phonation position
(closure of glottis)

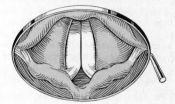

D Whispered speech

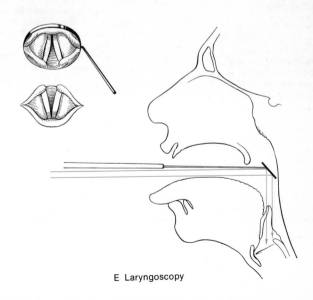

E Laryngoscopy

Position of the Larynx

Motility. The larynx is situated in the visceral space of the neck which merges into the connective tissue areas of the head and the chest. The position of the cervical viscera changes with movement of the head and the neck. The viscera of the neck are moveable against each other. *Movements of the larynx in the longitudinal axis of the body* occur during swallowing (deglutition), and it can be elevated by 2–3 cm during phonation and increased respiration. When the head is raised and the cervical spine is stretched the larynx rises by more than the height of 1 cervical vertebra. When the head and the neck are bent (flexed) the cricoid cartilage drops as far down as the opening into the chest cavity; its excursion does not exceed 4 cm.

Position. In the male standing with the head held upright, the hyoid bone **A 1** lies at the level of the 3rd–5th cervical vertebrae. The upper margin of the thyroid cartilage lies 1 segment lower and the lower margin of the cricoid cartilage lies at the border between the cervical and thoracic spine.

The low position of the larynx is characteristic of the adult human (see Speech). In babies the larynx lies higher.

The larynx is covered in front by the *superficial* **A 2** and *pretracheal* **A 3** layers of the *cervical fascia.* It is only a few millimeters below the skin. The *laryngeal prominence (Adam's apple)* **ABC 4**, the *cricoid cartilage* **A 5** and the *cricothyroid ligament* can be felt. In this position the larynx is passively attached to the base of the skull by means of the thyrohyoid membrane and the hyoid bone, and it is also anchored to the thorax by the tension of the elastic structures of the trachea and the bronchial tree. Its position is actively determined by the superior and inferior hyoid muscles and the inferior constrictor of the pharynx. **A 6** Epiglottis, **A 7** sternum (breast bone), **A 8** posterior wall of the esophagus in front of the prevertebral layer of the cervical fascia.

Effects of Age. The larynx in the *newborn* lies 3 vertebrae higher than in the adult and immediately below the hyoid bone. The upper margin of the epiglottis extends as far as the palate, the oral part of the larynx is short and its laryngeal part is all but absent. About half the rima glottidis in the newborn is still formed by the intercartilaginous part and altogether it is only 1/3 of the length of the adult. At the first growth period of the larynx the range of the voice is increased by about 3 notes in the first year and by 1½ octaves in the 2nd year. During *puberty* laryngeal growth leads to sex differences: in the male the rima glottidis becomes almost twice as long (about 1.5 cm) as before; compare **B** larynx in the male, **C** in the female. **C 9** Thyroid gland.

Clinical tips: If the upper airway is obstructed and there is a danger of choking, a new life-saving airway can be created by splitting the cricothyroid ligament and the layers of the skin and fascia in front of it, *coniotomy* **A 10**, or by cutting the trachea above or below the isthmus of the thyroid gland **A 11** *(upper* **A 12** *and lower* **A 13** *tracheotomy).*

Vessels and nerves. The superior laryngeal artery stems from the superior thyroid artery. It perforates the thyrohyoid membrane and passes to the larynx. The inferior laryngeal artery arises from the inferior thyroid artery and ascends dorsally to the larynx, see p. 52. Lymph vessels from the vestibule and the ventricle of the larynx run to the upper deep cervical lymph nodes, the lymph vessels from the vocal cords and the laryngeal cavity drain into the pretracheal nodes (lower deep cervical lymph nodes).

The superior laryngeal nerve (internal branch passes through the thyrohyoid membrane to the mucous membrane, the external branch passes lateral to the larynx to the cricothyroid muscle). Inferior laryngeal nerve supplies motor fibers to all the inner laryngeal muscles. See Vol. 3.

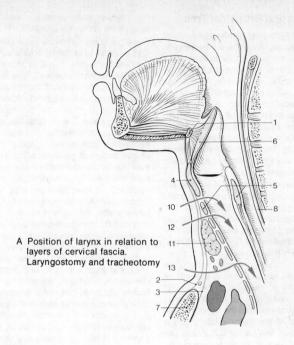

A Position of larynx in relation to layers of cervical fascia. Laryngostomy and tracheotomy

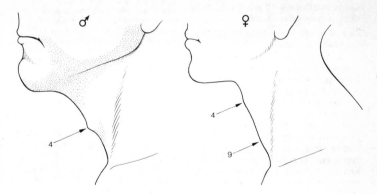

Form and position of larynx

B In an old man

C In a young woman

Trachea and Bronchial Tree

The 10–12 cm long **trachea** (windpipe) extends as an elastic pipe between the larynx and the bronchi. By its structure it maintains the air flow. The anterior lateral wall consists of 16–20 horseshoe-shaped hyaline *cartilages* **BCD1**, which are connected by the *annular ligaments* **B2**. In the posterior wall, *paries membranaceus* **BCD3**, the cartilages are closed into a ring by connective tissue and muscle. The transverse diameter of the trachea is 1.3–2.2 cm and exceeds its sagittal diameter by $\frac{1}{4}$.

The trachea bifurcates A4 into the left **ADE5** and right **AD6** *principal bronchi* forming an angle of 50–100 degrees. The right principal bronchus almost continues the course of the trachea, the left one turns more sideways. Because the lungs, which are of different size, pull with different force on the 2 principal bronchi, the bifurcation lies somewhat to the right of the mid-line at the level of the 4th–5th thoracic vertebra. Aspirated foreign bodies usually get into the right principal bronchus. At the tracheal bifurcation a sagittal spur, the *carina* **D7**, points upward. It separates the air stream during inspiration. Each principal bronchus divides into *lobar bronchi, three on the right* **A8, 9, 10** and *two on the left* **A8, 10**. **Large numbers 1–10** are the numbers of the *segmental bronchi* **A**.

Fine structure of the trachea. The wall consists of 3 layers.

The middle layer, a *fibrocartilaginous membrane,* is composed of the tracheal cartilages **BCD1** and the *annular ligaments.* The perichondrium radiates into the annular ligaments. The connective tissue contains collagen and elastic fibers and their longitudinal framework helps to keep the trachea open. The middle layer in the *paries membranaceus* consists of connective tissue and the transverse *trachealis muscle.* The latter can narrow the lumen of the trachea by about $\frac{1}{4}$. The external layer, the *adventitia,* consists of loose connective tissue and permits movement against the surrounding organs.

The inner layer, the *mucous membrane,* **C11** adheres to the perichondrium of the tracheal cartilages, but it is moveable over the membranous wall. It is in this layer that contraction of the trachealis muscle produces longitudinal folds in between which open the seromucous *tracheal glands.* The mucosa has a ciliated pseudo-stratified, *columnar epithelium,* containing goblet cells. Mucus and dust particles which penetrate as far as the trachea are shifted back into the pharynx within hours. In the epithelial cell layer there are also *brush border cells* which have no kinocilia (goblet cells after secretion of mucous, sensory cells?, immature cells) and *endocrine cells,* which occur either singly or in the form of *neuroepithelial bodies;* they are included in the APUD-system (see p. 148) and form *serotonin* and peptides *(bombesin, calcitonin, enkephalin).* The bodies which are innervated are thought to be *chemoreceptors sensitive to hypoxia.* *Clara cells* in the epithelium of bronchioles are notably light in colour, their secretion dissolves mucous and cell detritus etc., by proteolytic enzymes and other substances and so prevents obstruction of the airways.

In *straining* and before a *cough,* when the intrathoracic pressure is increased, the thoracic part of the trachea is constricted and the larynx pushed forward. After reopening of the glottis, i. e., after a cough, the larynx descends again, and the bifurcation ascends by about 5 cm due to the elastic shortening of the trachea and relaxation of the diaphragm. In *inspiration* the bifurcation descends by a maximum of 1 segment. The angle of the bifurcation narrows when the diaphragm descends and widens when it is elevated; the difference is 5–16 degrees.

Position. The aortic arch **E12** lies across the *left* principal bronchus **ADE5**. Below the aortic arch and in front of the left main bronchus the pulmonary trunk **E13** divides into the two pulmonary arteries **E14**. The esophagus **E16** runs behind the trachea **E15**. The azygos vein passes over the *right* principal bronchus, see p. 50.

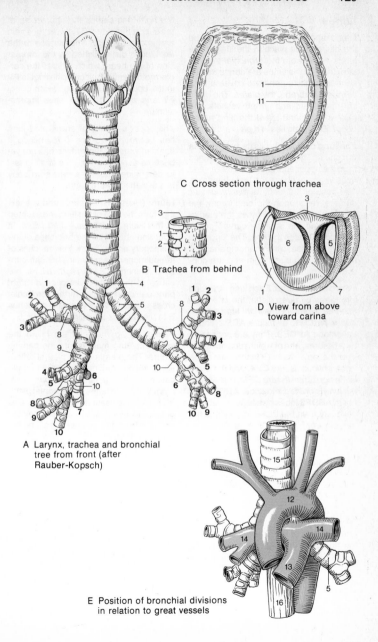

C Cross section through trachea

B Trachea from behind

D View from above
   toward carina

A Larynx, trachea and bronchial
   tree from front (after
   Rauber-Kopsch)

E Position of bronchial divisions
   in relation to great vessels

Lungs

Gas exchange between inhaled air and the blood takes place in the lungs. The lungs incorporate the *bronchial tree*. All structures of the lungs are connected by connective tissue and are covered by a serous membrane, the *pleura*. The bronchi, the blood and lymph vessels and nerves enter and leave the lung at the *hilum*. Each lung lies in a *pleural cavity*.

The **lung** completely fills its space within the chest. The *apex* of the lung extends from the upper aperture of the thorax and rises ventrally above the first rib in front. The base of the lung, its *diaphragmatic surface,* rests upon the diaphragm; the flat *medial surface* **EF** faces the mediastinum, and the strongly curved *costal surface* **CD** faces the ribs. The lungs are always adapted to the changing shape of the thorax and diaphragm and are constantly expanded.

Lobes of the lungs. Each lung is divided into lobes by deep clefts, the *interlobar fissures*. Normally the *right* lung has an upper **ABCE1**, a *middle* **ACE2** and a *lower* lobe **ABCE3**. They are separated by 2 fissures, one of which runs obliquely from the top at the back to the base at the front, *oblique fissure* **CE4**, and the other is the *horizontal fissure* **CE5**. The smaller *left lung* is divided into *upper* **ABDF1** and *lower* **ABDF3** lobes by a fissure that runs obliquely from the top at the back to the base in front, *oblique fissure* **DF6**. The upper lobe reaches the diaphragm on the left. Between the right upper and lower lobes is the middle lobe. The volume of the left lung is only about ¾ of that of the right lung. The fissures are lined by the visceral pleura.

The **surface** of the lung in a young person has a faint pinkish color. With age, pollution of the air we breathe gives the lungs a slate-gray, mottled and streaky discoloration. The surface of a lung that has been fixed in situ reflects all the unevenness of the mediastinal and costal walls. The mediastinal surface of the *right* lung carries the impression of the subclavian artery **E7**, the azygos vein and esophagus **E8**. The surface of the *left* lung is distinguished by the impression of the heart, which causes the cardiac notch and formation of the lingula **F9** in the upper lobe. The subclavian artery **F7** and aorta **F10** also leave impressions.

The specific gravity of an aerated lung lies between 0.13 and 0.75, hence it floats in water. A lung which has not yet become aerated sinks – the *floating test* to determine whether a new born baby has breathed before death.

Hilum (Hilus). The bronchi and vessels that form the root of the lung are situated at the hilum as follows – *Left hilum:* in front and above are the branches of the pulmonary artery, below the branches of the pulmonary vein; behind lies the principal bronchus **F11**. *Right hilum:* because of the early division of the right principal bronchus, a further bronchial cross section, the *"epartial"* bronchus **E11**, lies across the pulmonary artery.

Pulmonary ligament. At the hilum the visceral pleura merges into the parietal pleura. The margin of the pleural reflection stretches out into a point, a reduplication of the pleura, *"mesopneumonium"*, which extends in the frontal plane from the mediastinum to the lung. This *pulmonary ligament* **EF13** separates the anterior from the posterior half of the inferior paramediastinal portion of the pleural cavity.

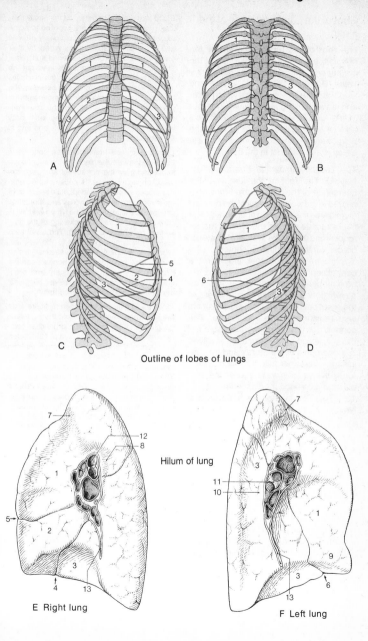

Outline of lobes of lungs

Hilum of lung

E Right lung

F Left lung

Root of the Lung and Base of the Heart

The **root of the lung** is formed by the *principal bronchi,* the main trunks of the *pulmonary artery* **B1,** the *pulmonary veins* **B2,** the *bronchial nerves* and *veins,* and the lymph vessels which enter and leave the lung at the *hilum.* The root of the lung is stabilized by the bronchi and by the flexible elastic tube of the pulmonary arteries, the blood pressure in which is about ⅛ that in the aorta.

The *bronchopericardial membrane (v. Hayek).* The root of the lung and the bifurcation of the trachea are firmly connected to the posterior wall of the pericardium by means of strong fiber bands. The fibers radiate across the wall of the inferior vena cave **A3** into the posterior border of the tendinous center of the diaphragm **A4.** Together they form a central connective tissue membrane, the bronchopericardial membrane **A5.** The adventitious connective tissue layers of the great pulmonary vessels and bronchi also radiate into this membrane. The connective tissue membrane secures the connexion of the structures involved in movements of the thorax and the diaphragm. Together with the bifurcation it forms the *partition wall* between the middle and posterior mediastinum, see p. 138. **A6** The inferior tracheobronchial lymph nodes.

The azygos vein **B7** passes over the root of the *right* lung from the posterior to the anterior mediastinum; the aortic arch **B8** passes over the *left* root from the anterior to enter the posterior mediastinum. **B9** Left subclavian artery, **B10** thyrocervical trunk, **B11** vertebral artery, **B12** common carotid artery, **B13** brachiocephalic trunk, **B14** superior vena cava, **B15** upper lobes of the lungs, **B16** left cardiac atrium, **B17** lower lobes of the lungs, **B18** coronary sinus, **B19** inferior vena cava.

Blood vessels and nerves. The connective tissue, bronchi and pleura are supplied to some extent by branches of the pulmonary arteries and veins, but to a much larger extent by their own vessels, the bronchial arteries and veins. The bronchial branches arise mainly from the aorta. They divide together with the bronchial tree. There are anastomoses between the bronchial and pulmonary arteries beneath the pleura and between small bronchi. The bronchial veins receive blood via the pulmonary veins and some mediastinal veins and end in the vena cava.

Lymphatics arise from the loose connective tissue underneath the pleura, in the interlobular septa and from the periarterial and peribronchial connective tissue. Because of the negative pressure prevailing there, fluid seeps from the alveoli into the peribronchial and subpleural connective tissue and into the lymph vessels (a consequence of this is blackening of the lung surface by inhaled coal dust). The lymphatics convey the lymph indirectly to the tracheobronchial, tracheal and mediastinal lymph nodes. The regional lymph nodes near the hilum (hilar glands, or pulmonary and bronchopulmonary lymph nodes) are visible against the air-filled lung tissue in radiographs.

Innervation: Efferent and afferent fibers from the vagus nerve and sympathetic trunk (stellate ganglion and 2nd–5th thoracic ganglia; compare neuroepithelial bodies, see p. 128). Afferent fibers of the vagus conduct impulses from stretch receptors in the lung which inhibit the respiratory center in the medulla oblongata. The cough reflex can only be elicited as far down as the bifurcation of the trachea and not in the lobar bronchi. Afferent fibers in the sympathetic nerves are part of a reflex arc through which nociceptive reactions may be mediated. Efferent fibers from the vagus nerve produce contraction of the bronchial muscles, and efferent fibers from the sympathetic trunk produce contraction of the pulmonary vessels, see Vol. 3.

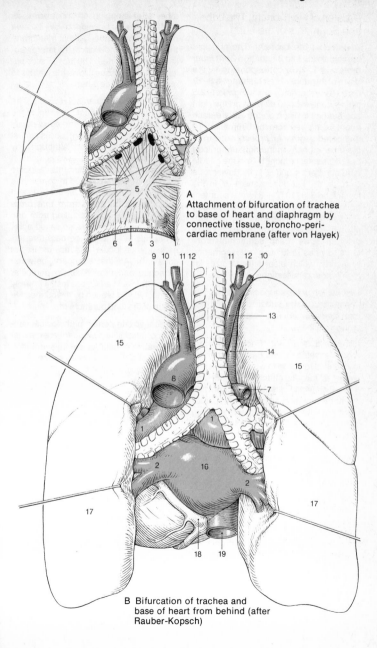

A Attachment of bifurcation of trachea to base of heart and diaphragm by connective tissue, broncho-pericardiac membrane (after von Hayek)

B Bifurcation of trachea and base of heart from behind (after Rauber-Kopsch)

Division of the Bronchi, The Lobes and Segments of the Lung

Division of the bronchi. The principal bronchi divide into 3 right and 2 left *lobar bronchi* **A1, 2, 3,** corresponding to the lobes of the lungs. Their diameter is 8–12 mm. The first division on the right is 1–2.5 cm, and the second division on the right and the first division on the left are each about 5 cm away from the bifurcation. 10 numbered *segmental bronchi* arise from the lobar bronchi: in the right upper lobe are segmental bronchi 1–3, in the middle lobe numbers 4 and 5, in the lower lobe 6–10; in the left upper lobe 1–5, in the lower lobe 6, 8–10 (7 + 8 form a segment on the left; **1–10** numbered according to *Boyden* **A**). The segmental bronchi supply cone-shaped pulmonary segments of different sizes. Each segmental bronchus divides into two subsegmental branches, which also differ in size; for their further ramifications see p. 136.

Distribution of the connective tissue. The connective tissue subdivides the lungs into *lobes, segments, lobules* and still smaller units.

Lobe and Segment. Each *pulmonary lobe* is enveloped by a *basal membrane* of connective tissue to which the visceral pleura is loosely attached without any folds. Connective tissue septae from this membrane extend at various depths toward the hilus. They subdivide the lobe into coneshaped bronchoarterial *segments* **1–10** (numbered like the segmental bronchi). In the center of each wedge lies a segmental bronchus. Each lobe of the lung is mostly covered by the pleura but not the segment.

Lobule. The surface of the lung, except in the interlobar fissures, shows polygonal areas. They are not completely delimited, their edges are 0.5–3 cm long and they may appear especially prominent because of their pigment infiltration. These areas delimit the basal part of the lobules, the *lobuli pulmonis,* which cannot be completely separated from each other after stripping off the pleura, as each lobule is enveloped by its own boundary membrane. At this membrane the elastic fiber network of the interalveolar tissue is inserted. The loose interlobular connective tissue allows the lobules to move against each other.

Details of the lobular subdivision. The lobular formation is most apparent at the periphery of the lung (marked capacity for deformation of the lung), and is almost absent from the center. Although any given lobule does not always contain equivalent structural elements, a lobule can usually be considered to contain the *terminal* and *respiratory bronchioles* and *alveoli* which originate from that *bronchiole.* The divisions resulting from one terminal bronchiole are also called *acini.* An acinus lacks septal borders (the terminology of lobule and acinus is not constant). At the boundary between bronchus and bronchiole, in particular where the wall cartilage is lost, it is firmly connected and braced by the connective tissue of the lung parenchyma.

Clinical tips: In certain pathological conditions, interstitial pulmonary edema (accumulation of fluid) or emphysema (collection of air) can develop in the *loose interlobar connective tissue.*

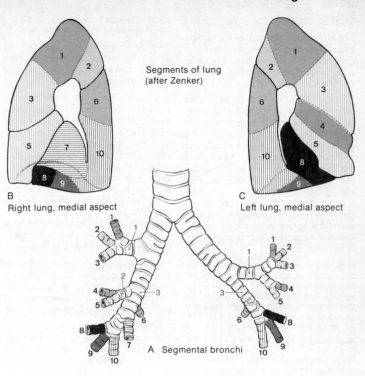

Segments of lung
(after Zenker)

B
Right lung, medial aspect

C
Left lung, medial aspect

A Segmental bronchi

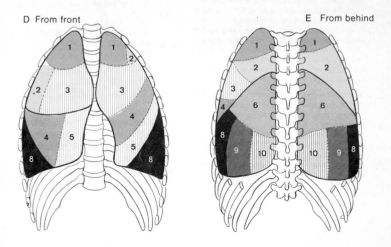

D From front

E From behind

Fine Structure of the Lung

The microscopic structure of the respiratory and vascular systems changes in the course of their subdivisions as the total cross sectional area of the branches increases.

Bronchi. They divide 6–12 times until they eventually reach a caliber of 1 mm. The adventitia contains elastic fibers in progressively increasing numbers, the cartilages become irregular plates and also contain elastic fibers. The height of the epithelium with its cilia and goblet cells diminishes, but seromucous bronchial glands are still present. A tubular layer of circular muscles appears beneath the mucosa. When these muscles contract longitudinal folds form and the lumen of the bronchus is narrowed, e. g., in bronchial asthma.

Bronchioles and terminal bronchioles. From the small bronchi **A1** *bronchioles* **A2** arise and from each of these 4–5 *terminal bronchioles* **A3** develop. Their walls are free of cartilage, but they are kept open by elastic fibers. The goblet cells disappear. **A4** Visceral pleura, **A5** acinus, **A6** branch of a pulmonary vein.

Respiratory bronchioles BC7 are 1–3.5 mm long and about 0.4 mm wide. They stem from the terminal bronchioles. They have a cuboidal epithelium and their walls are occasionally interrupted by thin walled sacs, the *alveoli,* which are followed by *alveolar ducts* **C8.** The walls of the latter consist only of alveoli which end in the *alveolar sac* **C9.** *Alveoli* **C10** measure 0.06–0.2 mm or more in diameter. About 200 alveoli are attached to each terminal bronchiole. Both lungs together contain about 300 million alveoli with a total surface area in the mean inspiratory position of 100–140 m².

Vascular system. The branches of the *pulmonary artery* follow the ramifications of the bronchial tree **A1, 2, 3** and the *arterioles* **B11** follow the *respiratory bronchioles. Precapillaries* accompany the alveolar ducts and the *capillaries* surround the alveoli **BC12, C10.** Between each precapillary artery and postcapillary vein there are 4–12 capillary loops: short ones *(resting capillaries)* for permanent perfusion, and longer ones *(active capillaries)* for circulation when the O₂ requirement is increased. The *postcapillary veins* **B13** lie in the interlobar and intersegmental tissue. Each vein carries blood from several

arteries. The veins **A6** converge in the center of the lung toward the arteriobronchial trunk.

Alveoli. Gas exchange takes place in the alveoli. The microscopic structure of the *blood-air barrier* can only be recognized under the electron microscope.

The alveolus has a flat *cellular* covering which in places is stretched into thin less than 0.1 μm thick plates sitting on a *basal lamina* **E14** – "small alveolar cells" (pneumocytes, Type I) **DE15.** In between these cells lie larger, taller cells without cell processes – "large alveolar cells" (pneumocytes Type II) **D16.** They are thought to produce a phospholipid film of an antiatelectatic factor. *Alveolar macrophages* **D17** (monocytes from the blood) take up dust *("dust cells")* or, after a hemorrhage (with vascular congestion) hemoglobin *("heart failure cells").*

The *capillary wall:* The uninterrupted *endothelium* **DE18** lies on a *basement membrane.* In the fissure between the alveolar epithelium and the capillary wall both basal laminae **E19** fuse for long stretches.

The *blood-air barrier* 0.3–0.7 μm thick, consists of *alveolar epithelium* **DE15,** *basal lamina* **E19** and the *capillary endothelium* **E18.** Breaks between the alveoli, *alveolar pores* **C20,** occur regularly.

In the *alveolar septa* are networks of elastic fibers **D21** and connective tissue cells **D22.** Together with smooth muscle cells they form "basal rings" at the entrance to the alveolus. These structures cross each other at an angle. By contraction of the myofibroblasts associated with the alveolar walls, the lumen of the interalveolar capillaries is reduced. The lymph capillaries only begin outside the lobule. Permeable intercellular contact of the capillary endothelium **E23,** compact intercellular contact of pneumocytes **E24.**

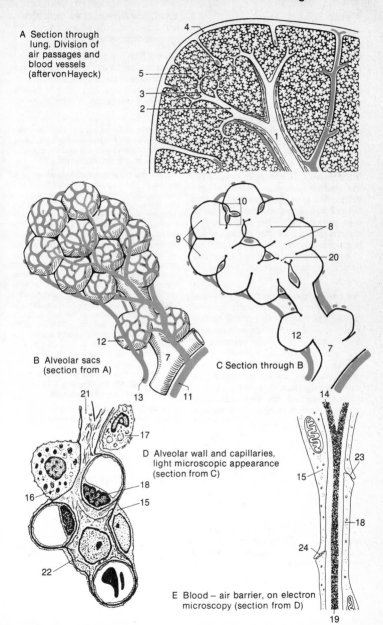

A Section through lung. Division of air passages and blood vessels (after von Hayeck)

B Alveolar sacs (section from A)

C Section through B

D Alveolar wall and capillaries, light microscopic appearance (section from C)

E Blood – air barrier, on electron microscopy (section from D)

Pleura

The *serous membrane* which invests the lung is called the **pleura.** We disinguish the *visceral* or *pulmonary pleura* that covers the lung from the *parietal pleura,* divided into costal, mediastinal and diaphragmatic pleura, that lines the wall of the lung cavities.

The **visceral pleura A1** forms a cover over the lungs. It consists of a single layer of epithelium, a fibrous layer of collagenous and elastic fibers and a subpleural layer in which lie the lymphatics and blood vessels. The *negative pressure* in the interpleural space causes a shift (transudation) of fluid from the lung into the interpleural spaces. This fluid carries dust particles from the alveoli to the visceral surface of the pleura, which leads to discoloration of the lung surface. The visceral pleura does not take up any particulate matter from the interpleural space.

The **parietal pleura,** particularly the *costal part* **A2** is firmly attached to the substrate. The development of the fibrous layer varies. Over the ribs and pericardium collagen fibers are predominant and over the diaphragm elastic fibers. The parietal pleura over the dome of the pleura is fixed by connective tissue strands to the deep layer of the cervical fascia and to the upper thoracic opening. Subpleural lymph vessels run in strands in the intercostal areas and over the diaphragm. The parietal pleura can absorb particulate matter, fluid and air from the pleural cavity. **AB3** Pericardium.

Pleural cavity (mediastinum). The visceral and parietal pleura, which join at the hilus **CD,** surround the completely closed pleural cavity **BD4** (see pericardium and peritoneal spaces). It contains a few milliliters (cm^3) of fluid. The parietal pleura over the ribs and that covering the diaphragm form a fold of variable width, the *costodiaphragmatic recess* **ABCE5**; the width depends on the respiratory position. This also applies, to a lesser degree,

to the parietal pleura of the mediastinum and the chest wall, where it forms the *costomediastinal recess* **D6.**

Clinical tips: The serous fluid in the pleural cavity may be increased in inflammatory disorders *(pleural exudate).* It contains protein and may lead to pleural adhesions (reduction of pulmonary expansion).

Innervation: While the visceral pleura is insensitive to pain, the parietal pleura is highly sensitive with afferent fibers contained in the intercostal nerves and the phrenic nerve.

The *position of the* **dome of the pleura** and of the **hilum.** The 3 scalenus muscles form a tent over the *pleural dome* **A7.** The brachial plexus and the subclavian artery pass through the *"scalene gap"* between the anterior and middle scalenus into the lateral cervical region. The subclavian vein runs between the anterior scalenus and the clavicle. These two vessels and the cervicothoracic ganglion of the sympathetic system (in front of the head of the first rib) lie on the pleural dome. The phrenic nerve, which runs on the surface of the anterior scalenus from above laterally to downward medially, touches the dome of the pleura on its medial aspect. The root of the right lung lies behind the superior vena cava **BD8** and partly behind the right atrium of the heart. The phrenic nerve descends ventrally and the vagus nerve dorsally to the hilus. The aorta **CDE9** and the esophagus **CDE10** cross in the posterior mediastinum. Compare mediastinum **B11** pulmonary trunk, **BC12** aortic arch, **C13** bifurcation of the trachea, **C14** azygos vein.

Planes of the section: for **B** see **D15,** for **C** see **D16,** for **D** see **BC17. E** View of the dome of the diaphragm.

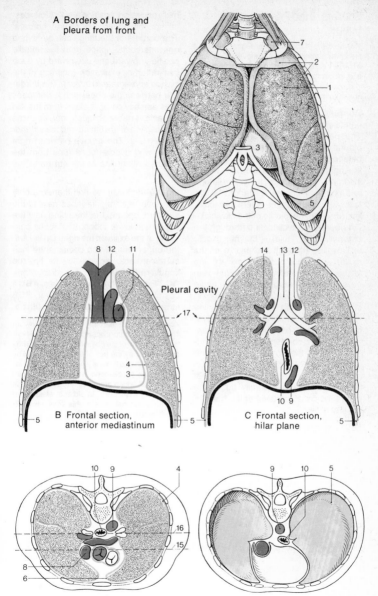

A Borders of lung and pleura from front

B Frontal section, anterior mediastinum

Pleural cavity

C Frontal section, hilar plane

D Horizontal section, hilar plane

E Pleural and pericardial coverings of diaphragm

Borders of the Pleura and the Lungs

The *borders of the pleura* are immovable but the lower *border of the lungs* move with breathing. They can be demonstrated by radiography and by percussion (air-containing lung tissue) and can be related to the intersections of the ribs with the meridional lines of the trunk.

Borders of the pleura. (Regions for percussion and auscultation). *Right side:* The parietal pleura extends from the pleural dome **A1**, 3 cm above the first rib behind the sternal angle, then medial to the right margin of the sternum to the insertion of the 6th rib. Its further lateral and posterior course is marked by the following intersections: medioclavicular line (through the middle of the clavicle) – 7th rib, anterior axillary line (through the anterior axillary fold) – 8th rib, middle axillary line (through the middle of the axilla) – 9th rib, posterior axillary line (through the posterior axillary fold) – 10th rib, scapular line (through the lower angle of the scapula with the arm hanging down) – 11th rib, paravertebral line (parallel to the vertebral column) – lower margin of the 12 rib. The *costodiaphragmatic recess,* corresponding to the variable origin of the lumbar part of the diaphragm, may extend paravertebrally up to 2 cm below the 12 rib. The pleural margin runs paravertebrally up to the pleural dome at the level of the head of the first rib.

Left side: The borders of the pleura take the same course on both the left and right sides. The only difference is that the left anterior border leaves the sternum at the level of the 4th rib and descends in a curve *(incisura cardiaca)* to the intersection of the medioclavicular line and the 6th rib. **A2** Heart.

Borders of the lungs. At the apex of the lung, behind the sternum and next to the side of the spine *(costomediastinal recess)* they correspond approximately to those of the pleura. On the other hand, the lower borders of both lungs in the mid-respiratory position are 1–2 intercostal spaces above the pleural borders. The borders of the lungs can be shifted one intercostal space from this middle position upward and downward by deep inspiration or expiration (opening of the *costodiaphragmatic recess).* The border line between the upper and lower lobes runs dorsally on both sides from the 4th thoracic vertebra obliquely forward to the intersection of the 6th rib and the anterior axillary line. The groove between right upper and middle lobes runs from the middle axillary line at the 4th rib to the sternum.

On the *right* side we find therefore only the upper and middle lobes next to the sternum, and next to the spine only the upper and lower lobes. All 3 lobes converge in the axilla on the right. On the *left,* the upper lobe, which contains the incisura cardiaca, extends by its tip, the lingula pulmonalis, to the diaphragm. **AC3** Costodiaphragmatic recess, **ABC4** costal arch, **A5** margin of the liver, **BC6** lung, **B7** spleen, **B8** course of the 9th rib.

Age differences: the baby and infant show an *"habitual inspiratory position"* with a more horizontal position of the ribs, but later in life the ribs are lowered by the pull of the muscles and by gravity. With age there is a general tendency for the organs to descend and to reduce joint excursions, which produces a reduction of the respiratory excursions. In old age, elasticity of the lung diminishes and this may result in emphysema.

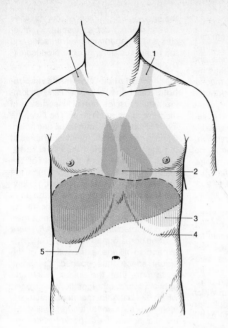

1
1

Borders of
lungs and pleura

2

3
4

5

A From front

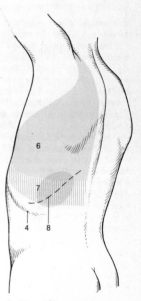

6

7

4 8

B From left

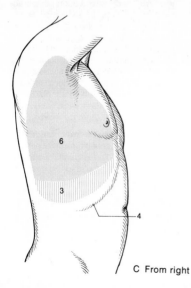

6

3

4

C From right

Mechanics of Respiration

During *respiration* there is an alternating increase and decrease in thoracic volume. Two components, the *ribs* and *diaphragmatic respiration,* work together to produce an ingenious 'pump' mechanism. The diaphragm may be regarded as a 'piston' in the 'pump cylinder' of the thoracic cavity. It is pushed in and pulled out, thus pushing air out of the trachea *(expiration)* or pulling it in *(inspiration).* Throughout this process the diameter of the thorax continuously alters. In expiration **B1, 2, 3** it becomes smaller, and during inspiration it enlarges **A1, 2, 3.** Apparatus of movement, see Vol. 1.

Mechanics of rib respiration. When none of the respiratory muscles are acting, the elastic thoracic cage lies in the median position of equilibrium, the *resting respiratory position.* This is influenced by the pull of the lungs and the abdominal muscles. The position of respiratory rest is the position from which *inspiration* starts **A.** During inspiration the thoracic cage is actively lifted – as the ribs are positioned obliquely – and enlarged. The epigastric angle is also enlarged. The sternum, which is attached to the ribs by cartilage, produces a 'parallel displacement' of the ends of the ribs. In quiet respiration, after inspiration the elastic thoracic cage passively returns to its resting position. It is only during forced *expiration* **B** that the thoracic cage in addition is actively lowered against the elastic force. During artificial respiration, compression of the thorax may produce a forced expiration, and the elastic recoil of the thorax to the resting position then produces an inspiration. During quiet rib respiration, the external intercostal and the posterior serratus muscles produce inspiration, and the internal intercostal muscles in the region of the bony ribs are the muscles of expiration. The intercostal muscles hold the intercostal spaces tense against the external air pressure and the negative intrathoracic pressure. In forced rib respiration, the muscles of the shoulder girdle ("accessory muscles of respiration") act as inspiratory muscles and the abdominal wall muscles and the latissimus dorsi muscle produce expiration.

Mechanism of diaphragmatic respiration. The abdominal viscera and abdominal wall muscles are of importance in diaphragmatic respiration. The liver, the 'core' of the 'piston', moves up and down. During *inspiration* **A,** the muscle fibers of the diaphragm shorten: the longer fibers, which originate from the back of the diaphragm **A4,** shorten more than the anterior fibers. The central tendon is lowered and the distance between the parts of the diaphragm which ascend to the dome and the chest wall becomes larger, the *costomediastinal recess* **A5,** more marked dorsally than ventrally. The lungs expand to fill the extra volume of the virtual space thus opened, the basal parts more than the apices. A left and right anterior diaphragmatic muscle band which lie beneath the pericardium attached to the central tendon, pull down the diaphragm like a pocket under the plane of the xiphoid process **A6.** The heart **AB7** is displaced during diaphragmatic respiration. In *expiration* **B** the diaphragm and the liver move into the thoracic cavity due to contraction of the muscles of the abdominal wall and the pull of the lungs. The excursion of the dome of the diaphragm **AB8** varies from 1.5–7 cm according to the depth of respiration. In quiet respiration about 75% of the change in intrathoracic volume is produced by diaphragmatic movement.

Costodiaphragmatic mechanism. In the adult both mechanisms of respiration work together. The essential condition for effective rib respiration is that the diaphragm contracts and is not pulled into the thoracic cavity by lung suction. Effective diaphragmatic respiration requires stable, tense intercostal spaces.

Respiratory position of the thoracic cage and the diaphragm on the left anteriorly and
on the right laterally.
Photograph and X-ray superimposed

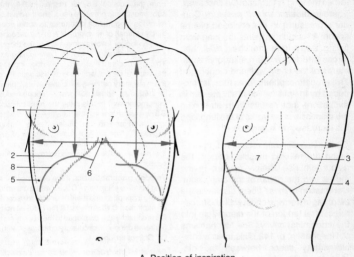

A Position of inspiration

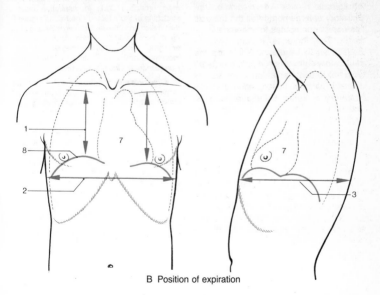

B Position of expiration

Respiratory Dynamics

Respiration. Air pressure exerted through the respiratory pathways **A2** forces the lung **A1** against the chest wall. There is *subatmospheric pressure* (Donder's pressure) in the enclosed pleural space **A3** which prevents the lung from falling away from the chest wall, and forces it to follow the respiratory movements of the chest wall and diaphragm. The subatmospheric pressure is produced by the pull (which is still present in expiration) that results from stretching the elastic fiber networks of the lung, and it is increased by inspiration.

Costodiaphragmatic recess **A4**. In expiration the relaxed muscle fibers of the diaphragm rise steeply upward in the thorax into the dome of the diaphragm. Contraction of these fibers displaces the diaphragm from the chest wall and forces the central tendon of the diaphragm into the abdominal cavity. See Mechanisms of Respiration, p. 142. This frees a complementary space between the diaphragm and the chest wall, the costodiaphragmatic recess **A4**: *Inspiration*. Inspiratory enlargement in all the thoracic diameters contributes to opening the recess. The increase in lung volume causes air to be drawn in through the respiratory pathways = *respiratory (tidal) volume*, forced inspiration = *inspiratory reserve volume*. In *expiration* the *tidal volume* is breathed out and forced expiration further displaces the *expiratory reserve volume*. This only leaves the *residual volume of air* in the lung.

The *costomediastinal recess*, which lies on both sides behind the sternum and on either side of the vertebral column, also serves for expansion of the lung. The inspiratory enlargement of the lung increases caudally and ventrally. In the thoracic type of respiration the upper lobes are better ventilated than during abdominal respiration.

Pathological Changes in Lung Dynamics

Pneumothorax. If the pleural space is opened from the outside **B5**, air rushes in **B6**, the subatmospheric pressure rises, and the elasticity of the lung makes it contract to about one-third (minimal) of its volume **B7**. In this case the lung no longer follows movement of the thorax and the diaphragm = *external pneumothorax*. This also occurs if there is a tear in the lung and the visceral pleura, and air is able to pass from the respiratory pathways into the pleural space = *internal pneumothorax* (spontaneous pneumothorax). In this case the mediastinum **B8** is pulled toward the healthy side during inspiration and to the diseased side during expiration *("mediastinal flutter"; danger of circulatory failure). After closure of the tear, air in the pleural space is reabsorbed within a few days.*

Tension pneumothorax. If the opening in the lung acts like a flap valve **C9**, air will be drawn in during inspiration and it will be unable to escape on expiration **C10**. As a result the mediastinum **C8** is considerably displaced to the healthy side (extreme danger of circulatory embarassment). **C7** Collapsed lung.

Bronchial obstruction. Mediastinal displacement may also be caused by obstruction of a main bronchus **D11**; in inspiration elastic stretching of the affected side fails to equalise the tension and the mediastinum is displaced to the diseased side **D8**. During expiration, poor emptying of air from the diseased side results in movement of the mediastinum to the healthy side (danger of circulatory failure).

Positive pressure respiration. In failure of the respiratory apparatus, e. g., paralysis due to curare given during surgery, or resulting from bilateral pneumothorax, breathing may be maintained by positive pressure respiration, in which the volume of the lungs is alternately increased and decreased by deliberately forcing air into and out of the respiratory pathways. This is also the principle of *mouth-to-mouth respiration* in the emergency treatment of respiratory paralysis.

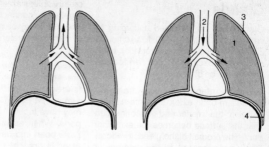

A Normal

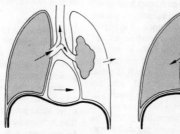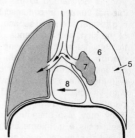

B Pneumothorax

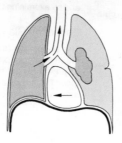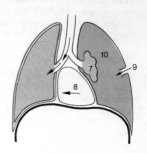

C Tension pneumothorax

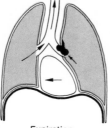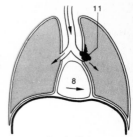

D Bronchial narrowing (after Pernkopff)

Expiration   Inspiration

Glands

Glands can be classified according to various criteria, see p. 150. If the criterion is the presence of an excretory duct, **exocrine** (excretory) and **endocrine** glands can be distinguished.

Exocrine glands A grow from the epithelium of an external or internal surface into the interior and the connection with the surface becomes the excretory duct. The product of the gland, its **secretion,** is discharged to the external or internal surface at circumscribed sites, e. g., sweat, scent and sebaceous glands of the skin, and the glands of the gastrointestinal tract. **Endocrine** (ductless) **glands** may originate from a surface epithelium with loss of their connection with the surface, e. g., the thyroid gland with its follicle formation, **B,** and the parathyroid gland which lacks follicles **C.** They may also arise from connective tissue cells, e. g., intercellular gland. The secretion of an endocrine gland, the **hormone,** passes via the intercellular connective tissue into the blood and lymph vessels and through the **blood circulation** it reaches all parts of the body.

Exocrine Glands

The epithelia of **exocrine glands** show polar differentiation. At the base they absorb basic substances (sugars, amino acids) from the intercellular fluid of the connective tissue, which is derived from the blood capillaries, and at their free superficial surfaces they secrete into external ducts, i. e. transport of substances takes place in a directed way.

Most glandular epithelia expel their secretions as described on p. 70 by exocytosis. The secretion is mostly formed *continuously* (e. g. sweat glands) or *intermittently* as a result of special stimulation (e. g. salivary glands), or periodically (e. g. scent glands).

Exo-epithelial glands are usually referred to by the term gland. They are secretory epithelial cells which are grouped together by connective tissue *as organs.* The gland may be microscopically small or many centimeters in size.

According to the *form of the terminal part of the gland* (= the secretory part) there are **acinous** (berry-shaped), **alveolar** (sac-shaped) and **tubular** (tube-shaped) *terminal parts,* and "mixed glands" which may have tubular and acinar or alveolar parts. When several secreting terminal parts open into one excretory duct, the gland is *branched;* if the excretory duct itself is divided then the gland is *composite.*

Three types of glands may be distinguished by the light microscope according to the method of *extrusion* of the secretion, *eccrine, apocrine* and *holocrine* glands. Basically glands may be distinguished by the chemical nature *of their secretions.* Although this is definitely the most important means of characterisation; in many cases it cannot be achieved by staining. It is supplemented by the observation of typical cellular structures and by examining the excretory duct.

Endocrine Glands

The **endocrine (ductless) glands,** are the *hypophysis* (pituitary gland) **D1,** the *epiphysis* (pineal body) **D2,** the *thyroid gland* **D3,** the *parathyroid glands* **D4** the *adrenal cortex* and *adrenal medulla* **D5,** the *islets of the pancreas* **D6,** and parts of the *gonads* **D7** (reproductive glands). They produce *"glandular hormones".* Parts of the diencephalon act as endocrine glands. Hormone-producing cells also occur in other organs, e. g., in the epithelial lining of the gastrointestinal tract ("system of gastrointestinal endocrine cells"). These **diffusely distributed endocrine organs** will be discussed with the organs in which they occur.

A Exocrine gland

B Endocrine gland,
follicle formation

C Endocrine gland
without follicles

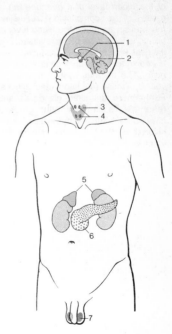

D Survey of major
endocrine glands

Endocrine glands are *organs of the endocrine system* which only produce hormones. On the other hand, organs which principally fulfil other functions but at the same time contain a hormone producing cell system are also known as *endocrinologically active organs*. The most extensive of the latter is the gastro-intestinal tract.

Hormones. Minute amounts of hormones coordinate metabolic processes in cells and organs by *activating enzymes*. The amount of a hormone which is discharged into the organism is important and can usually be calculated from its level in the blood. Hyper- or hyposecretion of a hormone leads to typical disease patterns. Hormones are vital substances. As many hormones are not specific to any particular species, deficiency of a hormone in man can often be made good by substituting an animal hormone. Certain hormones are also manufactured synthetically. The stability of the *blood level of a hormone* is the result of a complicated regulating mechanism so that if the blood level of hormone falls, more of it is released, and if it rises less hormone is produced. This is known as a *feedback* control mechanism. The nervous system is also involved in these regulatory mechanisms. Furthermore, endocrine glands act and react upon each other. Although each hormone is distributed by the blood circulation throughout the body, only particular cells, tissues or organs with a specific "receptor" for it react to each individual hormone. The effect of a hormone is often detectable morphologically. *Hormones* may be subdivided into different chemical types, the *amines* (e. g. adrenalin, noradrenalin, dopamine, melatonin, serotonin), the *steroids* (e. g. mineralocorticoids, glucocorticoids, sex hormones), the *peptides* (e. g. regulatory hormones of the hypothalamus, gastrointestinal hormones) and *proteins* (e. g. gonadotrophins, growth hormone).
Endocrine cells may produce more than one hormone.

When there are several hormones produced in one cell as is frequently observed in peptide-synthesising cells, they may be split off enzymatically from a *common precursor (pre-prohormone, prohormone)* and form a "peptide family"; e. g. **1** the family of the pro-opiomelanocortin derivatives. However they may **2** arise from *different precursors;* the endocrine cell may activate the programs for production of more than one peptide family. **3** Hormones produced by an endocrine cell may belong to *different groups of substances*. This is particularly true of peptides and amines. *Endocrine* cells **4** secrete into blood vessels and may **5** influence neighbouring cells in a paracrine manner.

Hormones of all chemical types are also produced as active substances by neurons of the central and peripheral nervous systems.

The neurons release active substances either **6** as *neurotransmitters* (or neuromodulators), or as **7** *neurohormones* into the blood vessels of a neurohemal region (see median eminence p. 156).

APUD-system. The origin of aminergic and peptidergic endocrine cells is not completely understood. A theory of unitary origin suggests that they are uniformily able to produce peptides with hormonal characteristics and simultanously to produce biogenic amines and/or to take up and concentrate the precursors of biogenic amines, i. e. amino acids and to decarboxylate them to amines. For this reason this cells were called "*Amine and/or amine precursor uptake and decarboxylation*"-cells, *APUD*-cells by *Pearse* (1968). Many workers believe that all these cells arise from the neural crest (see Vol. 3). *Pathologically,* there is increasing evidence that APUD-cells play an important role; hormone producing tumours, "apudomas" with characteristic symptoms may arise from them.

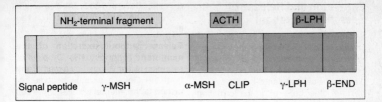

1. Pro-opiomelanocortin molecule.

   Vertical lines:
   basic amino acid pairs, splitting-off active peptide hormones.

   MSH: melanocyte stimulating hormone; CLIP: corticotropin-like intermediale lobe peptide; LPH: lipotropic hormone; END: endorphin.

2. Simultaneous occurrence in cell of different peptide precursors

Somatostatin	+ Enkephalin
Substance P	+ Enkephalin
Corticotropin	+ Enkephalin
Corticotropin	+ Vasopressin
Vasopressin	+ Dynorphin
Oxytocin	+ Enkephalin
Oxytocin	+ Cholecystokinin
TRF*	+ Somatostatin
TRF	+ Somatotropin
TRF	+ Substance P

   (* TRF-thyrotropin-releasing factor)

3. Simultaneous occurrence in cell of monoamines and peptides

Noradrenaline	Somatostatin
	Enkephalin
	Neurotensin
	Vasopressin
	Oxytocin
Dopamine	Enkephalin
	Cholecystokinin (CCK)
	Prolactin
Serotonin	Substance P
	Thyrotropin-releasing factor (TRF)
	Calcitonin
	Neurotensin

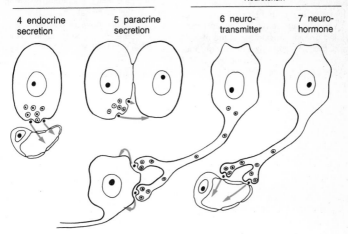

4 endocrine secretion   5 paracrine secretion   6 neuro-transmitter   7 neuro-hormone

Formation and Release of Secretion (Incretion)

Secretion is the production and discharge of cell-specific substances which have been synthesized for the specific purpose of release. Secretion takes place in the cells of glands but it may also occur in other cells, e. g., the formation and release of precursors of connective tissue fibers by connective tissue cells. Secretion may be triggered by vegetative (autonomic) nerves.

Formation: Protein-containing secretions ('protohormones' in endocrine glands) are produced by the general method of cellular synthesis of proteins. On a gene, i. e., on the DNA of a chromosome, a "matrix" messenger of ribonucleic acid (messenger-RNA or m-RNS) is formed. This substance migrates from the nucleus **A1** into the cytoplasm to the surface of ribosomes where protein production takes place. Ribosomes consist mainly of stable ribosomal RNA. The ribosomes and the lamellae of the endoplasmic reticulum (ER) form the granular ER (ergastoplasm) **A2**. Another type of RNA, transfer RNA, takes up amino acids, which have been activated by enzymes after their transfer from the blood vessels **A3** and converts them according to the genetic messenger of the m-RNA into protein molecules, in accordance with the information supplied by the DNA of the gene. Newly synthesized proteins pass via the cisterns of the granular ER into the vacuoles of the Golgi apparatus **A4**, where they grow into larger droplets of secretion (hormone) **A5**. The (stained) visible "basophilic" (with an affinity for basophilic dyes) protein-producing structures, which are visible by light microscopy, namely the nucleolus **A6** from which ribosomal-RNA and the ergastoplasm **A2** originate, are much magnified in Fig. **A**. Steroid hormone on the other hand occurs in the agranular ER.

Release: Several modes of release of a secretion (hormone) from the cell can be distinguished.

Release without excretion of the membrane (crinocytosis). Droplets of secretion still enveloped by a membrane from the Golgi apparatus become attached to the internal surface of the cell membrane (plasmalemma) **C7**. Where both membranes are in contact, the membrane enveloping the droplet becomes incorporated into the cell membrane and an opening into the extracellular space is created for the discharge (excretion) of the secretion (hormone) **C8**. Thus, secretions, such as hormones, that are discharged by crinocytosis have lost their covering membrane; microscopically, this is known as **"eccrine"** secretion **D**.

Release with excretion of the membrane is well recognized when milk fat is released from the epithelial cells of the mammary gland. The fat granules (droplets) at first arch from the surface of the cell **B9** and the protruded cell membrane forms a cover around the fat globule. After separation of the globule from the cell the secretion is enveloped by a membrane; microscopically **"apocrine"** secretion **E**. This mode of discharge probably occurs also in other glands, when, together with the cell membranes a cytoplasm is discharged.

Release with destruction of the cell. In some glands the formation of secretion is so excessive that eventually the entire cell becomes completely filled and perishes in the process. The lamellae of the ER and the Golgi apparatus disappear and pyknosis and disintegration of the nucleus ensues; microscopically **"holocrine"** secretion **F**.

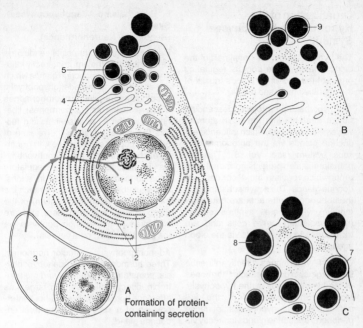

A Formation of protein-containing secretion

Modes of release, scheme from an electron micrograph
A, C Release without loss of cell membrane, eccrine secretion crinocytosis

B Secretion with loss of cell membrane, apocrine secretion

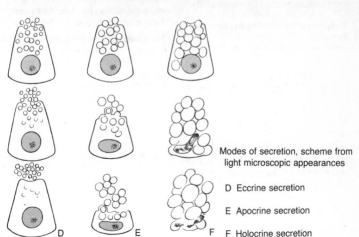

Modes of secretion, scheme from light microscopic appearances

D Eccrine secretion

E Apocrine secretion

F Holocrine secretion

Organization of the Hypothalamo-Hypophyseal System

The hypothalamus, the basal part of the diencephalon, is the primary regulating organ of the endocrine glands (nuclei of the hypothalamus, see p. 154).

The hypothalamus projects descending neural efferents into the brain stem to viscero-motor nuclei, which influence endocrine glands via the autonomic nervous system (see. Vol. 3). The hypothalamus also projects hormonal efferents which regulate the secondary endocrine glands. The active substances in these hormonal efferents are neurohormones which are synthesized in (neurosecretory) nerve cells and are transported to their targets via the blood stream.

Only a few hormones directly influence their targets as do the effector hormones of the hypothalamus or the hypophysis. Most of the hormones of the hypothalamus and hypophysis influence peripheral endocrine glands indirectly via the adenohypophysis, as regulating neurohormones of the hypothalamus, or as glandotropic hormones of the adenohypophysis. The hypothalamus and hypophysis are a functional entity.

I. Hypothalamo-Neuro-hypophyseal System
(Effector Neurohormones)

Hypothalamus. Oxytocin and vasopressin are hypothalamic hormones which influence their targets directly as effector neurohormones, i. e. they act without interposition of the adenohypophysis. Transported in axons these neurohormones reach the neurohypophysis where they are released into the blood.

The neurohypohysis is only a storage and release site for oxytocin and vasopressin, production of neurohormones does not occur in the neurohypophysis proper.

II. Hypothalamo-Adenohypophyseal System
(Regulatory Neurohormones)

Representing the principal endocrine center, the hypothalamus indirectly influences peripheral endocrine glands which are secondary to the adenohypophysis: hypothalamic regulatory neurohormones act as releasing factors (hormones), liberins, or inhibit as release-inhibiting factors (hormones), statins, the release of adenohypophyseal hormones. In general, each hormone of the adenohypophysis is associated with a regulatory neurohormone. Transported in axons, the regulatory neurohormones reach the median eminence (infundibulum) of the diencephalon, and from there they are conveyed to the adenohypophysis via the blood stream (portal vessels).

Adenohypophysis – effector hormones. Three hormones of the adenohypophysis, somatotropin, prolactin and melanotropin, directly influence their targets as effectory hormones, i. e. without interposition of a secondary peripheral endocrine gland.

Adenohypophysis – glandotropic hormones. The remainder of adenohypophyseal hormones are glandotropic hormones which induce growth and hormone production and release by secondary peripheral endocrine glands.

The feedback from the peripheral endorine glands to hypothalamus and hypophysis mainly occurs via the blood stream; in addition, neural feedback pathways are known.

Review of the important hormonal efferents of the hypothalamo-hypophyseal system in man

(Using the Nomenclature of Peptide Hormones of the Commission on Biochemical Nomenclature 1974)

Hypothalamic Hormones	Adenohypophyseal Hormones	Peripheral Endocrine Gland and Principal Action
A. Releasing factors and release-inhibiting factors	**I. Glandotrophic hormones**	
	1. Gonadotrophic hormones	
Folliberin (follicle stimulating hormone-releasing factor = FSH-RF)*	a) *Follitropin* (follicle stimulating hormone = FSH)	Stimulates ripening of egg follicles and spermatogenesis
Luliberin (luteinising hormone releasing factor = LH-RF [LRF])	b) *Lutropin* (luteinising hormone = LH; = interstitial cell stimulating hormone = ICSH)	Interstitial cell (ovary and testis), stimulate ovulation and luteinisation of egg follicle and secretion of testosterone
	2. Non gonadotropic, glandotropic hormones	
Corticoliberin (corticotropin releasing factor = CRF)*	d) *Corticotropin* (adrenocorticotropin hormone = ACTH)	Stimulates growth and secretion of adrenal cortex
Thyroliberin (thyrotrophin releasing factor = TRF)	e) *Thyrotropin* (thyrotropic hormone = TSH)	Stimulates growth and secretion of thyroid gland
	II. Effector hormones	
Somatoliberin (somatotropin-releasing factor = growth hormone releasing factor = GH-RF)	a) *Somatotropin* (somatotropic hormone = STH; = growth hormone = GH)	Stimulates growth of body
Somatostatin (somototropin-release inhibiting factor)		
Melanoliberin (melanotropin-releasing factor = MRF)*	b) *Melanotropin* (melanocyte stimulating hormone = MSH)	Probably an endogenous anti-opioid in man
Melanostatin (melanotropin-release inhibiting factor = MIF)*		
Prolactoliberin (prolactin releasing factor = PRF)*	c) *Prolactin* (mammotropic hormone = PRL)	Stimulates proliferation and formation of milk gland secretions (in rodents maintains function of corpus luteum)
Prolactostatin (prolactin-release inhibiting factor = PIF; Dopamin)		
B. Effector hormones		
Oxytocin (oxytocin = OXT)		*Leads to contraction of sensitised smooth muscle* *Encourages water retention*
Vasopressin (= VP; = adiuretin = ADH)		

* The existence of these substances has been postulated on indirect grounds. Their chemical structure has not yet been proven.

I. Hypothalamo-Neurohypophyseal System

Hormones. *Vasopressin* raises the blood pressure and promotes reabsorption of water from the renal tubules. *Deficiency:* diabetes insipidus. *Oxytocin* sensitises smooth muscle; it stimulates the uterus into labour and causes milk to be expressed by contraction of the muscle cells of the terminal parts of the breast glands. *Impaired function* produces uterine hypoactivity in labour.

Hypothalamus

The perikarya of the neurosecretory neurons, nerve cells of the hypothalamo-neurohypophyseal system lie in the large-celled nuclear zones of the diencephalon, in the *supraoptic nucleus* **AB1** and the *paraventricular nucleus* **AB2**. The axons terminate in the neurohypophysis **AB3**. The neurohormones are bound to a carrier substance, *neurophysin*. Both are produced by cell organelles in the perikarya. The *neurosecretion* (= neurohormone + carrier substance) lies in the perikaryon **C4**, the axon nerve cell process **C5** and the axon endings **C6**. It passes along the axons out of the perikarya to the neurohypophysis and is there released (Bargmann and Scharrer).

The axons which carry the neurosecretion form the *hypothalamo-hypophyseal tract* **B7,** which ends in the neurohypophysis. Larger axonal swellings, which contain neurosecretion, are known by the older name of Herring bodies. The neurosecretory substance may be stored in the axonal endings. **D** It consists of membrane-bound *elementary granules* with a mean diameter of 100–300 nm.

Neurohypophysis

The neurohypophysis **AB3** consists of a network of glial cells (pituicytes) and glial fibers, a type of "connective tissue" of the nervous system. There are numerous capillaries but no nerve cells. In the neurohypophysis one of the neurohemal regions (see p. 156) neurosecretory substances leave the axonal endings and pass into the blood vessels **B8**.

Hypophysis

The bean-shaped **hypophysis** (*pituitary gland*) weighs about 0.6 g and lies in the sella turcica of the sphenoid bone, at the center of the base of the skull, see Vol. 1. It consists of a *glandular* part, the *adenohypophysis* **A9**, and a *cerebral* part, the *neurohypophysis* **A3**. The following areas may be distinguished (see also p. 178). The *sella turcica* is bordered on both sides by the cavernous sinus. The optic chiasm lies anterior to the hypophysis and may be damaged by pressure caused by tumours of the hypophysis (*bitemporal hemianopsia* see Vol. 3).

The *adenohypophysis* (= *anterior lobe*) may be divided into: a) *pars distalis* (prehypophysis) which is the largest part, b) *pars intermedia* ('middle lobe' which forms a small border zone with the neurohypophysis, c) *pars infundibularis* (pars tuberalis, "funnel lobe") which lies anteriorly on the hypophyseal stalk.

The *neurohypophysis* (= posterior lobe) is joined to the hypothalamus of the diencephalon by the *infundibulum* (hypophyseal stalk). The *infundibulum* and the *median eminence* may be considered part of the neurophysis in the wider sense.

Vessels of the hypophysis. The internal carotid artery (see p. 58) gives off lower *arteries* which arise within the cavernous sinus, and upper arteries which arise above it. They pass to the infundibulum and the infundibular process and amongst other things form special vessels. *Venous* return is via the cavernus sinus and possibly via the cerebral veins.

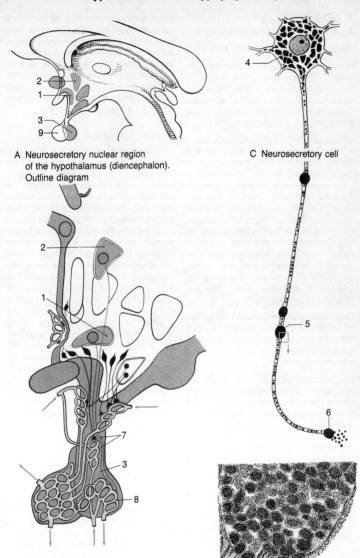

A Neurosecretory nuclear region
of the hypothalamus (diencephalon).
Outline diagram

C Neurosecretory cell

B Hypothalamo-neurohypophyseal system
(red cells) and hypothalamo-adenohypophyseal
system. Schematic view

D Neurosecretion, electron
microscope view

II. Hypothalamos-Adenohypophyseal System

Hypothalamus

Hormones. *Releasing factors* and *release inhibiting factors,* see review p. 153. *Malfunction* of these hormones leads to malfunction of the adenohypophysis and the peripheral endocrine glands which are dependent upon it.

The neurons of the infundibular nucleus in the mid-hypothalamic region control the release of *releasing factors and release inhibiting factors* by transmitter substances. These are also formed by the neuroendocrine cells of the hypothalamus **AB1,** but they are dispersed diffusely and do not form any nucleus-like cell groups. The *releasing and release inhibiting hormones* stimulate or inhibit the release of hypophyseal hormones, which themselves are mostly glandotropic hormones influencing the subordinate endocrine glands (thyroid, adrenal and sex glands).

The *releasing and release inhibiting hormones* are produced in the perikarya of the nerve cells **C** and pass in the axons of the *tubero-infundibular tract* **B2** to the *median eminence.*

The neuronal cell bodies containing the regulatory neurohormones *luliberin* (LRF, LHRH), *somatostatin* (SRIF) and *thyroliberin* (TRF) lie scattered in the periventricular zone, the perikarya of each individual hormone being situated in a different region of the "hypophysiotropic region". In contrast, perikarya producing the cortocotropin releasing factor corticoliberin (CRF) are concentrated in the *paraventricular nucleus.* The *nucleus infundibularis* contains scattered cells synthesizing prolactin inhibiting factor *prolactostatin* (PIF) (= *dopamine*) together with the neurons in which *corticotropin* (ACTH) is produced (and which may also contain *prolactin*), and those for *somatoliberin (somatokrinin).* The infundibular nucleus, a small celled, well de-

fined nucleus in the wall of the infundibulum, receives neural afferents from other parts of the brain, and regulates the release of regulatory hormones on the *median eminence.* The unmyelinated efferent fibers of these systems a directed towards the median eminence and each of them forms a relatively discrete bundle within the tuberoinfundibular tract.

Median Eminence

The **median eminence B3,** which includes the greater part of the wall of the infundibulum, is the *neurohemal region* where the neurohormones pass into the capillaries. The capillaries of the median eminence, the mesh of capillaries which penetrates the hypophyseal stalk radially from the outside, are surrounded by extended *perivascular connective tissue spaces.* The axonal endings of the neurohormonal nerve cells open into them and release the neurohormone into the perivascular space. Then, the neurohormones, pass into the blood stream through the "portal" veins **B4** ("portal" because like the portal vein of the liver, between two capillary beds they run between the capillary beds of the median eminence and the adenohypophysis) into the adenohypophysis **AB5,** where they stimulate or inhibit the release of hypophyseal hormones. The neurohormone appears in the form of variable-sized *vesicles with a dense core* in the axons and the axon endings.

Regulatory circuits. The perikarya and processes of neurohormonal cells have synapses (see Vol. 3) and are therefore under the influence of the central nervous system. The production and release of neurohormones is therefore under humoral control (through the blood vessels of the hypothalamic nuclear region, red arrow to 1), and also under control by different levels of the central nervous system (yellow arrow to 1, e. g., influence of psyche on ovarian circle, tactile stimulation on the breast to release milk etc.).

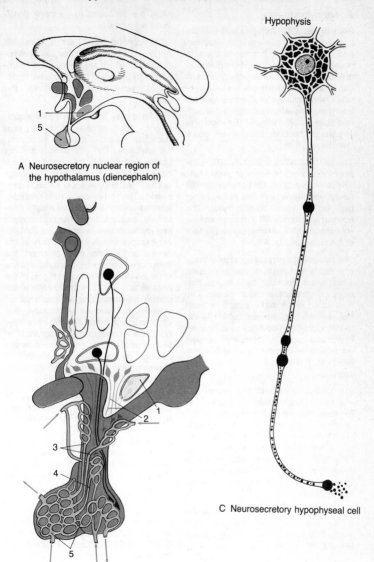

A Neurosecretory nuclear region of
the hypothalamus (diencephalon)

B Hypothalamo-adenohypophyseal system (red cells)
and hypothalamo-neurohypophyseal system:
schematic view

C Neurosecretory hypophyseal cell

Hypophysis

Adenohypophysis

Hormones. *Glandotropic* and *effector* hormones, see review p. 153. *Malfunction* of glandotropic hormones leads to malfunction of their dependent peripheral endocrine glands. *Inadequate production of somatotropin:* hypophyseal dwarfism. *Overproduction of somatotropin:* hypophyseal gigantism or, after full growth has ceased, acromegaly. (Enlargement of parts of the body like the ear, nose, chin, hands, feet).

The *pars infundibularis* of the *adenohypophysis* covers the hypophyseal stalk. The *pars intermedia* **AB 1** borders on the neurohypophysis and the other, larger part is the *pars distalis* **AB 2**. The **neurohypophysis AB 3** merges into the *infundibulum* **A 4,** which is part of the floor of the *diencephalon* **A 5.**

The **adenohypophysis** consists of irregular cords and nests of *epithelial cells,* which are surrounded by a sparse network of connective tissue fibers. The thin-walled *sinusoidal capillaries* spread out between the epithelial cells. The entire hypophysis is covered by a *connective tissue capsule.* The portal vessels enter the adenohypophysis near the pars tuberalis. The veins form a plexus in the capsule.

The *glandular cells* of the hypophysis may be stained differentially by a number of methods. A division into three groups of cells according to their staining characteristics mirrors the chemical nature of the hormones produced in these cells. The usual staining methods divide the cells into *chromophilic* (acidophilic or basophilic) and *chromophobic* ("neutrophil") cells.

The *proteohormones* STH and prolactin (PRL) are found in the so-called *acidophilic* cells, which may be stained with orange G **C 6.**

The *proteohormone* ACTH and the *glycoprotein hormones* TSH, FSH, LH and MSH are produced in *basophil (PAS-positive)* cells **C 8.**

Chromophobe cells **C 7** are *degranulated* (empty) cells of all types, or *stem cells* or *"star cells"* which are surrounded incompletely by groups of gland cells with long, thin processes which subdivide the gland.

The cells may be identified microscopically by the hormones which they produce with the aid of immunohistochemical methods. The gland cells are not strictly separated to their cell type, they have a scattered arrangement, though each all group does show local concentrations. The *somatotropin* (and *prolactin*) forming acidophilic cells lie mostly posteriorly and laterally; the *corticotropin* and *melanotropin* (or *lipotropin*) containing basophilic cells occur mainly in the central and anterior parts of gland; the *thyrotropin* producing basophilic cells are common in the antero-central part of the pars distalis and, basophilic cells which contain *gonadotropins (follitropin* and *lutropin*) are found more laterally. The *chromophobe,* possibly *stem cells* have no characteristic position. *Degranulated* chromophobe cells may be observed amongst all cell types and branches of chromophobic stellate cells penetrate everywhere.

The differently stained cells may be characterised electron microscopically by their content of membrane-bounded granules (vesicles with a dense core), the size of which varies from 140–600 nm according to the hormone they contain and according to their functional state. The cells may be further distinguished by the shape and position of their granules and the different appearance of their ergastoplasm and Golgi apparatus. The production and secretion of the hormone is by exocytosis. Specific hormone identification may be done immunohistochemically.

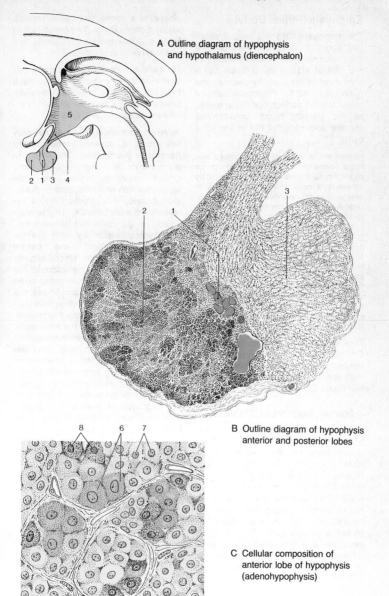

A Outline diagram of hypophysis and hypothalamus (diencephalon)

B Outline diagram of hypophysis anterior and posterior lobes

C Cellular composition of anterior lobe of hypophysis (adenohypophysis)

Epiphysis (Pineal Body)

The **epiphysis AB1** is a piriform structure, about 12 mm long, in the roof of the third ventricle. It lies above the anterior two colliculi **AB2** of the midbrain and is connected to the brain by pedicles, the *habenulae* **AB3**. Phylogenetically it arises like the parietal eye of the reptile, as the parietal organ (phylogenetic change from receptor cell to secretory cells).

Hormones. In amphibia the epiphyseal hormone, *melatonin,* causes contraction of the pigment cells, the melanocytes, with resultant lightening of color. It acts, therefore, as an antagonist to the melanotropin of the adenohypophysis. In the rat, melatonin which is produced from serotonin by an enzyme only present in the epiphysis, restricts the release of gonadotropin hormone and through this restrains development of the gonads. Particular types of pubertas precox (early sexual maturation) may be due to a subfunctional pineal gland. The effect of light on their development is also important. These and other observations suggest a relationship with circadian rhythms. In submammals (birds) the epiphysis is important in the coordination of hormonal processes in the hypothalamus.

Fine structure. In the lobular structure **C5**, vessels, connective tissue **C4** and parenchymal cell islands may be seen. With age, degeneration and cyst formation occur. **C6** Calculi.

Adrenal (Suprarenal) Glands

Each **adrenal** consists of two endocrine glands of different origin and with different functions: the *adrenal cortex* and the *adrenal medulla.* The adrenals weigh about 5 g each and sit like caps on the upper pole of the kidneys. Each is separated from the kidney by adipose tissue, but is enclosed within the renal fatty capsule. The right adrenal gland **D7** is more triangular and the left one **E8** more semilunar in shape. At the back each adrenal gland has a *hilum* where the veins and lymph vessels leave, whereas arteries and nerves enter the glands at many sites. The adrenal gland is enclosed in a connective tissue capsule which radiates into it. The organ, rich in lipids, is visible through the capsule as a yellowish shining structure.

On the cut surface of the in parts less than 1 cm thick fresh organ it is possible to distinguish the broader yellowish-brown *cortex* **E9** from the smaller, greyish-red colored *medulla* **E10.**

Adrenal hormones. The **adrenal cortex** produces several **corticosteroid hormones (corticoids).** The most important can be summarized in three groups: *mineralocorticoids,* (e. g., aldosterone) which regulates sodium-potassium balance (Na retention if K is excreted) and water balance. The *glucocorticoids* (e. g., cortisone) diminish carbohydrate utilization by cells (so the blood sugar rises) and control gluconeogenesis. The glucocorticoids can also produce a reduction in the number of lymphocytes circulating in the blood and inhibit the phagocytic action of granulocytes and monocytes, thus suppressing the symptoms of inflammation. Glucocorticoids also aid the nonspecific resistance of the organism to stress situations (adaptation to hunger, thirst and change of temperature). *Hyperfunction* (or excessive drug administration) leads to *Cushing's syndrome* with a moon face and obesity of the trunk, and *hypofunction* to general weakening of resistance which is ultimately lethal. Masculinizing hormones *(androgens)* are produced during the synthesis and breakdown of corticosteroids. *Overproduction* of androgens in women leads to their masculinization: *adrenogenital syndrome.* Female sex hormones are also produced in small quantities.

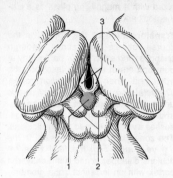

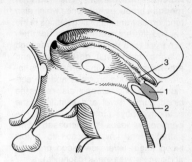

A Position of epiphysis from above and behind in relation to diencephalon and mesencephalon

B Position of epiphysis in relation to 3rd ventricle (sagittal section through diencephalon)

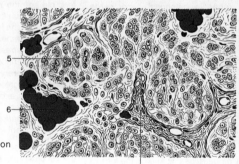

C Microscopic section through epiphysis

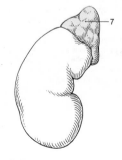

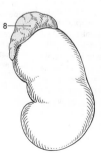

D Right adrenal gland

E Left adrenal gland, transverse section on right

(see also p. 163)

Adrenal hormones. The **adrenal medulla** produces hormones which also function as transmitter substances in the second efferent neurones of **sympathetic** nerves (see Vol. 3 its cells correspond to sympathicoblasts), **noradrenalin** and **adrenalin.** Both of these hormones increase the blood pressure and the cardiac output, and adrenalin also raises the blood surgar level and stimulates the adenohypophysis to relaese ACTH. *Hyperfunction* leads to increased sympathetic tone, *hypofunction* to a low blood sugar level, low blood pressure and general weakness (cachexia).

Adrenal Cortex

The *adrenal* **cortex A1** accounts for 80–90% of the total weight of the organ. It contains, apart from lipoids, lipofuchsin and other substances, and large amounts of vitamin C. **A2** Adrenal medulla.

Fine structure. The parenchyma of the adrenal cortex consists of strands and nests of epithelial cells between which connective tissue, blood vessels and nerves penetrate radially from the capsule into the medulla. During the course of life the epithelial cells of the cortex are arranged in zones which are transformed several times. Three layers are seen at the time of puberty. The glomerular zone *(zona glomerulosa)* **B3** lies in the form of cell nests externally beneath the capsule. The *zona fasciculata* **B4** follows toward the center and forms a broad layer of parallel strands of epithelial cells. This zone occupies the largest part of the cortex. Its big polygonal cells contain lipoid droplets which are dissolved by the usual histological fixatives to leave only holes. Lipochromes stain the cortex a yellowish-golden color. The *zona reticularis* **B5** adjoins toward the center. Its network of strands of cells is acidophilic; in old age the cells contain increasing amounts of lipofuscin granules. Even in freshly cut tissue the zone can be recognized with a magnifying glass as a pigmented band.

Transformation of the cortex. In the course of life the zones undergo changes, and shifts occur particularly at both borders of the zona fasciculata, the *outer* and *inner transformation zones. The fetal* and *newborn adrenal cortex,* under the influence of choriogonadotropic hormones from the placenta, possesses a well developed zona reticularis. The zona glomerulosa is absent. In *the 1st year after birth* the fetal zona reticularis disappears and a three-layered cortex with an indistinct zona glomerulosa and reticularis is apparent. In the *postpubertal period* the zona glomerulosa and reticularis become wider; the three layers are better defined in the female than in the male. In *postclimacteric* old age the zona glomerulosa and reticularis again become smaller, more abruptly so in the female than in the male – *regressive transformation.* If the organism is under *stress* ACTH is released in increased quantities, and this quickly results in temporary widening of the zona fasciculata, increased amounts of lipoid deposit and expansion of the zona glomerulosa and reticularis – *progressive transformation.* The two inner zones are dependent on the hypophysis (ACTH).

It is not possible to allocate with certainty the production of particular hormones to particular types or zones of cells with the exception of the mineralocorticoids which are in the zona glomerulosa and are not dependent on the hypothalamo-hypophyseal system but are stimulated by the renin-angiotensin-system. The precursors of the corticosteroids are probably produced by many cells of the cortex, but the last steps of synthesis are undertaken by specialized cell organelles of a restricted number of cells. The zona fasciculata is the most productive of all the layers.

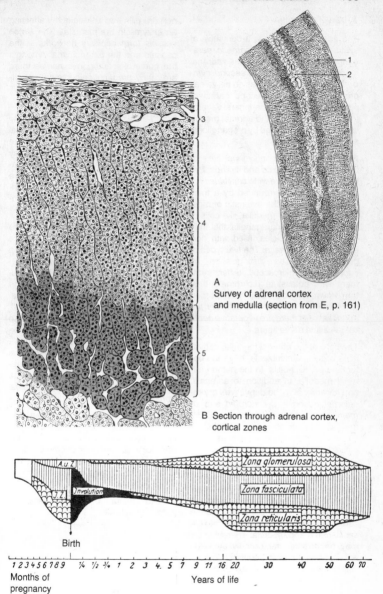

A
Survey of adrenal cortex
and medulla (section from E, p. 161)

B Section through adrenal cortex,
cortical zones

C Transformation of adrenal cortex at different ages (from Rotter)

A.u Z. = outer transformation zone
I. Z. = inner transformation zone

Adrenal Medulla

The **adrenal medulla** is a derivative of sympathoblasts, a *"sympathetic paraganglion"*. The *cells* of the adrenal medulla correspond to a *second sympathetic neuron,* which has no nerve cell processes. Like the second parasympathetic neuron in the peripheral vegetative nervous system, the adrenal medullary cells are innervated by preganglionic sympathetic fibers.

Fine structure. The polygonal cells of the medulla form nests and strands. Between them there are wide *capillaries* **A 1** and large *veins* surrounded by a thick muscle coat. It is characteristic of most cells of the adrenal medulla, like cells of other sympathetic paraganglia, that they contain cored vesicles, filled with *noradrenalin* or *adrenalin.* The two types of granules may be distinguished by fluorescence microscopy. After treatment with potassium dichromate they turn brown, and so the cells **A 2** are called *chromaffin (phaeochromic). Between the epithelial cells there are vegetative nerve cells* **A 3** and nerve fibers.

The cells of the adrenal medulla, derivatives of sympathicoblasts, have the potential of APUD-cells. In the human adrenal medulla, in addition to adrenalin and noradrenalin, the derivatives of *proopiomelanocortin* (β-*endorphin* and α-*melanotropin* and *lipotropin* and ACTH) may be observed as well as *enkephalins, substance P, somatostatin* and as *vasoactive intestinal polypeptide* (VIP).

Vessels and nerves. The *arteries* of the adrenal gland, the inferior phrenic artery and the renal artery (see p. 48), arise from the aorta. Wide, radially arranged capillaries run from the arteries into the cortex and blood flows from them into the capillaries of the medulla. There its flow may be arrested temporarily by *veins* with strong muscle coats (blocking veins). The direction of flow from the cortex into the medulla enables cortisol indirectly to induce the conversion of noradrenalin into adrenalin (by affecting an enzyme in the medulla). The *lymph vessels* form capillary networks in the capsule and the parenchyma. Amongst the numerous autonomic *nerves* there are many preganglionic sympathetic fibers (see Vol. 3) that enter the medulla.

Paraganglia

The small pea-sized knots of epithelial cell heaps which lie on the nerves are called "paraganglia". In the past they were subdivided into "sympathetic paraganglia", which lay on sympathetic nerve fibers, and "parasympathetic paraganglia" on nerve fibers which also contained parasympathetic nerves.

Sympathetic paraganglia. These are small *endocrine glands,* which produce noradrenalin and adrenalin and develop from sympathetic anlagen, which emigrate from the neural crest. Their structure and staining characteristics are the same as those of the adrenal medulla, the largest sympathetic "paraganglion". The sympathetic paraganglia in regard to the adrenal medulla are also called "extramedullary, chromaffin cell groups". They mostly occur in the retroperitoneal space. A sympathetic paraganglion about 1 cm long lies at the origin of the inferior mesenteric artery, the *abdominal aortic paraganglion* (Zuckerkandl).

The **"parasympathetic paraganglia"** associated with the glossopharyngeal and vagus nerves, the *carotid glomus* and *aortic glomus,* are *chemoreceptors* involved in regulation of respiration. The term "parasympathetic paraganglion" is no longer used.

Carotid glomus: **C 4** nerve fibers, **C 5** parenchymal cells.

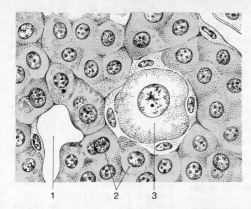

A Section through adrenal
medulla

1   2   3

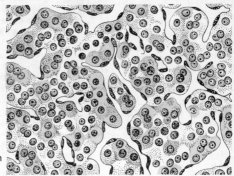

B Section through
retroperitoneal
sympathetic paraganglion
(after Watzka)

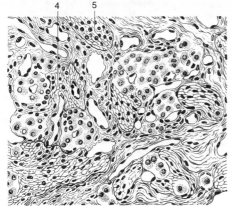

4   5

C Section through carotid
glomus (carotid body)
(after Watzka)

Thyroid Gland

The **thyroid gland** weighs 18–60 g and consists of two oval lobes **ABC1, 2**, which lie on either side of the trachea and the larynx. They are connected by a bridge, the *isthmus* **AC3**, at the level of the 2nd–4th tracheal cartilage. In 50% an appendix, the *pyramidal lobe* **A4** extends from the isthmus toward the hyoid bone. The gland is enveloped by a *fascia* **C5** and subdivided by connective tissue. Another tough *fibrous capsule* **C6** from the middle leaf of the cervical fascia is loosely connected to the thyroid gland in front and at the back to the trachea and to the sheath of the neurovascular bundle **C7** of the neck. In front of the thyroid are the sternothyroid and sternohyoid muscles and the middle layer of the cervical fascia **C8**. Between the organ capsule and the fibrous capsule posteriorly are the *parathyroid glands* **B9, BC10**.

C11 Skin of the neck, **C12** platysma, **C13** superficial layer of the cervical fascia and sternocleidomastoid muscle, **C14** deep layer of the cervical fascia, **C15** esophagus.

Hormones. The thyroid gland produces two hormones. **Thyroxine,** and its derivative *triiodothyronine,* stimulate cell metabolism and are essential for normal body growth. Somatostatin producing cells are involved in hormone release. In a case of *hyperfunction (hyperthyroidism,* "soft struma", goiter, *Grave's* or *Basedow's disease),* oxidative processes in cells are accelerated (loss of weight, rise of temperature, acceleration of the heart beat) and nervous irritation results. In *hypofunction (hypothyroidism)* metabolism, growth and mental activity are slowed down, and *myxedema* (swelling of subcutaneous tissues) develops. In congenital hypothyroidism (thyroid deficiency) dwarfism and idiocy *(cretinism)* ensue.

Calcitonin lowers the blood calcium level and promotes bone formation (antagonism to the action of the parathyroid hormone).

Fine structure. The gland consists of vesicular *follicles* **D16** and ducts where *thyroxin* is produced, and of loose connective tissue **D17**. The follicles are lined by a simple squamous to high cuboidal epithelium. Their lumen is filled with homogenous colloid. In the process of fixation marginal vesicles develop. Within the conglomeration of follicular epithelial cells and in the connective tissue are interspersed groups of lighter staining *parafollicular C-cells* **D18** which form *calcitonin.*

Hormone production and release. Thyroxine and triiodothyronine are produced in a stepwise fashion, then stored in the follicle bound to *thyroglobulin,* and released into the blood when required.

The protein *thyroglobulin* is synthesised in the basal part of the cell and is released apically into the lumen of the follicle. Circulating iodide (under the influence of thyrotropin) is rapidly taken up from the basal plasmalemma and oxidised to iodine. It is bound to thyroglobulin extracellularly in the colloid. In order to relaease the hormone, which is inactive in the follicle, under the influence of thyrotropin the epithelial cells take up the colloid at their apices. There, the apical lyosomes break the bonds between the hormone and the globulin and the hormone is released into the circulation by diffusion.

Functional phases of the follicle. While the secretion is being formed the epithelial cells are cuboidal **E19**, and during storage of the secretion they become flatter **E20**. During the phase of secretion the epithelium again grows taller **E19**; (in animal experiments within 30 minutes after injection of TSH). A follicle may shrink into a collapsed vesicle, but it can be refilled with colloid within a day. The thyroid gland is activated by cold and darkness, and is inhibited by warmth and light. The gland is enlarged during puberty and pregnancy. It gets smaller with age; starvation leads to degeneration.

Vessels and nerves. For the *superior thyroid artery* see p. 52. The *inferior thyroid artery* divides near the inferior laryngeal nerve. The *veins* drain into the jugular and left innominate veins. *Lymph vessels* drain into the deep and pretracheal cervical lymph glands. The *nerves* arise from the superior and inferior laryngeal nerves and from the third cervical ganglion of the sympathetic trunk.

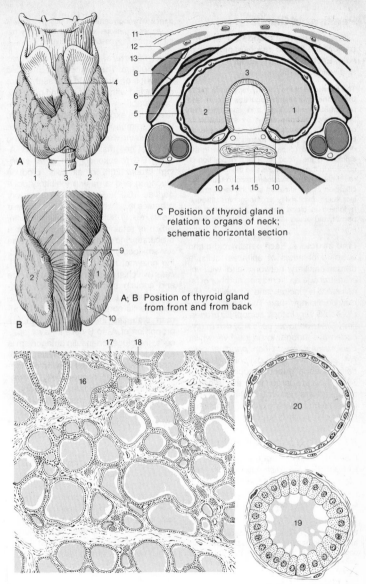

C Position of thyroid gland in relation to organs of neck; schematic horizontal section

A; B Position of thyroid gland from front and from back

D Section through thyroid gland

E Thyroid follicle above: filled, below: empty

Parathyroid Glands

The 4 **parathyroid glands A1** lie at the back of the thyroid inside its organ capsule. Each is lentiform, about 8 mm long and weighs 30–50 mg.

The **hormone** of the parathyroid glands, **parathormone**, (parathyrin) regulates calcium and phosphate metabolism, and stimulates the breakdown of bone tissue by stimulation of osteoclasts. *Hyperfunction* of the glands leads to increased phosphate excretion, breakdown of bone tissue (decalcification), a rise in the calcium in the blood level with deposition of calcium in the vessel walls and the formation of renal calculi. *Hypofunction* leads to faulty calcification of the skeleton and the teeth and to increased excitability of the nerves through reduction in blood calcium. Removal of the parathyroid gland leads to muscle spasms – *tetany*.

Fine structure. Each parathyroid gland consists of nests of epithelial cells, a dense capillary network, and with increasing age an increasing number of fat cells **A2**. Principal and oxyphilic cells may be distinguished. The (smaller) principal cells vary in their acidophilic properties. *Clear principal cells* **A3**, rich in glycogen are thought to be inactive, whilst *dark*, ergastoplasm rich *principal cells* **A4** are active. The *oxyphilic (Welsh) cells* appear at puberty (proportion about 3%), they are larger then the principal cells and are rich in mitochondria. It is possible that they are different functional states of a single type of cell.

Islet Cell Organ of the Pancreas

0.5–1.5 million **cellular islets of Langerhans** are distributed throughout the *pancreas*, more in the body and tail than in the head. They consist of endocrine cells and are known as the *islet organ*.

Islet hormones. The islets produce at least 2 antagonistic hormones: **Insulin** promotes glycogen synthesis in the liver and thereby lowers the blood sugar level. **Glucagon** leads to glycogenolysis in the liver and so raises the blood sugar level. Hyper- and hypofunction with regard to insulin are known. *Hyperfunction:* tendency to hypoglycemia *Hypofunction* is present in *diabetes mellitus* and is manifested by hyperglycemia, glycosuria and polyuria.

Fine structure. Each islet measures 100–500 μm. The islet cells are arranged in groups or strands. In places they are connected with exocrine gland cells **B5**, comparable with endocrine cells between the exocrine intestinal epithelia. Using immunohistochemistry and special staining techniques, the following types of cell may be distinguished. A-cells, **B6** (blackened with silver staining), about 20% of all cells, produce glucagon and a gastrin inhibitory polypeptide (GIP). B-cells, **B7** (secretory granules fixed with difficulty, cells appear clear), about 80% of cells, form insulin, which is retained in the granules as a precursor zinc complex. At the same time a γ-aminobutyric acid (GABA) synthesising enzyme, may be demonstrated. *D-cells* (which stain blue with azan) are rare and contain *somatostatin,* which regulates insulin release (inhibitory). In addition D-cells contain β-endorphin. In human diabetics there is a shift in the proportion of A- to B-cells in favor of A-cells; the glucagon-insulin antagonism is upset in favor of glycogenolysis. Many other peptide hormones may be demonstrated immunohistochemically. Corresponding to its developmental history, in the formation of the association of its exocrine and endocrine parts, the pancreas resembles a greatly compressed collection of modified epithelial cells.

Vessels and nerves. The islets are permeated by wide *capillaries* **B8**. A *microcirculation* exists between the individual islet and the surrounding exocrine tissue through small *insuloacinar portal vessels* (influence, for example of the D-cells on the activity of the exocrine pancreas). *Sympathetic fibers* stimulate the secretion of glucagon and inhibit that of insulin, which is stimulated by the *vagus nerve. Serotininergic nerve fibers* inhibit the release of insulin.

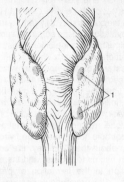

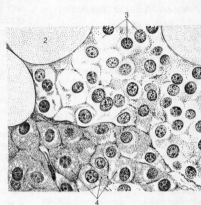

A Position of parathyroid
glands
Right: microscopic section
through gland

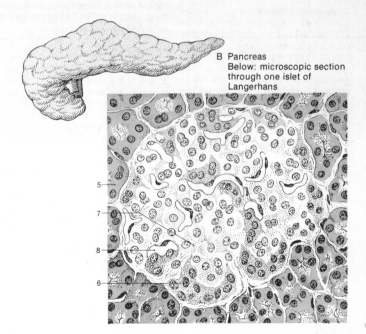

B Pancreas
Below: microscopic section
through one islet of
Langerhans

Gonads as Endocrine Glands

The sex hormones are formed in the gonads (reproductive glands), and to a lesser extent in the adrenal cortex (see pages 158/162).

Ovary as an Endocrine Gland

The regulation of endocrine activities is particularly conspicuous in the female sexual (menstrual) cycle. We distinguish the actions of the *hypophysis,* and indirectly the *hypothalamus,* on the ovary. Other actions result from the influence of the *ovary* on the mucosa of the uterus (womb), and so eventually (by retroaction) on the hypothalamus and the hypophysis.

Ovarian (menstrual) cycle. For further information on the phases of the menstrual cycle see p. 300. *Releasing factors* of the *hypothalamus* (see p. 153) cause secretion of the gonadotropic hormones of the anterior lobe of the hypophysis **A1.**

Desquamation-regeneration phase (Days 1–4). *Hypophysis:* a small rise in LH and FSH at the end of the previous cycle affects the ovary. *Ovary:* the corpus luteum of the previous cycle degenerates and there is a fall in the production of progesterone. The beginning of a rise in estrogen production in the cells of the *theca interna* of the growing follicle **A2** with effect on the mucous membrane.

Phase of proliferation ("estrogenic phase") (Days 5–15). *Hypophysis:* Steady level of LH and at first a small reduction in FSH. Around the 15th day a short-lived peak in LH production and a (smaller) FSH-peak. These hormones act on the ovary. *Ovary:* FSH stimulates the follicle to ripen and LH stimulates the production of **estrogens.** Immediately after the LH peak *ovulation* (follicle rupture) occurs **A3.** *Hormonal feedback:* the rise in estrogen level suppresses the formation of hypophyseal hormones. The estrogens act on the mucous membrane. *Secretory phase* ("luteal or gestagenic phase") (Days 15–28). *Hypophy-*

sis: reduction in the LH and FSH peak until the 28th day, when there is a small rise in LH and FSH release. The hormones act on the ovary. *Ovary:* LH produces the conversion of the empty follicle after ovulation into the *corpus luteum* **A4,** with simultaneous inhibition of estrogen production, and starts the secretion of **progesterone,** the hormone which is produced by the corpus luteum. At the same time, the ripening and ovulation of other follicles is inhibited. *Hormonal feedback:* the high concentration of progesterone restricts the formation of gonadotropic hormones in the hypophysis. At about the 22nd day, the corpus luteum begins to regress and there is a reduction in the formation of progesterone. (If pregnancy ensues, the trophoblast takes over its production = chorionic gonadotropin, progesterone production remains high and the corpus luteum remains intact = corpus luteum of pregnancy). Toward the end of the cycle, the low level of progesterone stimulates the production of FSH. The hormones act upon the mucous membranes.

Testes as Endocrine Glands

Leydig's interstitial cells **ABC5,** the producers of the male sex hormones, **androgens,** are arranged in groups situated around the blood vessels **C7** in loose connective tissue between the convoluted seminiferous tubules (tubuli seminiferi contorti) **AB6.** Small amounts of female hormones are also supposed to be produced. Androgens have a local effect on sperm maturation, as well as a general one in which they promote development of the genitalia and of the secondary sex characteristics. The *interstitial cells* are stimulated by LH and *sperm maturation* **AB6** by FSH from the hypophysis (see p. 154).

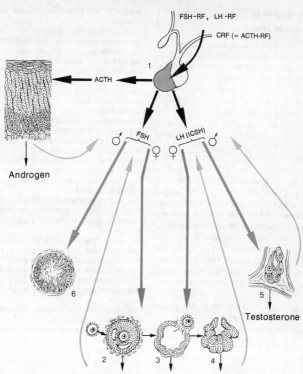

FSH-RF, LH-RF

CRF (= ACTH-RF)

ACTH

Androgen

FSH

LH (ICSH)

Estrogen Estrogen Progesterone

Testosterone

A Sex hormones, dependence on hypothalamus and hypophysis and their effects on them (menstrual cycles)

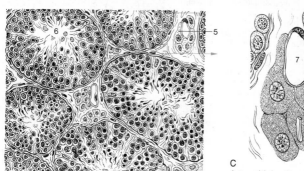

B Section through testis

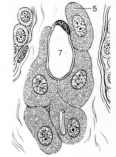

C
Interstitial cells of testis (section from B)

System of Gastrointestinal Endocrine Cells

The endocrine cells of the gastrointestinal tract are together known as the system of gastrointestinal endocrine cells. They are found in the layer of secretory and absorptive epithelial cells and as representatives of the *APUD-system* they produce *peptide hormones* and monoamines. The cells **A1** are oval or pyramidal in shape smaller than mucosal epithelial cells and have extensive contact with the basal lamina and the adjacent capillaries. The opposite secretory granules of endocrine cells lie basally ("basal granulated cells"). The apices **B2** of some of the cells reach the lumen of the gut **B3** ("open type") and are directly exposed to stimuli from the intestinal lumen (e. g. changes in the piti-value), whilst others do not reach the lumen ("closed type"). The hormones produced in the cells **B4** in response to a stimulus **B5** secreted basally by exocytosis **B6.**

Distribution **C** and effects of the active substances of endocrine cells.

EC-cells, enterochromaffin cells, produce a brown reaction product on fixation with potassium chromate due to oxidation and they are found to be "argentaffin" by reducing silver nitrate to form a black reaction product. They produce *serotonin,* which induces contraction of the musculature of the blood vessels and the intestinal wall etc. Occurrence: stomach, intestine, pancreas, bronchi.

ECL-cells, "enterochromaffin-*like*", resemble EC-cells but can be histochemically distinguished from them. They produce *histamine,* a hormone which increases the local permeability of capillaries. Occurrence: only stomach.

EG-(A-) cells are *glucagon producers* (enteroglucagon) like the A-cells of the islet organs. Occuring in stomach, small and large intestine, *D-cells* functionally correspond in their locally inhibitory effect to somatostatin producers in the appropriate organ.

G-cells produce *gastrin,* which enhances water, electrolyte and enzyme secretion in the stomach, duodenum and pancreas, and inhibits reabsorption of water in the small intestine.

S-cells produce *secretin,* which stimulates the release of pepsin and enhances the secretory activity in the intestine, pancreas and gall bladder.

Vasoactive intestinal polypeptide (VIP) which is structurally related to secretin, GIP and glucagon, produces dilatation of smooth muscle (vasodilatation, sphincter control), stimulates intestinal secretion, production of bicarbonate in the pancreas, enhances bile flow, lipolysis, release of several hormones and inhibits release of gastric acid.

Cholecystokinin (CCK) (pancreozymin). Under the influence of fatty acids, amino acids and low pH in the intestinal lumen it is released from the intestine into the blood; in the stomach it stimulates secretion of gastric acid and peptides, enhances stomach and small intestinal motility, bile flow of the gall bladder and the production of pancreatic juice rich in enzymes. It gives the feeling of satiation.

Gastrin inhibitory peptide (GIP) acts antagonistically to gastrin (inhibition of hydrochloric acid secretion) and thus shortens the gastric phase of digestion.

Bombesin (gastrin releasing peptide, GRP) stimulates the release of gastrin, PP and motilin, and thereby increases gastric acid secretion. Occurrence: stomach, duodenum.

Motilin stimulates intestinal motility.

Pancreatic polypeptide (PP) is of uncertain function.

Neurotensin (NT) produces hyperglycaemia on release into the blood after a meal.

Enkephalin inhibits the action of somatostatin. *Occurrence: stomach-intestinal tract, particularly the antrum.*

Endocrine and paracrine secretion.

Many hormones of the gastrointestinal endocrine cells produce their effects, via the blood stream. However, other hormones have such a short half-life in the blood (less than one minute), that a (prolonged) action via the blood stream **B6** is unlikely (somatostatin, SP, VIP, neurotensin). A more local action has been hypothesised either in the extracellular space **B7,** or via locally limited vascular systems. This has been called "paracrine" secretion.

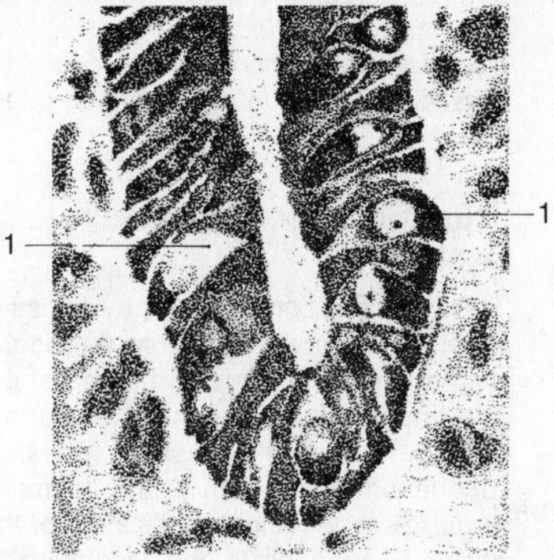
1
1

A Basal granulated cells of human duodenum

B Endocrine cells, open type

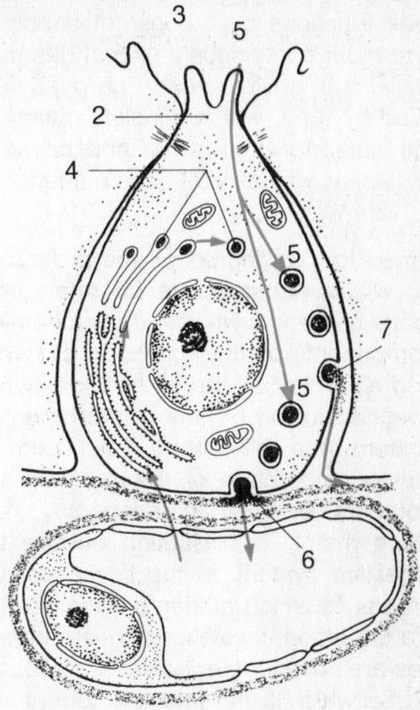
3
5
2
4
5
7
5
6

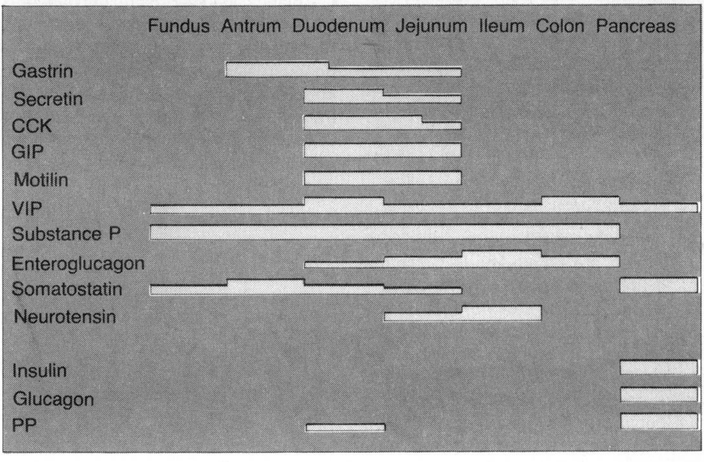
Fundus
Antrum
Duodenum
Jejunum
Ileum
Colon
Pancreas
Gastrin
Secretin
CCK
GIP
Motilin
VIP
Substance P
Enteroglucagon
Somatostatin
Neurotensin
Insulin
Glucagon
PP

C Distribution of a selection of gastrointestinal endocrine cells, human (from Heitz 1980)

174

Digestive System

The body maintains its structures and their functions by a constant *supply of energy*. If continuously exchanges matter with its environment. Energy is supplied by food, which consists mainly of the *nutrients protein, fat* and *carbohydrates,* as well as vital trace nutrients, e. g., vitamins.

Digestion comprises all the processes by which nutrients are liberated from food, broken down into their chemical components by the actions of enzymes and absorbed by the body. Digestion is confined to the organs of the digestive system, the alimentary tract and its glands. Liberation of energy from the nutrients or their components takes place mainly by oxidation outside the digestive system, in the tissues of the organs to which nutrients have passed via the blood circulation. These processes are called *intermediate metabolism,* part of which is the "internal respiration", which must be distinguished from the "external respiration", i. e., the movement of air to the alveoli via the respiratory organs. An essential prerequisite for intermediate metabolism is carefully regulated salt and water metabolism. Thus, several organ systems contribute to the intermediate metabolism: the digestive, respiratory and circulatory systems, and, in a coordinating capacity, the endocrine glands and the nervous system.

Role of the Foregut in the Digestive System

The digestive organs, like the organs of respiration, are situated partly within the *head* and *neck,* and partly within the trunk. The anterior part *(foregut)* of the digestive system comprises the oral cavity 1 with numerous small and 3 pairs of large salivary glands, and the middle and lower parts of the pharynx (throat) 2 to the beginning of the esophagus (gullet) 3. In the foregut the food is taken in by means of the *lips, teeth* and *tongue,*

chewed, lubricated with saliva and swallowed in single bits. The digestion of starch starts within the mouth, e. g., if soaked with saliva, bread tastes sweet. The *taste* and *olfactory organs* control the chemical composition of food. The *tonsils* serve for defense against infections.

Trunk Section of the Digestive System

This section comprises the digestive tube (alimentary canal) from the beginning of the esophagus to the end of the gut (anus). It is divided into the *esophagus* 3, *stomach* 4, *small intestine* (duodenum 5, jejunum 6 and ileum 7), *large intestine* (colon), consisting of the cecum 8, appendix 9, and ascending 10 and transverse 11 and descending 12 colon, the sigmoid colon 13 and the rectum 14. The esophagus is only a tube for transport. The break-down of nutrients takes place partly within the stomach and is completed in the small intestine, where the separate constituents of the nutrients are absorbed. As in the foregut, there are a large number of small and the two large digestive glands, the *liver* 15 and the *pancreas* 16, which produce digestive juices and release them into the small intestine. The unabsorbed part of the food is thickened in the colon by water extraction and transformed into feces by fermentation and decomposition. The feces are then moved to the end of the intestine (anus).

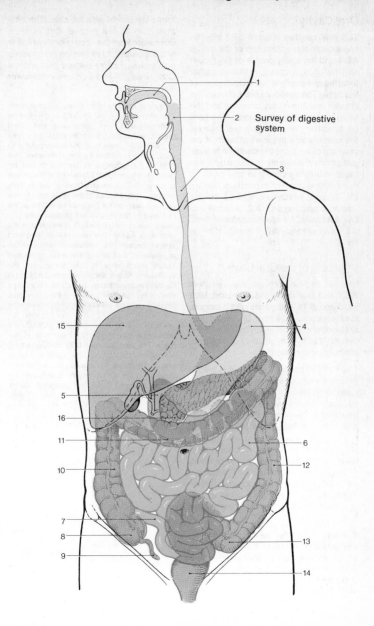

Survey of digestive system

Oral Cavity

The **oral cavitas** *(cavum oris)* may be divided into the *vestibulum of the mouth* **AB1** and the cavity proper of the mouth *(cavitas oris propria)*, which together form the oral cavity in the widest sense of the term. The vestibule lies between the cheeks and lips on the one side and the teeth and alveolar processes of the jaws on the other side. The true oral cavity is the space inside the rows of teeth. If the jaws are closed and the dentition is complete there is no communication between the vestibule and the true oral cavity. If the mouth is opened its posterior border, the *fauces,* which are formed by the posterior palatine arch **A2,** become visible. **A3** Uvula, **A4** anterior palatine arch, **A5** palatine tonsil, **A6, 7** frenulum of the lip.

Vestibule of the Oral Cavity

Median section through the *vestibule* **AB1** and *oral cavity* **B8. B9** Lips, **B10** hard and **B11** soft palate, **B12** tongue, **B13** pharynx, **B14** root of the tongue, **B15** entrance into the larynx, floor of the mouth (**B16** mylohyoid, **B17** geniohyoid, **B18** digastric muscles).

The *lips* and *cheeks* form the very elastic external wall of the vestibule, a muscle plate (orbicularis oris and the buccinator muscle) which in some places is firmly attached to the skin of the face. Mimetically, the skin follows the movements of the muscles. Inside, the muscle plate is loosely covered by the mucous membrane of the mouth. Where the opening of the mouth is bordered by the lips the skin of the face and mucous membrane of the oral cavity join each other across an intermediate zone, the red area of the lips.

The lips, *labia oris,* are connected laterally at the corners of the mouth. The upper lip reaches to the base of the external nose and to the *nasolabial fold* **C19,** and the lower lip to the chin-lip groove *(mentolabial fold)* **C20,** which forms the border with the chin. The *philtrum* **C21** is the groove that extends downward from the nose over the midline of the lip, and in its extension the upper lip possesses a berry-shaped prominence **C22,** which fits into a groove in the lower lip.

Skin zone, red area of the lip and mucosal zone. The *skin zone* of the lip **D23** is covered by epidermis with hairs, sebaceous and sweat glands (see p. 336). Its border with the *red area of the lip* **D24** projects as the *margin of the lip* **D25,** which is formed by the extroverted part **D26** of the sphincter of the mouth, the *orbicularis oris muscle* **D27.** This structure serves to curve the lip inwards and to aid its tight closure. The red area of the lip is not sharply demarcated from the mucosal zone **D28.** In the red area of the lip **D24** cornification and pigmentation of the epithelium diminish and the skin dries easily. The red color of the blood shines through the epithelium because the capillaries of the high papillae of the connective tissue reach near to the surface of the epithelium. A dark discoloration of the blood in O_2 deficiency, cyanosis, is apparent in this red area of the lips. The papillae of the connective tissue secure the interlocking of epithelium and connective tissue. At the border between the red area and the mucosa in 50% of adults are sebaceous glands which appear only at puberty. The *mucosal zone* **D28** continues as a pocket into the gums of the upper and lower jaw. In the center the mucosa is connected to the gum by the *frenulum of the lip* **A6, 7.** In the mucosal zone there are pinhead sized, seromucous lip glands (glandulae labiales) **D29.**

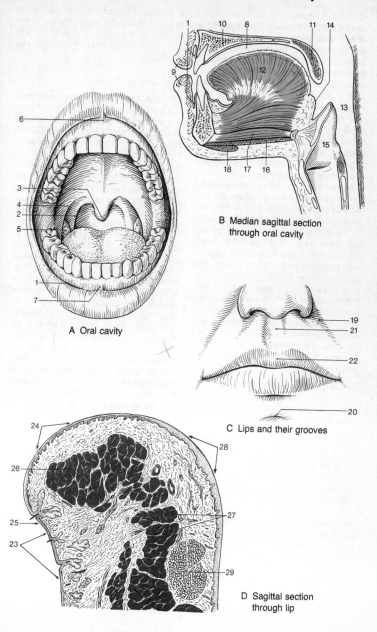

A Oral cavity

B Median sagittal section through oral cavity

C Lips and their grooves

D Sagittal section through lip

Cheeks (buccae). The *buccinator muscle* **AB1** forms the muscular basis of the cheek. It extends from the pterygomandibular raphe and the neighboring parts of the upper and lower jaw to the angle of the mouth. The muscle plays a part in sucking and chewing and can be formed into a cheek pouch. Outside the buccinator muscle, at the anterior margin of the masseter muscle, is the *adipose body of the cheek* (Bichat's fat plug). The fat plug helps to stiffen the cheek during suction. At the level of the 2nd upper molar tooth the buccinator muscle and the mucosa are obliquely perforated by the excretory *duct of the parotid gland* **B2** which opens into the vestibule **A3** (the parotid duct). At the back, a vertical fold, an extension of the row of teeth, ascends to the soft palate. It is formed by soft tissue in front of the ramus of the mandible (lower jaw) and it can easily be seen and palpated in the open mouth. The mucosa of the cheek is movable. Two rows of seromucous glands of the cheek, the buccal glands, continue the line of the glands of the upper and lower lip laterally. They extend into the muscles. Small sebaceous glands are also present. **A4** Tongue, **A5** palate, **B6** pterygoideus lateralis, **B7** pterygoideus medialis, **B8** orbicularis oris.

Gums (gingiva). The alveolar processes of the upper and lower jaw (maxilla and mandible) are covered by the *oral mucosa*. Its connective tissue firmly grows together with the periosteum. However, in the transition zone between the cheek or the lip and the gums there is only a loose connection between the mucosa and the periosteum, and here inflammatory processes may cause the mucosa to bulge out; in this case the nasolabial or the chin-lip groove is often flattened. The epithelium of the mucosa at the external surface of the alveolar processes, known as the *external border epithelium*, consists of tall cells and is firmly interlocked with the connective tissue of the mucosa by tall papillary processes. It may cornify in places. The *inner border epithelium* reaches above the upper margin of the alveola to the neck of the tooth and covers the *periodontum* from above, see p. 182.

Oral Cavity Proper

The **oral cavity** in the narrower sense **A9** lies behind the teeth and extends to the *isthmus of the fauces* in the region of the back of the tongue, see p. 176. The muscular floor of the mouth is formed by the *mylohyoid muscles* **AC10**, which run from the mylohyoid line of the mandible to a medial raphe and to the hyoid bone **C11**. Above the floor of the mouth near the median plane lie the two *geniohyoid muscles* **AC12**. The *sublingual gland* **A13** lies on either side between these muscles and the mandible. The paired *genioglossus muscles* **AC14** arise above the floor of the mouth in the center of the inner side of the mandible. They constitute the bulk of the body of the tongue. The anterior belly of the *digastric muscle* extends below the floor of the mouth on both sides from the region of the lesser horn of the hyoid bone to the digastric fossa of the mandible. On either side the *submandibular gland* **A16** lies between the muscle and the mandible. The *roof* of the oral cavity is formed by the hard and soft palate. The cavity is mainly filled by the tongue. **A17** Antrum (maxillary sinus), **A18** platysma.

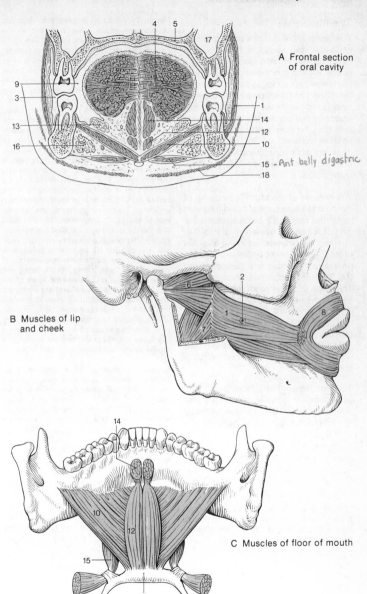

A Frontal section of oral cavity

- Ant belly digastric

B Muscles of lip and cheek

C Muscles of floor of mouth

Teeth

The **teeth** *(dentes)* form an arch which separates the vestibule from the oral cavity proper. Human teeth lie close together without spacing and are individually shaped according to their specialized functions. Chisel shaped incisors are for biting, long canine teeth which are firmly rooted in their anchorage, are for tearing and gripping, the premolars with their broader crowns are for grinding and crushing, while the molars do most of the chewing with the broad grinding surface of their crowns.

The nucleus of the **tooth** is made up of *dentin* **B1**. Teeth consist of the following parts: the *crown* **A2** extends above the gingiva **B3** and is covered by enamel **B4**. The *root* **A5** is anchored in the alveolus of the jaw by connective tissue fibers, the periodontal membrane *(desmodontium)* **B6**. The root is covered by *cement* **B7**. The edge of the enamel meets the cement at the neck of the tooth, *collum* **A8**. The *root apex* **A9** is perforated by the root canal **B10**, which leads into the pulp cavity *(cavitas dentis)* **B11**, filled by the *pulp* of the tooth. The pulp of the crown with its horns is distinguished from the pulp of the root.

The *root canal* before closure is a simple, wide tube. It can be subdivided into smaller canaliculi **C12** interconnected by transverse anastomoses. The apical end of the root canal can give off lateral branches, or it can divide into divergent branches that perforate the apex of the root like a sieve, *"apical ramification"* **C13**. This may make root canal treatment very difficult.

Fine structure. *Dentin, enamel and cement,* the hard substances of the tooth are very much like bone. Comparative composition of bone, dentin and enamel:

Dentin B1 (substantia eburnea) is produced by *odontoblasts* which are comparable to the bone-forming osteoblasts, but unlike them odontoblasts are not embedded in the material they produce. They lie adjacent to the inner surface of the dentin, and their processes (Tomes' fibers) lie in the dentin canaliculi and extend as far as the enamel-dentin of the cement-dentin border. Another difference from bone is that dentin does not contain any blood vessels. In a section through a tooth near the border with the enamel, there are rhomboidal or serrated *interglobular spaces* **B14**, areas of deficient calcification of the ground substance. The dentin of the deciduous teeth (milk teeth) is thinner and less resistant than that of the permanent teeth. The **enamel B4** (substantia adamantina) is the hardest substance in the human body. It consists of small, convoluted, unilaterally furrowed columns, the *enamel prisms*, about 5 μm in diameter. They are bound together by a sparsely calcified ground substance. The prisms are mainly arranged radially. The **cement B7** (substantia ossea) is a network of *bony* substance with few cells and with collagen fibers linking it to the dentin and the alveolar wall **B15**, where the fibers of the periodontium (Sharpey's fibers) are anchored. The **pulp of the tooth:** the loose connective tissue of the pulp **B11** contains blood vessels, and medullated and non-medullated nerve fibers. New odontoblasts arise from the connective tissue of the pulp and they are responsible for the formation of *secondary dentin* later in life.

Marginal epithelium: The stratified, non-keratinized squamous epithelium of the gingiva covers the alveolar process as the *outer marginal epithelium* **B16**. As the *inner marginal epithelium* **B17** (pocket epithelium) it stretches over the margin of the alveola to the neck of the tooth and covers the periodontium **B6** from above. Formation of a pocket more than 1 mm deep at this site is pathological (periodontopathy).

	Inorganic matter	Organic matter	Water
Bone (4 yrs)	48%	25%	27%
Bone (adult)	62%	25%	13%
Dentine	69.3%	17.5%	13.2%
Enamel	96%	1.7%	2.3%

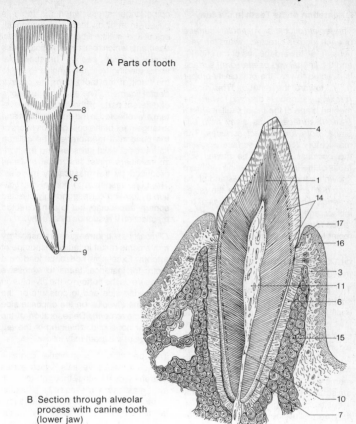

A Parts of tooth

B Section through alveolar
process with canine tooth
(lower jaw)

C Variability of root canals: casts of
first and second premolars
(after Keller)

Orientation of the Teeth in the Jaw

The *labial (buccal or vestibular)* surface of each tooth is directed toward the vestibule (i. e., toward the lips or the cheek), and the *lingual* and *palatal* (oral) surface is directed toward the oral cavity proper (i. e., toward the tongue). The *mesial (proximal)* surface is directed toward the median plane of the facial skull, its *distal (lateral) surface points away from this plane; both are contact surfaces.* The masticatory (chewing) surface is called the *occlusal surface.* The *mesial surfaces* of the incisor and canine teeth are known as the *facies medialis* and of the premolar and molar teeth as the *facies anterior,* whilst the *distal surfaces* of the incisor and canine teeth are called the *facies lateralis* and of the premolar and molar teeth the *facies posterior.*

Structures Supporting the Teeth in Position

The **periodontium** provides a springy suspension of the teeth in the bony alveoli **B** of the alveolar processes of the jaw, *gomphosis.* The periodontium consists mainly of collagenous fibers which extend between the periosteum of the alveolar wall **DE 1** and the cement **DE 2** to which they are attached. The periodontal membrane and the alveolar wall, form the structures which keep the teeth firmly in position and, together with the cement, they are called the **parodontium.** They develop together during eruption of the teeth.

The course of the fibers in the periodontal membrane **DE 3.** Most of the fibers run steeply downward from the alveolar wall **DE 1** toward the apex of the root, *oblique ligament;* they are exposed to the axial stresses of mastication. Toward the alveolar margin the fibers run more horizontally or radially, the *horizontal ligament,* and the fibers connect adjacent teeth in the interdental space. Finally, fibers ascend from the alveolar margin toward the neck of the tooth. These fibers

come under stress when the tooth is pulled. In addition there are tangential, spiral and netlike fibers, the *tangential ligaments* which resist mechanical stress in any direction, see **C.** The superficial fibers running from the gingiva (gum) to the neck of the tooth form the *circular dental ligament.* The separate functions of different parts of the masticatory system are reflected in the development and arrangement of the collagenous fibers of the peridontal membrane. The oblique fibers resist axial stresses and the horizontal fibers resist horizontal stresses produced by the masticatory muscles. The bone responds to normal mastication by forming correspondingly directed spongy trabeculae, but to a misdirected mastication it responds by atrophy.

Clinical tips: Abnormalities in masticatory movements result in faulty anchorage of a tooth. Excessive horizontal load on a tooth, for instance, leads to excessive pressure on the bottom of the alveolus on the loaded side and to pressure on the rim of the alveolus on the opposite side (tilting movement). Degeneration of the alveolar bone and loosening of the anchorage of the tooth may follow.

The periodontal membrane contains glomerular blood vessels, which act as hydraulic cushioning thought to curb axial masticatory pressure. Its pulsation is transmitted to the tooth. The periodontal membrane is supplied by tactile nerves *(pressure sensation)* and contains lymph vessels. The lymph flow is directed toward the tongue and cheeks into regional submandibular lymph nodes (from the maxilla via the lymph vessels of the cheek).

DE 4 Dentin, **D 5** pulp cavity.

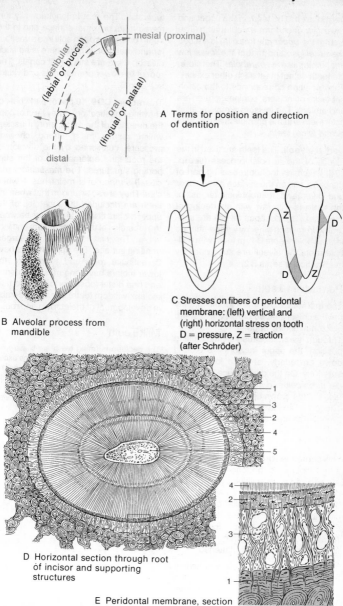

mesial (proximal)

vestibular
(labial or buccal)

oral
(lingual or palatal)

distal

A Terms for position and direction
of dentition

B Alveolar process from
mandible

C Stresses on fibers of peridontal
membrane: (left) vertical and
(right) horizontal stress on tooth
D = pressure, Z = traction
(after Schröder)

1
3
2
4
5

D Horizontal section through root
of incisor and supporting
structures

E Peridontal membrane, section
from D

4
2

3

1

Dental arch. The teeth in the upper and lower jaw each form a dental arch. The arch of the *upper jaw* (maxilla) is shaped like an *ellipse* and that of the *lower jaw* (mandible) like a *parabola*. Therefore, the teeth do not meet each other exactly. Their position corresponds to the different tasks of incisors, canines, premolars and molars. The incisors are also called *front teeth* and the premolars and molars *lateral (side) teeth*.

Dental alveoli. The teeth are fixed in the alveoli of the alveolar process (see p. 176) which, as the tooth-bearing part of the bone, sits on the surface of the upper and lower jaws. It atrophies after loss of the teeth (senile jaw with shortening of the face). The osseous alveoli are separated from each other by wedge-shaped *interalveolar septa* **B 1**. In teeth with multiple roots the alveoli are subdivided by inter-radicular septa **B 2**.

Permanent Teeth

The **incisors CD 3, 4** are for biting, they have a chisel shaped crown with a sharp horizontal *cutting edge*.

Through wear, because of their biting position, the upper incisors are ground down posteriorly, and the lower teeth anteriorly. On the oral surface is a tubercle *(tuberculum dentis)*. The lateral surfaces of the crown are almost triangular. The root of an incisor is long, conical and slightly flattened at the side. The upper incisors **CD 3** are broader than the lower ones **CD 4**.

The **canine teeth CD 5, 6** do the tearing and gripping. As the longest teeth they are secured against tilting by a long root, particularly the maxillary canines, embedded in the canine pillar of the facial skeleton. The crown has two cutting edges converging at an acute angle *(chewing point)*. The root is single, strong, long and laterally flattened.

The **premolars CD 7, 8** carry out chewing movements. They have a *masticatory (occlusal) surface* and a crown with two tubercles. The root is longitudinally furrowed at its proximal surface and in the upper premolars often split into a vestibular and oral root. Where there is no such division there are still 2 root canals. The root of the lower premolars is undivided, see also **B**.

The **molars CD 9, 10** perform the bulk of chewing (mastication). They lie in or near the direction of the masticatory muscles, which results in a strong chewing pressure, cushioned by the division of the roots and enlargement of the supporting structures. The *masticatory (occlusal) surface* of a molar has 4 tubercles. They are so arranged that when the teeth are occluded the tubercles of the upper molars fit into the grooves between the tubercles of the lower molars and vice versa. The first molar has the largest grinding surface. The upper molars have 2 vestibular roots and 1 oral root, the lower molars have one mesial (anterior) and one distal (posterior) root. The third molars (wisdom teeth) vary considerably in development of their crown and root.

Milk Teeth

The **deciduous (milk) teeth** are bluish-white and translucent like porcelain. With the exception of the first molars they resemble the permanent teeth. The divergent course of the roots of the milk molars is of practical importance, since the permanent tooth germs develop between their roots.

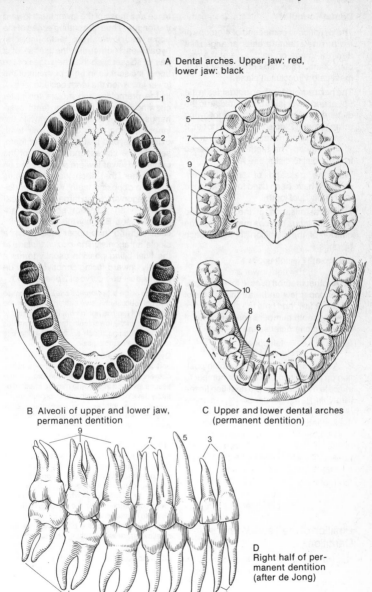

A Dental arches. Upper jaw: red, lower jaw: black

B Alveoli of upper and lower jaw, permanent dentition

C Upper and lower dental arches (permanent dentition)

D Right half of permanent dentition (after de Jong)

Dental Formula

The dentition is composed of 4 *groups of teeth* in the 2 dental arches, arranged like a mirror-image on each side of the median plane. They face each other in the masticatory (occlusal) plane.

The **permanent teeth** on both sides in the upper and lower jaw from the middle to the distal end of the row have the following arrangement: 2 incisors, 1 canine tooth, 2 premolars and 3 molars, i. e., a total of $4 \times 8 = 32$ teeth. The teeth are numbered from medial to distal, see **A.**

With the passage of time, different methods have been used to characterise each individual tooth.

1. Characterisation of the medial and occlusal planes by lines may indicate the quarter of the dentition to which each belongs.

Example: $\underline{1}$ = left upper 1
$\overline{3|}$ = right lower 3

2. Another method uses + for the middle of the upper jaw and − for the middle of the lower jaw, and puts the sign before or after the tooth number in each quarter of the jaw to characterise the position.

Example: +1 = left upper 1
3− = right lower 3

3. A new method of showing tooth position on paper is to use the first number to indicate the quarter of the dentition (in the remaining jaw, right upper = 1, left upper = 2, left lower = 3, right lower = 4) and the second digit in the number to indicate the number of the tooth (from mesial to distal). In the milk dentition, the quadrants of the bite, taken in the same order use digits 5–8.

Example:
21 = left upper 1
43 = right lower 3

Position of the Teeth Within the Dentition

Neutral bite *(scissors bite)*. Normally (in **orthognathism**) the crowns of the upper teeth are slightly oblique toward the ves-

tibule and those of the lower teeth toward the tongue. Thus, the cutting edges of the upper and lower front teeth (incisors) bypass each other like the blades of a pair of scissors, so that the edges of the upper incisors lie in front of those of the lower jaw when the teeth occlude, see **C.** In the premolars and molars the outer chewing edge of the upper teeth overhangs that of the lower teeth, while the inner chewing edge of the lower teeth extends beyond that of the upper teeth, see **D.** The corresponding teeth of the upper and lower jaw are so displaced on each other, that each tooth articulates with two opposing teeth, with the principal antagonist (over a large contact area) and with an adjacent secondary antagonist, see **B.** Only the first lower incisor and the third upper molar have a single antagonist. The *occlusal plane* in the final biting position usually forms a slightly upward facing concave curve on each side – see curve of *Spee,* see **B 1.**

Edge to edge bite *(forceps bite).* In rare cases (radical peculiarity) the cutting edge of the upper front teeth is not in front but lies on top of the edges of the lower teeth; the bite resembles that of a forceps. Such an edge-to-edge bite may also be pathological.

Dysgnathism (anomaly of the position of teeth and jaw) is due to faulty development of the chewing apparatus, i. e., all the organs and tissues involved in chewing may be affected by faulty development: teeth, periodontium, maxilla, mandible, joints of the jaws, masticatory muscles, mimetic muscles, tongue. In **prognathism** *(distal bite)* the upper jaw is displaced anteriorly. In *progenism* (anterior displacement of the lower jaw) the chin is unusually prominent and there is a reversed front tooth overbite. It is usually an inherited abnormality and may be visible even during the milk dentition. Complete antero-occlusion *(overbite)* is usually also an inherited abnormality, in which the upper front teeth completely cover the lower ones in the final bite (recession of the lower jaw, *deep bite).* If the abnormality of the bite is very marked, swallowing, nasal breathing and speech may be affected.

B 2 Styloid process, **B 3** condyle of the mandible, **B 4** zygomatic bone, **B 5** maxilla, **B 6** mandible.

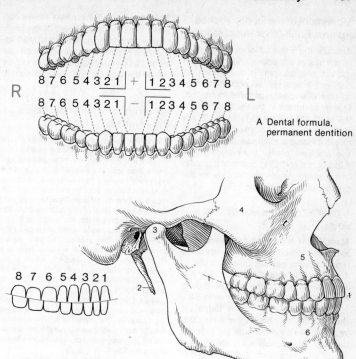

R   8 7 6 5 4 3 <u>2 1</u> + <u>1 2 3</u> 4 5 6 7 8   L

8 7 6 5 4 3 <u>2 1</u> − <u>1 2 3</u> 4 5 6 7 8

A Dental formula, permanent dentition

8 7 6 5 4 3 2 1

B Biting position in orthognathism, right half of permanent dentition (left schematic, after Faller)

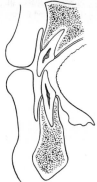

C Position of middle incisors (antagonists) in orthognathism (orthognathic dentition)

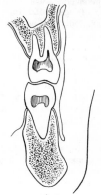

D Position of second molars (antagonists) in orthognathism

Movement of the Dental Arches Against Each other (Articulation)

Articulation is the movement of the dental arches of the upper and lower jaws against one another. It is important to distinguish between *full occlusion, side-to-side bite* and *protrusive occlusion.* During occlusion in the resting position the teeth meet (face-to-face) in the occlusal (masticatory) plane. If an antagonist is missing a tooth can grow beyond the occlusal plane. During life physiologic wear (grinding down) of the teeth takes place, which contributes to the preservation of occlusion.

Movements in the Mandibular Joint

Mandibular joint. The two mandibular joints must be considered as parts of the masticatory apparatus (see also Vol. I). The mandible (mandibular *condyle* **B1**) and the temporal bone (the mandibular fossa **B2** and the articular tubercle **B3** – *joint socket*) take part in its formation. Between the two bones lies a disk **B4**, which grows together with the articular capsule. The capsule is the site of insertion of the lateral pterygoid **B5**. The disk divides the joint into two, usually completely separate, chambers. In the median plane, the transverse axes of the two condyles form an angle of 150–160° which is open anteriorly. The articular capsule is wide and is reinforced by strong ligaments. It permits extensive movements during which the disk changes its position.

If the *mouth is passively open* (in sleep or after death) a rotating (hinge) movement takes place "in situ", the axis is not displaced.

During *active opening of the mouth* (chewing, talking) an essentially rotatory (hinge) movement takes place in the lower chamber and a sliding movement in the upper chamber; the condyle moves forward on the articular tubercle **BD3**.

Thus the *axis of the hinge moves forward.* Grinding movements of the teeth involve alternating, predominantly unilateral active movements in the mandibular joint **E.** Movements of the condyle can be felt by the finger in the external meatus **D6.**

Movements of the Tooth Inside the Alveolus

During the act of chewing the tooth moves slightly in various ways inside the alveolus. It is pressed into the alveolus in its longitudinal axis and is tilted against the vestibule or against the oral cavity around a transverse axis, which lies roughly in the center of the root and runs in the direction of the dental arch. Thus, the tooth acts as a two-armed lever and pressure and traction zones develop between the tooth and the alveolar wall. The tissue at the entrance of the alveolus may be exposed to very heavy pressure by the lever. In addition there is a slight lateral tilting movement, which is increased if the adjacent tooth is absent. In a full row of teeth each tooth is supported by its adjacent tooth via contact points. Loss of these contact points if a tooth is missing causes increased mobility of the tooth and damages the tissue at the entrance to the alveolus. Finally, even under normal conditions, a minimal rotatory movement can develop around the longitudinal axis.

For anatomical reasons human teeth have the greatest resistance against axial masticatory pressure. If, because of deficient articulation (e. g., due to loss of teeth), unphysiologically strong horizontal pressures arise, the periodontium may be damaged and the dentition is at risk.

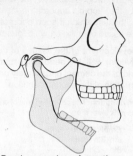

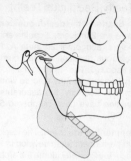

A Passive opening of mouth
(in sleep, after death) (after Töndury)

C Active opening of mouth
(after Töndury)

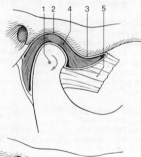

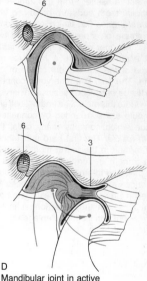

B
Mandibular joint in passive opening
of mouth. Top: closed, bottom:
opened

D
Mandibular joint in active
opening of mouth.
Top: closed, bottom: opened
(after Braus)

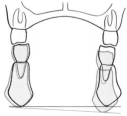

E Grinding movement (oblique
movement of lower against upper
jaw; after Strasser)

Milk Teeth (Deciduous Teeth)

In each quarter of the **deciduous dentition** there are 2 incisors, 1 canine and 2 milk molars, altogether 20 teeth **A.** For the deciduous dentition Roman numerals are used **BC**. The "left upper 1" tooth of the milk dentition may also be written as ⌊1 or +1. Recently, the quadrants of the dentition have been numbered 5–8 (see p. 186).

Development of the teeth. The deciduous and permanent teeth develop in two phases. In the 2nd fetal month the arched dental lamina grows from the epithelium of the upper and lower jaws **E1** into the connective tissue. From this lamina 10 epithelial tooth buds are produced, the *enamel organs* **D2,** which later assume the shape of a bell. The bell has two walls: the inner wall, the internal *enamel epithelium* **D3,** is like a mold for the shape of the subsequent crown of the tooth. The pulp **D4,** embryonic connective tissue with blood vessels and nerves, grows into this mold. The enamel bell and the pulp are surrounded by connective tissue which contains many cells, the *tooth follicle.* The first hard substances are formed in the fourth month. Dentin and cement are formed from the cells of the pulp, the internal enamel epithelium produces the enamel. These processes go on in almost the same fashion in both dentitions, but they take longer in the permanent dentition. The anlagen of the permanent teeth develop in the 6th intrauterine month. They can be damaged in early life, e. g., during forceful extraction of milk teeth. **E5** Meckel's cartilage, **E6** tongue, **D7** epithelium of the oral cavity.

Eruption of the Teeth

Shortly before eruption of a tooth the crown is fully developed, while the root is still growing. It is elongation of the latter that causes the tooth to erupt. The site of the gum where a tooth is going to erupt becomes at first rather swollen and dis-

colored bluish-red. Later, the white apex of the tooth appears beneath the epithelium which it soon perforates and the tissue over the crown degenerates.

Development of the tissues of the bony alveolar wall, the periodontal membrane and of the cement ends largely only after eruption of the second dentition.

1st dentition. The order and time of eruption of the milk teeth are

	Eruption of the teeth	Order
I	6th–8th month	1
II	8th–12th month	2
III	15th–20th month	4
IV	12th–16th month	3
V	20th–40th month	5

2nd dentition. The two milk molars are eventually joined by 3 further molars of the permanent dentition which do not erupt at the same time as the other teeth of that group. The incisors, canines and the "milk molars" are replaced by teeth of the 2nd dentition, see **E.**
Order and time of eruption of the permanent teeth:

	Eruption of the teeth	Order
1	6th–9th year	2
2	7th–10th year	3
3	9th–14th year	5
4	9th–13th year	4
5	11th–14th year	6
6	6th–8th year	1
7	10th–14th year	7
8	16th–30th year	8

The *times of eruption* can vary considerably. During eruption of a permanent tooth the root of the milk tooth is absorbed, until eventually only the crown of the milk tooth still remains connected to the gum and it can then easily be removed.

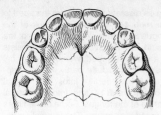

A Row of milk teeth in upper and lower jaw

$$V\ IV\ III\ II\ \underline{I|} + \underline{|I}\ II\ III\ IV\ V$$
$$V\ IV\ III\ II\ \overline{I|} - \overline{|I}\ II\ III\ IV\ V$$

B Schematic representation of teeth in milk dentition

$$V\quad IV\quad III\quad II\quad I$$

C Arrangement of the right half of the bite in the milk dentition

5 years

8 years

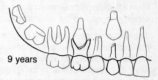

9 years

11 years

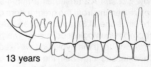

13 years

F Eruption of permanent teeth (2nd dentition) (after Korkhaus)

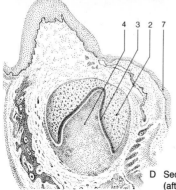

D Section through tooth anlage (after Rauber-Kopsch) (Section from E)

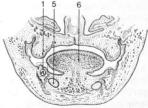

E Dental anlage (dental lamina): frontal section, 3 months fetus

Tongue

The **tongue** *(lingua),* is instrumental in chewing and sucking and carries sense organs for taste and touch, as well as being involved in speech production. Its major regions are the *root, dorsum* (back) and the *apex* (tip).

Muscles of the Tongue

The *extrinsic muscles of the tongue* arise from the lower jaw (mandible), hyoid bone and the styloid process of the skull. Their fibers radiate into the tongue and form a 3-dimensional network in its body, which merges into the *intrinsic muscles.*

The *extrinsic muscles of the tongue.* The *genioglossus* **AB1,** the strongest of the tongue muscles, is paired and arises from the center of the inner (proximal) surface of the mandible (spina mentalis). It fans out from the apex of the tongue through its body and extends as far as the root. Its lowest fibers, which run toward the base of the tongue pull the tongue forward. The other fibers pull the tongue toward the floor of the mouth. The *hyoglossus* **A2** arises as a thin 4-sided plate (lamina from the greater horn of the hyoid bone **A3**) and reaches laterally from the genioglossus into the margin of the tongue. If the hyoid bone is fixed, the hyoglossus can pull the tongue backward. The *styloglossus* **A4** arises from the styloid process and extends along the side of the tongue anteriorly to its apex. This muscle can pull the tip of the tongue backward and move the whole tongue upward and toward the back. The muscles arising from the mandible and hyoid bone can transmit movements of these parts to the tongue (see Vol. 1).

A5 Stylohyoideus, **A6** thyrohydeus, **A7** lesser horn of the hyoid bone, **A8** geniohyoid, **AB9** mylohyoid, **AB10** anterior belly of digastric muscle.

The intrinsic (inner) muscles of the tongue consist of the following muscle tracts on each side: the *longitudinales superior* **B11** and *inferior* **B12** *linguae* run as discrete fasicles from the apex to the root, one close to the back of the tongue and other close to the floor of the mouth. The *transversus linguae* consists of a well developed system of diagnoal fibers. They radiate partly into the septum **B13,** partly into the dorsal aponeurosis **B14** and into the lateral margin of the tongue. Some of the fibers traverse the septum. The *verticalis* muscle consists of fiber bundles running in the free part of the tongue from the upper to the lower surface of the tongue. **B15** Sublingual gland. **B16** Secretion through the body of the hyoid bone.

The main task of the *intrinsic muscles* is to change the shape of the body of the tongue. Usually one of the 3 muscles acts as antagonist to the other 2, contraction of the latter enforcing relaxation of the antagonist: thus, if the transversus and the verticalis contract, the longitudinalis relaxes and the tongue becomes slim and long; if the longitudinalis and transversus contract the verticalis relaxes and the tongue becomes short and thick: and, if the longitudinalis and verticalis contract the transverse fibers relax and the tongue is shortened, flattened and broadened. If, in a case of paralysis of half of the tongue (hypoglossal paresis), the tongue is protruded, then the transverse and vertical muscles of the healthy side push the ipsilateral longitudinalis muscle forward. Since the longitudinalis muscle of the contralateral, paralyzed side cannot be protruded similarly and does not offer any resistance, the healthy side is turned toward the diseased side, the apex of the tongue deviates toward the paralyzed side.

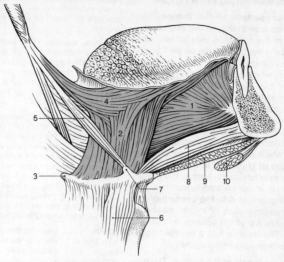

A Muscles of tongue

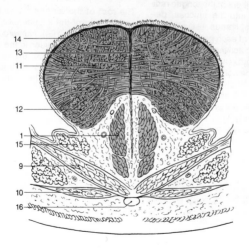

B Transverse section through
muscles of tongue

Mucous Membrane of the Tongue

Inferior surface of the tongue. The mucous membrane on the inferior surface of the tongue is only loosely connected to the body. In the center it forms the *frenulum of the tongue* **A1**, a narrow ribbon, which runs toward the gum of the lower jaw. The well developed *apical veins of the tongue* **A2** on both sides of the frenulum have a bluish appearance and shine through the transparent mucosa. More laterally is a fringed fold, the *plica fimbriata* **A3**, a rudiment of a sublingua which exists in some animals. The sublingual gland lies in the floor of the oral cavity beside the tongue. It is hidden beneath a narrow longitudinal fold, the *sublingual fold* **A4**. The opening of the submaxillary duct is in the *caruncula lingualis* **A5** at the anterior end of this fold. The mucosa has a stratified squamous, non-keratinized epithelium.

Dorsum (back) of the tongue. The mucosa of the dorsum is immovably fixed to the tough *aponeurosis* of the tongue, the connective tissue layer of the mucosa being interlocked with the epithelium by means of tall papillae. Strands of the intrinsic muscle fibers of the tongue radiate like bristles into the mucosa, particularly into the fibrous tissue which forms the center of each papillae. These muscle fibers actively deform the aponeurosis linguae. A shallow medial groove divides the dorsum of the tongue into left and right halves; the V-shaped *terminal sulcus* **B6** separates the *dorsum* **B7** from the root of the tongue **B8**. At the apex of the V lies the *foramen caecum,* a blind hole **B9**, the origin of the anlage for the thyroid gland. On the dorsum of the tongue there are numerous papillae.

Filiform papillae. The filiform papillae, also called papillae conicae, are distributed across the dorsum of the tongue. They are small elevations of the epithelium directed toward the pharynx that have become keratinised. The filiform papillae are not so strongly developed in man as in many animals in which they give the tongue a rough surface. In humans they are mostly sensitive to touch. The connective tissue core contains numerous sensory nerve endings.

Fungiform papillae. The reddish fungiform papillae are 0.5–1.5 mm high and are mainly distributed around the margin and at the apex of the tongue. They are more numerous in the newborn than in adults. In the infant they carry a greater number of *taste buds;* gustatory glands (taste glands) are absent.

Vallate papillae are 6–12 wart-shaped gustatory papillae **B10** which project only a little above the surface of the tongue. They are 1–3 mm in diameter and are arranged in the form of a V in front of the root of the tongue and of the terminal sulcus. Each papilla is surrounded by a trench (vallum) **C11** which is lined by an epithelium containing 3–5 rows of *taste buds* **C12**, see Vol. 3. Serous gustatory glands **C13** (v. *Ebner*) lead into the bottom of the trench. Their secretion washes away flavors.

Foliate papillae. These are transverse mucosal folds **B14** at the posterior lateral margin of the tongue. They contain *taste buds* in the epithelium and serous gustatory glands, which open into the depth of the mucosal folds.
The 4 *qualities of taste* – sour, salty, bitter and sweet – are detected at different sites on the tongue. There are, however, no recognizable differences either by light or electron microscopy between the corresponding taste buds.

The root of the tongue shows the rugged surface of the *lingual follicles* (*see Lingual Tonsil,* p. 104). Numerous small mucous glands are found in the connective tissue and between the muscle fibers. Three mucosal folds stretch from the root of the tongue to the surface of the epiglottis.

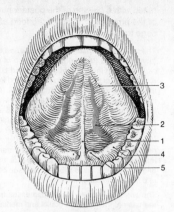

A Mucous membrane of tongue,
seen from below

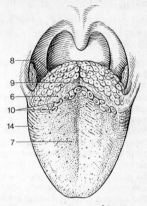

B Mucous membrane of tongue
seen from above; papillae
of tongue

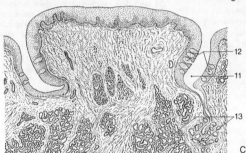

C Vallate papillae

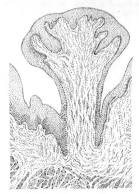

D Fungiform papillae

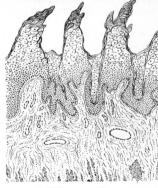

E Filiform papillae

Palate

The anterior ⅔ of the roof of the oral cavity is formed by the *hard palate,* the posterior ⅓ by the *soft palate.*

The **hard palate** *(palatum durum)* is formed by the palatine process of the maxilla and by the horizontal plates of the palatine bones, see Vol. 1. The bone is covered by periosteum and a mucous membrane, which anteriorly is firmly fixed to the periosteum and which continues into the gingiva (gum). In the middle of the hard palate there is a longitudinal ridge, the *raphe palati* **C1.** It is fixed to the palatine suture by connective tissue and ends anteriorly in a small elevation. The mucosa on both sides of the raphe carries shallow transverse ridges **C2,** an area of small furrows, against which the food is pressed by the tongue. Further back, between the periosteum and the mucosa, is an area containing small mucous glands, the *palatine glands* **C3.** They produce mucus to lubricate food while it is being chewed.

The **soft palate** *(palatum molle,* also called velum palatinum) hangs down posteriorly from the hard palate like a sail. The *uvula* **A4** projects in the midline from the posterior border of the velum palatinum. If the velum palatinum is relaxed the uvula rests on the root of the tongue. Two folds on each side of the uvula (the palatoglossal and palatopharyngeal arches) diverge and run downward. In the groove between them lies the palatine tonsil. The anterior, the *palatoglossal arch,* extends to the lateral border of the tongue, and the posterior, the *palatopharyngeal arch,* to the wall of the pharynx. They form the narrowing of the pharynx, the *isthmus of the fauces,* the entrance to the pharynx, which can be closed by muscle contraction (sphincter action). The mucosa and glands extend to the soft palate, where they lie on a tough aponeurosis. The velum palatinum plays an important part in deglutition (see p. 204), when com-bined with a bulging of the posterior wall of the pharynx, it shuts off the food passage from the upper airway.

The *tensor veli palatini* **AB5** originates as a thin triangular plate (lamina) from the base of the skull between the spine of the sphenoid bone and the root of the pterygoid process and from the membranous part of the auditory tube. The muscle descends and ends in a tendon which turns around the pterygoid hamulus **AB6** and radiates horizontally into the palatine aponeurosis; it elevates and stretches the velum palatinum up to the horizontal plane during deglutition **B7.** In this way it opens the entrance into the auditory tube as the muscle pulls at its origin. Thus, it is responsible for *passage of air into the middle ear* on swallowing.

The *levator veli palatini* **AB8** also arises from the base of the skull, but dorsal and medial to the tensor muscle (the region of the mouth of the carotid canal), and from the lower margin of the cartilage of the auditory tube **A9** ("levator bulge"). The rounded belly of the muscle runs obliquely downward and forward toward the midline. It has a broad insertion into the palatine aponeurosis. The muscle pulls the velum palatinum upward and backward **B10.**

The paired *uvulae muscle* **A11** arises from the hard palate (posterior nasal spine) and the palatine aponeurosis. It extends behind the levator and radiates into the uvula which is shortened by its contraction.

The *palatoglossus muscle* **B12** arises at the palatine aponeurosis, lies in the *anterior* palatine arch, and at the lateral border of the tongue it radiates mainly into the transverse muscle of the tongue. Together with the latter the palatoglossus closes the isthmus of the fauces **B13.** The *palatopharyngeus* **AB14** is stronger than the former muscle and lies in the *posterior* palatine arch. It counts as one of the levator muscles of the pharynx **B15.**

A16 Salpingopharyngeus, **A17** pharyngeal tonsil, **A18** inferior nasal concha, **B19** entrance into the larynx, **B20** esophagus.

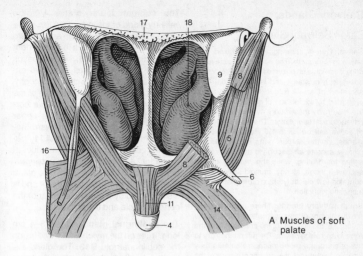

A Muscles of soft palate

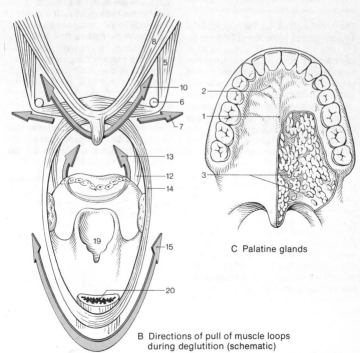

B Directions of pull of muscle loops during deglutition (schematic)

C Palatine glands

Salivary Glands

Large Salivary Glands

Saliva. During chewing food is mixed with saliva in the mouth. The saliva acts as a lubricant, it contains an enzyme (an amylase) for the conversion of starch and is bactericidal. The secretion of saliva is due to a reflex action prompted by stimulation of chemoreceptors in the mouth, by masticatory movements and by psychological stimulation. The daily production of saliva can be up to 1.5 liters. The *serous* glands, or the serous parts of the salivary glands, secrete an electrolyte and protein-rich saliva for diluting, while the *mucous* glands secrete a tough, stringy, mucoid saliva, which contains few electrolytes or protein and has a lubricating function.

Small salivary hlands. There are numerous small salivary glands with short ducts in the mucosa of the *lips*, **A1,** of the *cheeks* **A2,** of the *tongue* and of the *palate*. The nearer they are to the pharynx the more *mucous* elements they contain. In contrast, the *gustatory* glands of the taste buds (see p. 194) are *serous* glands.

The **parotid gland A3,** the largest of the salivary glands, lies in front of the ear on the ramus of the mandible and the masseter **A4** and has a process which penetrates deep around the posterior margin of the mendible. The gland extends only a little beyond the angle of the lower jaw and extends upward to the zygomatic arch. During movement of the mandible the interlobar fatty tissue allows the lobules of the gland to move against each other. The *parotid duct* **A5,** which is 3–4 mm thick and 5–6 cm long, emerges from the anterior surface of the parotid gland. It runs below the zygomatic arch, across the masseter and Bichet's fat pad (of the cheeks), pierces the buccinator in front of the masseter and opens eventually into the vestibule of the mouth at the *parotid papilla*, which lies at the level of the second upper molar tooth. Frequently a small *accessory gland* **A6** adjoins the duct. The parotid gland is covered by the parotid fascia, which continues into the fascia of the masseter, and the medial pterygoid.

The **submanidbular gland A7** (the salivary gland of the lower jaw) lies in the niche between the mandible and the two bellies of the digastric. It extends right down to the mylohoid **A8,** the hyoglossus **A9** and the styloglossus. The 5–6 cm long *submandibular duct* **A10,** accompanied by a hook-shaped process of the gland, passes around the posterior border of the mylohoid **A8** to its upper surface. From there it runs forward medially to the sublingual gland **A11,** frequently combined with the main duct of the latter, and opens on the *sublingual caruncula* **AB12,** see p. 194. The gland is enveloped by its fascial capsule and is also covered by a superficial layer of the cervical fascia and the platysma.

The **sublingual gland A11** is 3–4 cm long. It lies on the mylohoid **A8** and forms the sublingual fold **B13**. The gland extends laterally to the mandible and medially to the genioglossus. It consists of numerous small mucous glands, the *minor sublingual glands* **A14,** and of the predominantly mucous-secreting main gland, the *major sublingual gland* **A11**. Transport of the viscid mucus is facilitated by the shortness of the numerous ducts. Some of the minor glands open along the *sublingual fold* **B13** and others into the *submandibular duct* **A10**. The *main sublingual duct,* which comes from the major sublingual gland, runs with the submandibular duct to its orifice, the *sublingual caruncula* **AB12.**

The **gland at the apex of the tongue,** the *anterior lingual gland* **B15,** an almost purely mucous gland, lies on both sides at the apex of the tongue.

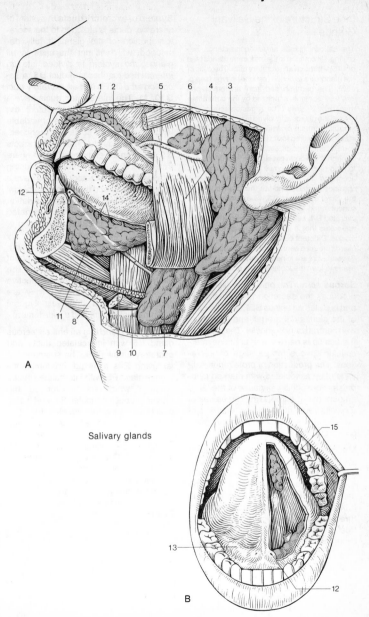

Salivary glands

Fine Structure of the Salivary Glands

The salivary glands are *exocrine* glands, secreting – in contrast to the endocrine glands (see p. 146) – through an excretory duct. The cells of the glands are exocrine *(eccrine)* in type (see p. 150). The berry-shaped acini are enveloped and their contents squeezed by the contractile *myoepithelial* cells. The ergastoplasm of the cells of the glands is involved in formation of the *serous*, protein-containing secretion. The materials required for its formation of the secretion **D1**, amino acids, etc., reach the gland cells through the blood vessels. The secretions are formed on the ribosomes of the ergastoplasm according to information specified by the chromosomes **D2** and are then transported to the *Golgi* apparatus **D3**, where they become concentrated and whence they are eventually excreted **D4**, see p. 150. It is mainly the Golgi apparatus that is involved in the production of *mucus*. Different cell groups in the acini of the glands produce serous and mucous secretions. Several acini are joined together by connective tissue to form the *lobules of the glands* **AB**.

Serous acini. The part of the gland that produces the secretion is berry-shaped and is called an *acinus* **BEFG5**. The cells of the glands **C6** are tall and cylindrical with central rounded nuclei. The lumen of the acinus is narrow, and contains intercellular spaces for passage of secretions. The production of protein-rich saliva is often associated with marked cytoplasmic basophilia, because of the large amount of ergastoplasm. The apices of the cells frequently contain precursors of the secretions, and therefore they often stain more deeply.

Mucous acini. The secretion-producing part of the gland is tubular – the *tubulus* **CFG7**. The gland cells are low, their flat nuclei are pushed toward the base of the cell by the secretion and the lumen of the tubule is wide. A honey-combed structure seen in histological preparations is due to the dissolution of the secretory granules. The mucous areas of glands are only faintly stained by the majority of staining techniques.

System of excretory ducts. A system of secretory ducts is attached to the secretory portion of the gland. If fully developed, which is not the case in all glands, the system is divided into an *intercalated* (connecting) duct **B8**, a *secretion tube* **BEF9** and an *excretory duct* **B10, 11**. The intercalated ducts have a flattened epithelium and therefore the smallest diameter. They are of variable length, and may be solitary or branched. In mixed glands the intercalated ducts **B8** are often full of mucus, i. e., they are transformed into mucus-producing tubules **CFG7**. Secretory tubules have a larger diameter and their simple cuboidal epithelium may show basal striation (hence the alternative name of "striated section"), which is caused by folding of the basal cell membrane and by mitochondria. These appearances suggest that there is a considerable amount of fluid passing through this section. The excretory ducts run in the connective tissue between the lobules. They are characterised by a stratified polygonal columnar epithelium and a wide lumen.

The *parotid gland* **E** is an entirely serous gland **E5** with intercalated ducts and secretory tubules **E9**. The *submandibular gland* **F** is a mixed, predominantly serous gland **F5**, with intercalated ducts, some of which are transformed into mucus-producing tubules **F7**, and it also possesses secretory tubules **F9**. The serous acini sit like half moons *(Giannuzzi or Ebner half moons)* compare **G6** on the mucus-containing tubules. The *sublingual gland* **G** is a mixed, predominantly mucous gland **G7**. Intercalated ducts and secretory tubules are almost entirely absent. The *anterior lingual gland* is classified as an almost pure mucous gland. **E12** Fat cell.

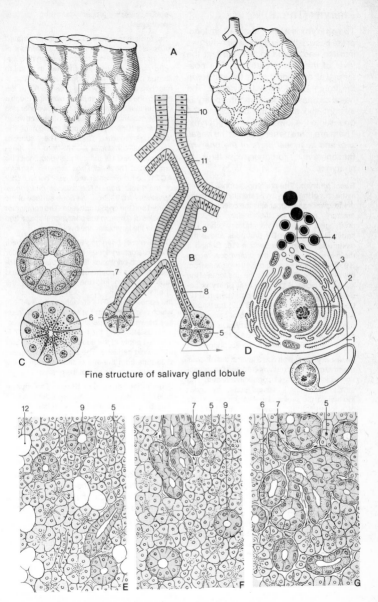

Fine structure of salivary gland lobule

Histologic sections through various salivary glands

Pharynx (Throat)

The **pharynx** is a tube, about 12 cm long, at the base of the skull **A 1**, which merges into the esophagus (gullet) **AB 2** at the level of the cricoid cartilage. The posterior wall of the pharynx is flat and lies in the frontal plane without any gaps. The nasal cavities **B 3** open into the pharynx in front and above, the oral cavity **B 4** opens into the lower end of the pharynx. There are three general areas: the *nasal, oral* and *laryngeal parts of the pharynx* (nasopharynx, oropharynx, laryngopharynx).

Nasal pharynx. The *pharyngeal tonsil* **B 6** lies in the roof of the pharynx. Bilaterally, at the level of the lower nasal concha, is the *opening of the auditory tube* from the middle ear, the pharyngeal ostium of the auditory tube. The cartilage of the tube **B 7** protrudes above the ostium as the *torus tubarious* **B 8**. Behind it there is a small groove, the *pharyngeal recess* **B 9**. The *levator veli palatini* **B 10** protrudes from below into the orifice of the tube (the torus of the levator muscles **B 11**). The *oral pharynx* begins at the level of the *velum palatinum* **B 12**. Both the *palatine tonsils* **B 13** and the *lingual tonsils* **B 14** are related to the oral portion. On both sides of the **laryngeal pharynx**, alongside the larynx is the *piriform recess* **B 15**, a groove that leads to the entrance to the esophagus.

The **pharyngeal wall** has 3 layers: the *mucous membrane, the muscularis* **B 16** and the *adventitia* which consists of connective tissue. The tunica muscularis is absent superiorly under the base of the skull and the wall there consists of a tough fibrous tissue membrane (*pharyngobasilar fascia* **AC 17**) by which the pharynx is attached to the base of the skull over a wide area.

The **mucous membrane** *(tunica mucosa)* is loosely connected with the tunica muscularis. A tough, elastic, longitudinally directed membrane broadening downward contributes to the reversible extensibility of the pharynx. The nasal portion is covered by the nasal epithelium and the oral and laryngeal portions of the pharynx have the same epithelium as the oral cavity. Mucous glands, the *pharyngeal glands,* provide the lubricating mucus. In the laryngeal portion at the transition to the esophagus, the mucosa of the anterior and posterior wall of the pharynx is separated from the skeleton of the larynx and the cervical spinal column by a venous plexus (upper narrowing of the esophagus).

The **tunica muscularis** consists of striated muscles, the *elevators* and the *constrictors* of the throat. The 3 *constrictores pharyngis* ascend posteriorly and are inserted in a tough fibrous band, the *pharyngeal raphe,* which is attached to the *pharyngeal tubercle* **A 1** at the base of the skull. The muscles overlap in such a manner that the lower parts of the superior and middle constrictor muscles are always covered on the outside by the upper margin of the following muscle. *Origins:* The *superior pharyngeal constrictor* **AC 18**: from the pterygoid process **C 19** and the pterygomandibular raphe **C 20**. The *medial pharyngeal constrictor* **AC 21**: from the greater and lesser horns **AC 22** of the hyoid bone. The *inferior pharyngeal constrictor* **AC 23**: from the outer surface of the thyroid and the cricoid cartilage. The constrictor muscles can narrow the pharyngeal space and elevate the larynx and the hyoid bone.

The *levatores pharyngis* are poorly developed. The *palatopharyngeus* **B 24** in the posterior palatal arch originates from the pterygoid process and the aponeurosis of the palate and runs downward into the dorsal wall of the pharynx where the fibers partly cross those of the opposite side. The *stylopharyngeus* **AC 25**, which originates from the styloid process, passes between the superior and middle constrictor muscles along their internal surface. It is inserted partly at the submucosa and partly at the cricoid cartilage. The *salpingopharyngeus* **B 26** runs from the end of the cartilage of the auditory tube to the wall of the pharynx.

Adventitial connective tissue. The muscle tube of the pharynx is covered by a thin fascia. It may be moved against the vertebral column because of a connective tissue space, the *retropharyngeal space.* This extends laterally into the parapharyngeal space. Both spaces are continuous with the mediastinum.

AB 27 Digastric muscle, **A 28** parotid gland, **A 29** submandibular gland, **A 30** border between the pharynx and esophagus, **AB 31** thyroid gland, **C 32** stylohyoid muscle, **C 33** cricothyroid muscle.

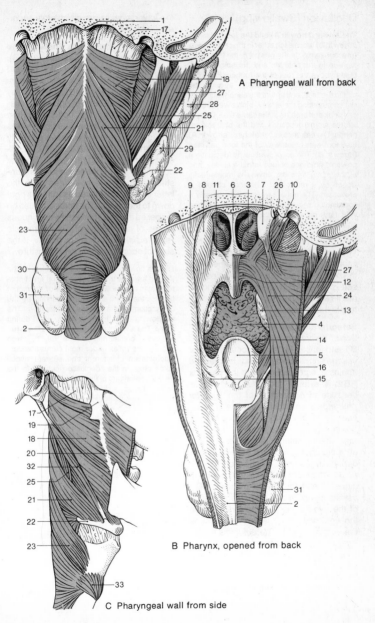

A Pharyngeal wall from back

B Pharynx, opened from back

C Pharyngeal wall from side

Deglutition (Swallowing)

The *airway* (arrow in **A**) and the *food passage* (arrow in **B**) cross each other in the *oral pharynx* (mesopharynx). In the newborn, the larynx lies high up in the pharynx and the epiglottis extends above the root of the tongue. Food bypasses the epiglottis laterally, through the piriform recess, and enters the esophagus without endangering the airway. In this way a baby can drink and breathe at the same time. Later in life the larynx descends and the pharynx ascends. The entrance into the larynx (arrow in **A**) therefore gets in the way of the food passage (arrow in **B**). The *act of swallowing* (deglutition) prevents entrance of food into the airway, and at the same time the airway is momentarily closed by a reflex and inhalation of food is prevented. The act of swallowing is actually a coordinated sequence of movements with one voluntary and two subsequent involuntary phases.

Voluntary inception of swallowing. The muscles of the *floor of the mouth* contract and the *tongue,* together with the bolus (of food), is pressed against the soft palate **AB1**. Subsequent movements are due to stimulation of the receptors in the mucosa of the palate.

Safeguarding the airway by reflex action. The *velum palatinum* is elevated **B1** and tensed by the tensor and levator veli palatini muscles, and is pressed against the posterior wall of the pharynx **AB2**. The latter *protrudes* like a torus because of contraction of the *superior pharyngeal constrictor muscle (Passavant's ring torus* **B2**) *separating the upper airways from the food passage* **C**. If the velum palatinum is paralyzed, e. g., after diphtheria, food will enter the nose during deglutition. Contraction of the *floor of the mouth* **AB3** by the mylohoid, digastric and thyrohyoid muscles **AB4** assists in visible and palpable *elevation* of the *hyoid bone* and the *larynx,* while the entrance to the larynx approaches the *epiglottis* **AB5**. The *epiglottis is lowered* by the root of the tongue with the help of the aryepiglottic muscles **B5** and the entrance to the larynx is (incompletely) closed. Simultaneously, the rima glottidis is closed and breathing stops. Thus,

the lower airways are completely separated from the food passage **C**.

Transport of the bolus through the pharynx and esophagus (arrow in **B**). The slit of the *pharynx unfolds* upward and forward when the larynx ascends. Then the *tongue* is pulled like a piston by the styloglossus and hyoglossus muscles and pushes the bolus over the fauces into the pharynx. The bolus slides through the piriform recesses primarily and partly also over the epiglottis. *Shortening of the pharyngeal wall* by its lower constrictor and contraction of the constrictors above the bolus can push the latter through the dilated esophagus right down to the cardia. The bolus can also be propelled into the stomach by continuous waves of contraction of circular muscle *(peristalsis),* even against gravity, if the subject adopts an appropriate posture.

Liquids flow in a groovelike flattening of the tongue towards the pharynx. In the erect posture rapid *contraction of the floor of the mouth* propels them into the cardia assisted by the contraction of the *tongue. Innervation:* The swallowing reflex remains active even during sleep. The afferent and efferent nerve fibers run in several cranial nerves (see Vol. 3), which secures the reflex under most circumstances. Coordination of afferent and efferent stimuli takes place in the "deglutition center" in the medulla oblongata of the brain.

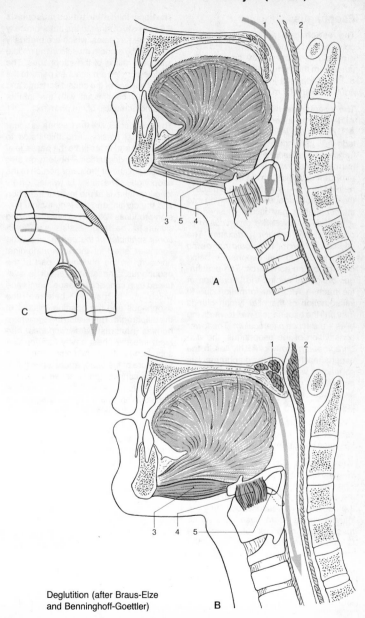

Deglutition (after Braus-Elze and Benninghoff-Goettler)

Esophagus

The **esophagus** *(gullet)* conveys the bolus of food to the stomach. In the adult it measures about 25–30 cm. The distance from the front teeth to the transition of the esophagus into the cardia in the adult is about 40 cm.

The esophagus has 3 *narrow places* **ABI–III.** The uppermost, the *sphincter* **ABI,** lies at the level of the cricoid cartilage **A1.** Its function is to close the entrance to the esophagus and it is the narrowest part with a width of only 14 mm. During deglutition the constriction is relaxed for 0.5–1 s. A thin-walled place in the posterior surface of the opening into the esophagus may become pushed out – a so-called *pressure diverticulum* or *pharyngoesophageal diverticulum.* The middle narrowing, the *aortic narrowing* **ABII,** is due to the esophagus being crossed by the aortic arch. The esophagus descends behind the bifurcation of the trachea **A2.** Scar formation due to inflammation of the hilar lymph glands may pull the esophageal wall towards the hilus – a *traction diverticulum.* The lower constriction of the esophagus, the *diaphragmatic narrowing* **ABIII,** lies in the esophageal hiatus of the diaphragm. It is connected with the complicated closure mechanism of the lowest 2–5 cm of the esophagus, which relaxes during deglutition.

The **layers of the esophageal wall** resemble those of the remainder of the intestinal tract, see p. 218.

The **mucosa** has a stratified, non-keratinized, squamous epithelium **C3.** The small numbers of mucous glands, the *esophageal glands,* increase along the course of the esophagus. They are situated in the submucosa. The strong *muscle tracts* of the mucous membrane **C4** run spirally. These muscles pleat the mucosa in the undilated esophagus into longitudinal folds ("reserve folds"). The *submucosa* **C5** adapts to these changes.

The upper third of the **tunica muscularis C6** consists of striated but autonomically innervated muscles, which are gradually replaced by smooth muscle in the middle and lower thirds of the esophagus. The muscle bundles are attached partly to the dorsal surface of the cricoid cartilage and are partly continuous with the inferior constrictor muscle of the pharynx.

The esophagus, like the trachea, is under *longitudinal tension,* and this helps to stabilize it and to facilitate the passage of food during deglutition. The tension also assists closure of the lower portion of the esophagus. The latter is twisted on its longitudinal axis at the level of transition into the cardia, and the longitudinal tension produces *"stretch closure"* (twisting closure). The latter assists the *"functional sphincter of the cardia",* i. e., the sphincter effect of the intra-abdominal pressure on the abdominal part of the esophagus. The stretch effect is sustained by a connective tissue membrane which is suspended between the esophagus and the opening (hiatus) of the diaphragm. This is also the site where hernias (ruptures: *paraesophageal hernias)* into the chest cavity can form. A submucous venous pad also contributes to closure of the esophagus. Since these vessels connect the portal vein with the inferior vena cava they can become swollen, varicose, and may bleed if the portal vein is obstructed (see p. 248).

The connective tissue of the **adventitia C7** contains smooth muscle fibers, which may run in bundles to the left main bronchus and into the left mediastinal pleura.

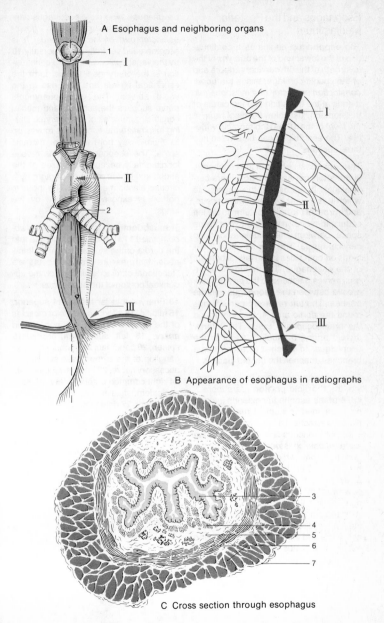

A Esophagus and neighboring organs

B Appearance of esophagus in radiographs

C Cross section through esophagus

Esophagus and the Posterior Mediastinum

The **esophagus** begins as a continuation of the lower end of the pharynx at the level both of the 6th cervical vertebra and of the cricoid cartilage; this is the *upper constriction sphincter of the esophagus*. It ends at the transition into the cardia of the stomach, just below the diaphragm at the level of the 10th–12th thoracic vertebra. The esophagus can be divided into three sections.

The uppermost section of the esophagus is the short **cervical portion** (pars cervicalis) **1**. It lies behind the trachea, to which it is bound by connective tissue, and in front of the spinal column to the left of the mid-plane, in the connective tissue space between the middle and deep cervical fascia. The thyroid gland **2** extends on both sides to the lateral borders of the esophagus. The recurrent laryngeal nerves **3** ascend to the larynx in the groove between the esophagus and the trachea. The left recurrent nerve winds round the aortic arch **4**, the right round the brachiocephalic trunk. The inferior thyroid artery, **5** situated lateral to the esophagus, to which it gives off some branches, reaches the thyroid gland **2**.

After entering the upper thoracic opening, the **thoracic portion** *(pars thoracica)* **6** of the esophagus extends into the posterior mediastinum. It runs first between the tracheal bifurcation **7** or the left principal bronchus and the descending aorta (*middle constriction of the esophagus* **6**), from which it also receives branches. Then the esophagus moves away from the spine and continues downward in a shallow arch toward the right, behind the left atrium of the heart (see radiograph, p. 207). Over a short stretch it approaches the right mediastinal pleura and then it passes through the esophageal hiatus of the diaphragm **8** while the descending aorta is inserted between the esophagus anteriorly and the spinal column (clossing with the esophagus). Where the esophagus passes through the diaphragm *(lower sphincter part of the esophagus)* it is accompanied ventrally **9** and dorsally **10** by the vagal trunks, which as a continuation of the pharyngeal plexus, form the esophagal plexus and give rise to the vagal nerves. The glossopharyngeal nerve supplies the superior and medial constrictor muscles of the pharynx. During inspiration the esophagus moves up to 7 cm away from the lower thoracic spine. The esophagus remains mobile longitudinally because it is fixed at the esophagial hiatus of the diaphragm by a circular elastic membrane. The thoracic portion is exposed to traction by the lungs.

The **abdominal portion** *(pars abdominalis)* **11**, 1–3 cm long, merges into the cardia of the stomach **12**. The intraabdominal pressure effectively acts like a "functional cardiac sphincter" on the abdominal portion of the esophagus.

13 Pharyngobasilar fascia, **14** superior, **15** middle, **16** inferior constrictor muscles of the pharynx, **17** left common carotid artery, **18** left subclavian artery, **19** spleen, **20** pancreas, **21** superior cervical ganglion of the sympathetic system, **22** accessory nerve, **23** internal jugular vein, **24** right common carotid artery, **25** sternocleidomastoid muscle, **26** right subclavian artery, **27** right lung.

Age changes. The esophagus of the newborn is relatively longer than that of the adult. It begins at the level of the 3rd or 4th cervical vertebra, i. e., with the pharynx in its low position. The distance of the cardia from the front teeth changes during growth (after *Bischoff*):

Age:	cm:
1 month	16.3
2 years	22.5
9 years	32.9
12 years	34.2
	and in the adult = 40 cm.

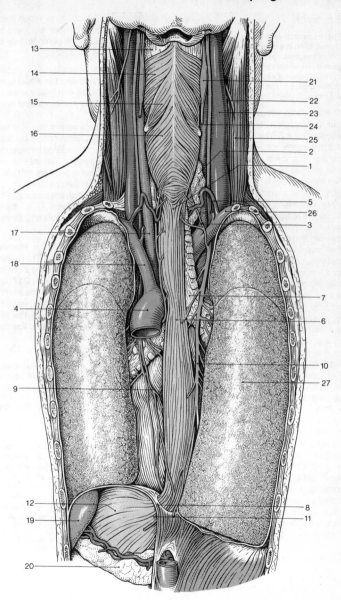

Dorsal view of esophagus in situ (after Rauber-Kopsch)

Stomach

The fragments of food (boli) are chemically broken down in the **stomach** (ventriculus, gaster) by the gastric juice to produce chyme. The gastric juice contains protein-digesting enzymes (pepsinogen), hydrochloric acid and mucus or mucin. The chyme is discharged from the stomach intermittently.

The *cardiac portion* **A1** of the **stomach** is the continuation of the orifice of the esophagus **A2**. The cupola of the fundus **A3** of the stomach arises on the left of the cardia. The *cardiac notch* **A4** (forming the *plica cardiaca,* inside the stomach) lies between the fundus and the esophagus. The main part of the stomach is its *body* **A5** which is continuous with the fundus. It merges into the *pyloric part* **A6** which is dilated to form the *antrum pyloricum* and then ends as the *pylorus* **A7**, the opening of the stomach into the duodenum. The stomach has an anterior and a posterior surface. The upper (medial or inner) border, the *lesser curvature* **A8** is indented at its lower third: the *angular notch* **A9** (forming the *plica angularis* inside the stomach where it marks the borderline between the body and the pyloric region). The lower (outer) border of the stomach, the *greater curvature* **A10,** has its most prominent bulge **A11** just opposite the *angular notch.* The *basic shape* of the stomach as just described is influenced by several factors, such as body posture, stomach contents, its muscle tonus and the influence of nearby organs. The capacity of the stomach is about 1200–1600 ml. The *mucosa* shows a few longitudinal folds along the lesser curvature (*"gastric pathway"* **B12**) and there are other folds running in an oblique or transverse direction (*"reserve folds"*) which demarcate mucosal niches (*"digestive chambers"*) where larger pieces of food may be held for a while.

Peritoneum

A serous membrane covering the peritoneum (see p. 22), enables the organs in the peritoneal cavity to move freely against each other. There is usually about 50 ml of fluid in the peritoneal cavity, which serves as a lubricant.

Peritoneum. It consists of a visceral layer **C13** which covers the viscera, and a parietal layer **C14** that lines the wall of the peritoneal cavity, *visceral and parietal peritoneum.*

The two layers are continuous and form the peritoneal sac which contains a capillary space. The connections between the visceral and parietal layers of the peritoneum, where organs are completely enveloped by the peritoneum (intraperitoneal position), are formed by thin peritoneal reduplications, i. e., tissue laminae which carry vessels and nerve to those particular organs.

These reduplications to the stomach **C15** are called ligaments. The *gastrosplenic ligament* **C16** passes from the greater curvature to the spleen **C17**; and its continuation to the wall of the trunk is called the *phrenicosplenic ligament* **C18**. From the lesser curvature to the liver **C20** extends the *hepatogastric ligament* **C19,** which continues downward as the *hepatoduodenal ligament,* see p. 236.

Fine structure. The peritoneal epithelium (mesothelium because of its origin from the embryonic mesoderm) consists of flat, simple, cuboidal cells with a border of microvilli, a sign of considerable absorptive activity. The subserous connective tissue is partly a loose, mobile layer, as in the mesenteries, and partly a firm connection between the serosa and the adjacent organs, as in the parietal peritoneum above the liver. Only the parietal peritoneum has a sensory innervation. Fluids, small particles, such as India ink particles and toxins from pathogens reach the lymph channels of the subserous connective tissue within a few minutes. The peritoneum reacts violently in inflammations and exudes a protein-rich fluid, so that adhesions may result.

C21 Inferior vena cava, **C22** abdominal aorta, **C23** pancreas, **C24** kidney, **C25** omental bursa, a loose peritoneal reduplication which permits the stomach some freedom of movement.

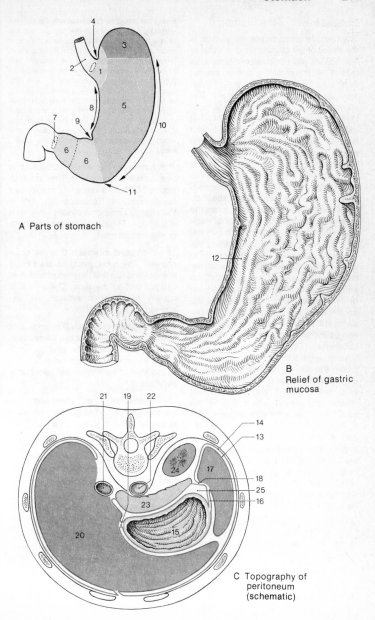

A Parts of stomach

B Relief of gastric mucosa

C Topography of peritoneum (schematic)

Muscle Layer of the Stomach

The **muscle layer,** *tunica muscularis,* is the motor of the stomach and consists of bundles of smooth muscle fibers. Like that of the intestine it is divided into an *inner layer* of *circular* **AB1** and an *outer layer* of *longitudinal* **A2, 3,** *fibers.* The stomach has in addition, a 3rd innermost layer formed by *oblique fibers* **B4,** that extend from the cardiac notch to the boundary **B5** between the body and the pyloric region. The latter, as well as the lesser curvature, does not contain any oblique fibers. The strong circular muscle layer of the *pyloric region,* which consists of two circular loops connected above by longitudinal muscle fiber tracts, forms the pyloric sphincter. The longitudinal muscle layer controls the length of the stomach; its fibers run partly along the greater curvature **A3** from the cardia to the pylorus. Some end at the lesser curvature **A2** and at the posterior and anterior walls of the stomach at the level of the angular notch **AB5.** Beyond the notch new longitudinal muscle fiber tracts **B6** extend across the pyloric portion of the stomach into the wall of the duodenum. The angular notch, therefore, is regarded as the boundary between the two portions of the stomach: an upper (proximal) portion, which mainly serves digestion, and a lower distal portion, concerned with emptying the stomach. Between the circular and longitudinal muscle layers lies the *myenteric-plexus* (of Auerbach) a vegetative nerve supply to the muscles. In contrast, the submucosa is supplied by the *submucosal plexus (Meissner's plexus)* which innervates the muscularis mucosae and the glands (*"intramural nervous system",* see Vol. 3).

Movements and motility of the stomach G.
As the stomach fills layer upon layer, its contents are accommodated without an increase in tension in the gastric wall. The pressure inside the stomach is equal to that in the rest of the peritoneal cavity. The layers of chyme adjacent to the mucosa are pushed away by contraction in the upper part of the stomach and deeper layers are brought into contact with the mucosa – *mixing movements* of the stomach. If the stomach is filled, peristaltic waves occur about every 3 minutes and take about 20 seconds to move from the fundus to the pylorus. The stomach is *emptied* under the influence of pressure conditions between it and the duodenum. Gastric motility is under the control of *tissue hormones* from the cells with basal granules (see p. 214) in the gastric and duodenal glands. In general the pylorus is slightly open and it becomes completely closed only when a peristaltic wave reaches the pyloric canal.

The lumen of the **empty stomach** is shaped like a narrow tube. The *"stomach bubble"* **CDE7** – swallowed air which can be demonstrated radiologically, lies in the slightly expanded fundus under the diaphragm. Three main types of stomach configuration occur:

The **J-shaped stomach C** is the commonest. The two curvatures are almost parallel and run first downward and then upward and to the right. The J-form is usually found when patients are examined in the erect position.

The **elongated stomach D** is atonic and is also J-shaped. It is more common in females than males. The angular notch forms an acute angle and extends down to the level of the 4th lumbar vertebra or even lower.

The **transverse type E** (steer horn) has no angular notch. The greater curvature is directed ventrally and has a high position, reaching the level of the 2nd lumbar vertebra. The transverse form often arises if the abdominal wall is tense, in the recumbent position, or if the abdominal organs need more space.

The *mucosal folds* can be brought into relief radiologically by filling the stomach with small quantities of *radio-opaque medium.* They appear coarsened in gastritis. Defects of the mucosa, e. g., in gastric ulcer or tumors, can also be detected in X-ray films. **CDE8** Duodenal bulb (see p. 216), **CDE9 pyloric region** (pars pylorica).

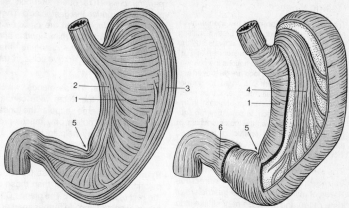

A Longitudinal and circular
muscle fibers

B Oblique muscle fibers

Radiologic appearances of stomach

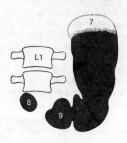

C J-shaped stomach

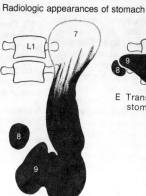

D Atonic, elongated J-shaped
stomach

E Transverse (steer horn)
stomach

F Mucosal relief

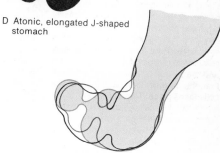

G Stomach movements
(after Braus-Elze)

Gastric Mucosa

The **mucous membrane of the stomach A1** (tunica mucosa) produces the protein-splitting enzyme pepsinogen, which is activated by the hydrochloric acid (HCl) of the stomach at a pH of 1.5–2 into *pepsin*. By its actions the connective tissue structures of food are destroyed and fat and muscle fibers are freed for further digestion in the small intestine. Other physiologically active substances originate in the mucosa of the stomach. In part, they regulate the secretory and in part the motor functions of the stomach. Another product of the gastric mucosa is "intrinsic factor" which makes possible the absorption of vitamin B_{12}.

The **surface of the mucous membrane** shows gross *mucosal folds,* numerous areas of millimeter size, the *gastric areas* (fine relief) and small point or slitlike little gastric pits, the *foveolae gastricae* **BD2,** which are only visible under a magnifying glass. In each pit several gastric glands *(glandulae gastricae)* **B3** are open. The mucosa and the foveolae are covered by a single layer of columnar epithelium **D4,** which stands out sharply against the pale esophageal epithelium at the cardia, when seen through a gastroscope. The epithelial cells liberate mucus which protects the epithelium against autodigestion. The protection does not persist after death when the mucosa undergoes digestion. The connective tissue of the mucosa is studded with tubular glands which reach as far as the *muscularis mucosae* **A5.** There are different glands in the cardia, fundus, body and pyloric region of the stomach.

The **glands in the fundus and the body of the stomach** are elongated, densely arranged and contain 3 types of cells. In the neck of the gland are mainly *mucoid cells* **BD6.** They resemble the epithelial cells of the mucosa, form neutral mucus and may show mitoses. It is from there that the regeneration of the epithelial cells takes place. The middle part of the gland has chief cells (peptic or zygomatic cells) and parietal cells (oxyntic cells). The *chief cells* **BD7** are basophil (ergastoplasm) and produce pepsinogen. The

parietal cells **BD8** often sit upon the tubuli and produce the gastric acid. They are more acidophil than the other cells and contain numerous mitochondria and intracellular secretory cisternae that arise from lamellae and vesicles in the interior of the cells.

Cardiac glands. These glands, which are situated in the 1 cm wide cardiac region, are the same shape as the fundal cells, but they contain only mucus-producing cells. They help to form an alkaline barrier zone between the stomach and the esophagus.

Pyloric glands. The foveolae **C9** of the pyloric region are deeper than those of the rest of the stomach. The glands **C10** are very convoluted, lie further apart and most of their cells produce mucus **C11.** Cells with "basal granules" **C12,** endocrine cells, can be demonstrated by special staining techniques (compare p. 146). Lymph follicles are common in the connective tissue of the mucosa. The body and antrum have a different pH; with the help of color indicators it is possible to demonstrate their surgically important boundary and to recognize it through a gastroscope.

Secretion of gastric juice. Two distinct phases in gastric secretion can be determined. *Nervous secretion* depending on impulses transmitted by the vagus nerve, is stimulated by sensory impressions, such as taste, smell and sight. This phase can occur even if the stomach is empty, but ceases after the vagus nerve has been cut. *Gastric (digestive) phase secretion,* which is stimulated by the ingestion of food, starts as soon as food enters the stomach. It is triggered off by a hormone (gastrin) which is produced by the cells with basal granules in the pyloric region in the intestinal phase.

The **muscularis mucosa A5** may influence the circulation in the mucosa by obstruction of the blood vessels. The larger blood and lymph vessels and the nerves of the mucosa *(submucous plexus)* run in the **submucosa A13. A14** Tunica muscularis of the stomach.

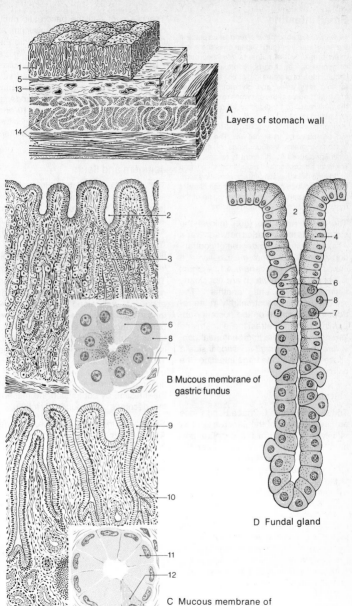

A
Layers of stomach wall

1
5
13

14

2

3

6
8
7

B Mucous membrane of
gastric fundus

9

10

11

12

C Mucous membrane of
pyloric region

2

4

6

8
7

D Fundal gland

Small Intestine

Digestion and absorption of food takes place in the small intestine (gut). Digestion is the enzymatic breakdown of nutrients into absorbable components, i. e., of carbohydrates into monosaccharides, of proteins into amino acids and of fat into fatty acids and glycerol. The most important source of the enzymes responsible is the pancreas. To be digested fat must first be emulsified by bile. The chyme is carried along through the intestine by mixing and transporting movements of the intestinal wall. The intestinal mucosa contains various types of epithelial cells specialized for different functions, e. g., some for absorption and mucus formation, and endocrine cells that stimulate pancreatic secretion and produce peristalsis of the gall bladder and the intestine. The connective tissue of the mucosa contains many lymph follicles.

The **small intestine** (gut) follows the stomach. Its length varies between 3 and 4 m, according to the degree of contraction of its longitudinal muscle layer. It consists of the *duodenum* **A1**, *jejunum* **A2** and *ileum* **A3**, which are not clearly delimited from one another. The duodenum lies almost entirely in the retroperitoneal space on the posterior abdominal wall. In contrast the jejunum and ileum form a mobile intraperitoneal *convolute (coils) of small intestine* inside a space framed by the large intestine (the colon).

Duodenum

The **duodenum** is C-shaped and winds around the head of the pancreas. It lies mainly to the right of the vertebral column. We distinguish the *superior part* **B4** at the level of the 12th thoracic − 1st lumbar vertebra, the *descending part* **B5** at the level of the 3rd–4th lumbar vertebra, and the *horizontal* and *ascending part* **B6**, which climbs across the midline to the *duodenojejunal flexure* **BC7** at the level of the 1st–2nd lumbar vertebra; the levels of the planes apply to the recumbent position during expiration. The beginning of the superior part is dilated – the *duodenal bulb*. It still lies intraperitoneally (hepatoduodenal ligament, see p. 236).

The bile duct and the pancreatic duct, open into the middle of the duodenum. At the transition of the duodenum into the jejunum is the *duodenojejunal flexure* **ABC7**, where the small intestine turns into the peritoneal cavity. At this site niches may be present in the peritoneum: the *superior* **D8** and *inferior duodenal recess* **D9**, the *retroduodenal* **D10** and *paraduodenal recess* **D11** into which the intestinal loops may stray and form internal herniae.

Jejunum and Ileum

The upper $^2/_5$ of the intraperitoneal section of the small intestine are called the **jejunum A2**, and its lower $^3/_5$ the **ileum A3**. They merge into each other without any definite border. If the abdominal cavity is opened the convoluted small intestine, which is movable at the mesentery, can be pushed aside, compare **A** and **C**.

The **mesentery** is fixed at its root, the *radix mesenterii* **C12**, to the posterior abdominal wall, along a line which extends from the duodenojejunal flexure obliquely downward and to the right as far as the opening of the small intestine into the colon. The radix mesenterii is 15–18 cm long, but the total length of the mesenteric attachment to the small intestine may exceed 4 m. With shortening of the intestine by contraction the mesentery becomes correspondingly folded.

Sections of the large intestine: caecum **AC13** with the vermiform process (appendix) **AC14**, the ascending **A15**, transverse **AC16**, descending **C17** and sigmoid colon **C18**, the rectum **C19**. Below the upturned sigmoid is the intersigmoidal recess **E20**.

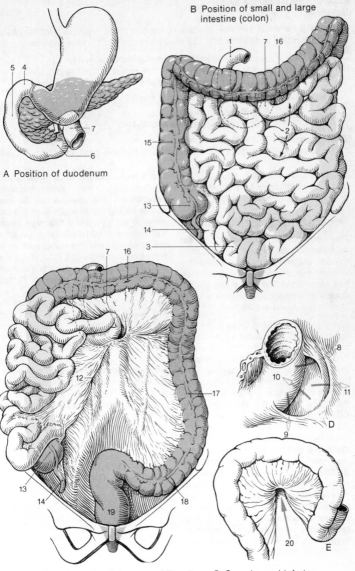

A Position of duodenum

B Position of small and large intestine (colon)

C Small intestine (jejunum and ileum) turned toward the left; transverse and descending colon visible

D Superior and inferior duodenal recesses

E Intersigmoidal recess

Layers of the Wall of the Small Intestine

The wall of the small intestine, like that of the rest of the intestinal tract (compare esophagus p. 206 and stomach, p. 212 et seq.), consists of the following layers:

Tunica mucosa { *Lamina epithelias* **A 1** (epithelial mucosal layer)
Lamina propria **A 2** (connective tissue mucosal layer)
Muscularis mucosae **A(C–F)3** (muscular epithelial layer)

Tela submucosa **A(C–F)4** (connective tissue mobile layer)

Tunica muscularis { *Stratum circulare* **A(C–F)5**
Stratum longitudinale **A(C–F)4**

Tunica serosa **A 7** *with tela subserosa* (peritoneum)
or: *tunica adventitia* (in retro- and extraperitoneal position; connective tissue layer)

In the intraperitoneal position, the mesentery **AB 8** carries the blood vessel and nerve supply for the intestine.

Mucosa of the Small Intestine

The considerable enlargement of the epithelial surface through mucosal *folds* (plicae), *villi,* and the *microvilli* seen in the electron microscope, facilitates digestion and absorption. The plicae and villi decrease in numbers and size in the lower parts of the small intestine.

The circular **Kerckring folds** *(plicae circulares)* **B 9** project about 1 cm into the lumen of the gut. They are formed by folding of the *mucosa* and *submucosa* and do not completely disappear if the intestine is stretched. They enlarge the surface of the intestine by about one-third. The **villi of the small intestine** **CDE 10** are 0.5–1.2 mm high and about 0.1 mm in diameter. They are leaflike or finger-shaped processes of the *epithelium* and the *lamina propria* of the mucosa. The muscularis mucosae **ACDEF 3** does not take part in the formation of the villi. There may be as many as 40 villi per mm², which gives a velvety appearance to the mucosa and increases its surface area 5–6 times. In between the villi lie the 0.2–0.4 mm deep **crypts of Lieberkühn,** *intestinal glands* **CDF 11,** which reach as far down as the muscularis mucosae.

The 3 regions of the small intestine are characterized by the following peculiarities: the **duodenum C** has numerous high folds which start 3–5 cm distal to the pylorus. It has high, often leaf-shaped villi **C 10** and its crypts **C 11** are shallow. Only the duodenum contains **Brunner's glands, C 12,** which are branching, clustered groups of glands in the submucosa. The **jejunum D** has numerous high folds and villi **D 10.** They diminish in number and height towards the ileum, while the crypts **D 11** get deeper. Thousands of isolated lymph follicles are distributed over the entire intestine. While the folds and villi **E 10** gradually disappear in the **ileum E,** the lymphoid tissue increases. It is concentrated on the side opposite the mesenteric attachment, mostly as 20 to 30 clusters, each of about 20 lymph follicles **E 13;** these *aggregated lymphoid follicles* are known as **Peyer's patches.**

The **colon F,** in contrast, has no villi, its crypts **F 11** are deep and lymph follicles **F 13** are present, particularly in the appendix (see p. 104).

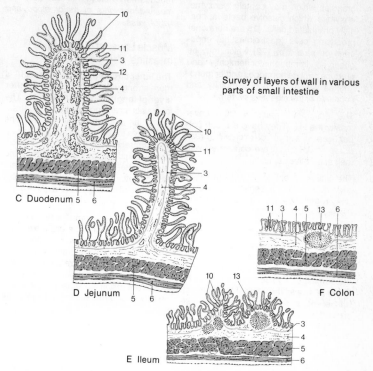
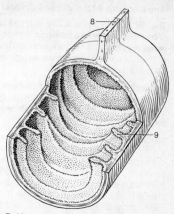
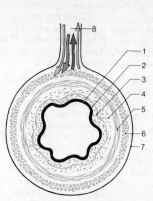

A Cross section through B: layers of intestinal wall (schematic)

B Upper part of small intestine (jejunum), partly cut away.

Survey of layers of wall in various parts of small intestine

C Duodenum

D Jejunum

E Ileum

F Colon

Fine Structure of the Villi and Crypts

The villi **A1** and crypts **A2** appear round or oval in **transverse section B**. In transverse sections the *villi* are *covered by epithelium* and the *crypts* have an *epithelial lining* (compare **C, D**).

The **epithelium of the mucous membrane** is one layer thick and consists of tall polygonal cells. We distinguish *secretory* and *absorptive* epithelial cells.

Various types of **secretory epithelial cells** can be distinguished: mucus producing *goblet cells* **CD3** which are quite common in the crypts as well as the villi: the mucus lubricates the contents of the intestine and thereby protects the mucosa; *granular Paneth cells* **D4**, which appear in small groups in the fundi of the crypts: they contain apical, acidophilic granules which may contain lysozyme, enzymes which dissolve bacteria; "basally granulated cells" **D5** are hormone-producing cells (gastrointestinal endocrine cells, see p. 172) which may be observed immunohistochemically and produce peptide hormones and/or monoamines which control the motility and secretion of the digestive system.

The *absorptive epithelia* **CD6** cover mainly the villi. They have a *brush border*, due to a layer of microvilli *(brush border cells)* which are concerned with their absorptive capacity.

The intestinal epithelial cells are only short-lived and are sensitive to injury. They **regenerate** from the epithelium of the crypts where cells undergo mitosis **D7**, and the daughter cells so produced are constantly shifted towards the summit of the villus from where old epithelial cells are shed into the intestinal contents. The migration of a cell from the crypt to the summit of a villus takes about 36 h. Every 2–4 days, the whole intestinal tract epithelium is replaced. It can be assumed that the enzymes of the shed epithelia play an active part in digestion.

The **connective tissue of the mucosa** is mainly reticular connective tissue containing lymphocytes and other cells of the *immunologic system*, as well as arterioles **C8**, capillaries **C9**, venules **C10**, lymphatics **C11** and terminal branches of the submucous venous plexus.

Muscle fibers from the **muscularis mucosae** radiate into the connective tissue of the villi and extend into the basement membrane of the epithelium. When the muscle fibers contract, which they do every 10–15 secs., the villi become shortened, their surface is folded and venous blood and lymph are squeezed into the vessels of the submucosa; *"villous pump"*. As soon as the vessels are refilled the villi become erect once more.

The **submucosa A13** carries the *submucous nerve plexus* as well as wide-meshed *networks of major blood and lymph vessels*.

Muscle Layer of the Small Intestine

The **inner circular muscle layer A14** is more strongly developed than the **outer layer of longitudinal muscle A15**. The layers act as antagonists so that contraction of the longitudinal muscle shortens and dilates the relevant portion of the gut, and contraction of the circular muscle narrows and elongates that part of the intestine. Between the two muscle layers lies the *myenteric plexus* (of *Auerbach*) consisting of nerve fibers and nerve cells. **A16** Peritoneal covering.

Movements of the small intestine. The contents of the intestine are mixed and moved by both *pendulum* and *segmentation movements*. Transportation takes place by *peristaltic waves*, i. e., orderly shifting zones of contraction. Peristaltic waves are *rolling movements* which pass quickly over large sections of the small intestine.

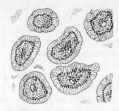

B Cross section – left, through villi; right, through crypts

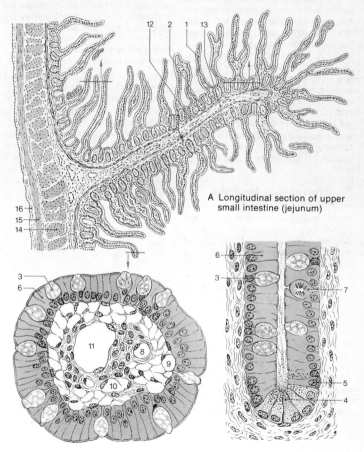

A Longitudinal section of upper small intestine (jejunum)

C Cross section through villus, section from A

D Longitudinal section through crypt, section from A

Blood and Lymph Vessels of the Villi

In the connective tissue of the villi a network of *capillaries* lies below the epithelium. It is supplied by one or more *arteries* which run undivided to the summit of the villus. A central *vein* returns the blood to the portal system; and there are *arteriovenous shunts* at the summit of the villus. The bulk of the material absorbed by the epithelium, namely amino acids, sugar, free fatty acids and glycerol, is carried in the blood via the portal vein into the liver. The *endothelium* **B1** of the blood capillaries is *fenestrated*. Inside the vascular coat is a lymph vessel (yellow), the *"central lacteal"*, which contains chyle – the lymph drained from the intestine. Fatty acids are resynthesized in the mucosa into triglycerides and are carried with the lymph stream via the thoracic duct into the blood. During the process of emulsification of fats in the intestine, whole droplets smaller than 0.5 µm are formed, which can also penetrate through the epithelium into the lymphatics.

Fine Structure of the Absorptive Epithelium

In the absorptive epithelium the **border cells** have a striated border 1.2–1.5 µm long, consisting of very densely packed *microvilli* **B2** (about 3000 of them on a single cell, about 200 million per mm^2 surface). They enlarge the surface area of the epithelium 30 times and thereby help to create a total surface area of the small intestine of some 100 m^2. The microvilli contain alkaline phosphatase for the absorption of carbohydrates. Lipases and esterases (fat absorption) are present in the microvilli and in the deeper zones. The absorption of materal from the intestine takes place mainly by *active transport* against the concentration gradient. Specific types of facilitated diffusion also play a part. Absorption by *membrane vesiculation* **B3** *(pinocytosis)* can be demonstrated with the electron microscope. Small vesicles of plasmalemma

containing material are detached from the plasma membrane of the cell. In the reverse *process (crinocytosis)* material is extruded into the intercellular space after the membrane of the vesicle has fused with the plasmalemma **B4** on the inner side of the cell. The intercellular space is completely shut off from the lumen of the gut by specialized cell contacts: zonulae adherens **B6** and *occludens* **B5**. Eventually material that has been absorbed penetrates the basement membrane **B7** and enters a blood or lymph capillary. **B8** *Golgi* apparatus, **B9** mitochrondium.

Fine Structure of the Mucus-Producing Epithelia

The part of a **goblet cell** pointing toward the lumen of the intestine is filled with *mucus particles* **C10,** which are surrounded by a membrane from the Golgi apparatus (compare **B8**). With increasing intracellular pressure the cell membrane bursts and the mucus enters the intestine. The cytoplasm and the nucleus **C11** are displaced toward the base of the cell. The cytoplasm surrounds the mucus particles like a goblet.

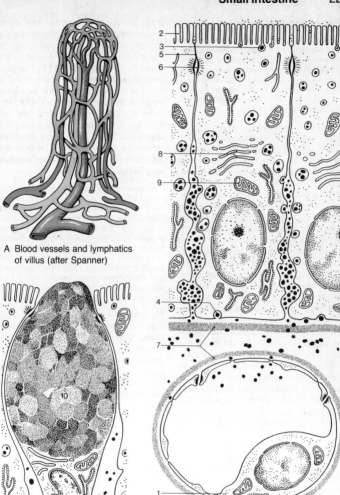

A Blood vessels and lymphatics of villus (after Spanner)

B Electron microscopic appearance of absorptive epithelial cell of intestinal mucosa (after W. Schmidt and Zetterquist) (schematic)

C Electron microscopic appearance of mucus-producing intestinal epithelial cell (schematic)

Large Intestine

Digestion and absorption are completed in the small intestine and the large intestine (colon) contains the indigestible residues of the foodstuffs which are being decomposed by bacteria (fermentation and putrefaction). The main task of the colon is the vital reabsorption of water and electrolytes which have entered the intestines with the digestive juices.

The **large intestine** is 1.5–1.8 m long. It begins with the ileocecal (colic) valve (see p. 226). The **appendix** (vermiform appendix; see p. 228) **A 2** hangs from the **cecum A 1** as a dilated intestinal pouch. The cecum is followed by the colon, which frames the convolutions of the small intestine. The **ascending colon A 3** runs close to the anterior abdominal wall on the right side under the liver; at the *right* is the *colic flexure* **B 4.** As the **transverse colon A 5** it runs in an arch along the anterior abdominal wall to the left upper corner of the abdominal cavity. At the *left colic flexure* **B 6,** at the level of the lower pole of the spleen, it turns at right angles into the **descending colon A 7** which, covered by the convolutions of the small intestine, passes downward and posteriorly along the left lateral abdominal wall. The attachment of the left flexure, the *phrenicocolic ligament,* forms the floor of the splenic niche. The bend of the colon at the left flexure represents an obstacle for the passage of the intestinal contents and increased peristalsis is required to overcome it. The **sigmoid colon AB 8** lies in the left iliac fossa and enters the false pelvis in an S-shaped loop. The **rectum A 9** starts in front of the 2nd-3rd sacral vertebra and ends at the *anus.* **A 10** Duodenojejunal flexure, **A 11** root of the mesentery, **A 12** ileum.

B The full colon effectively "floats" on the abdominal viscera and its position is higher than when empty.

Features of the large intestine. The external longitudinal muscles are compressed into three 1 cm wide longitudinal bands, the **teniae coli.** The *tenia libera*

D 13 is seen anteriorly, the *tenia mesocolica* **D 14** lies medial to the ascending and descending colon, and on the transverse colon it is related to the origin of the mesocolon **A 15**. The *tenia omentalis* **D 16** lies posteriorly and medially at the transverse colon below the origin of the great omentum **A 17**. The colon is sacculated (**haustra**) and has folds which protrude into the lumen, the **semilunar folds D 18**. There are small projecting appendages of fat from the subserosa of the colon, the **appendices epiploicae D 19**.

Peritoneum. The ascending and descending colon lie behind the peritoneum, i. e., only in front are they covered by peritoneum **D 20.** Their posterior wall is fixed to the posterior wall of the trunk. The cecum may be fixed to the posterior wall of the trunk or it may be mobile by means of a short mesocolon. The vermiform appendix and the transverse and sigmoid colon lie within the peritoneum and have a *transverse* and *sigmoid mesocolon.* The rectum is initially covered anteriorly by the peritoneum, but later lies outside the peritoneum.

Mucosa. There are no villi in the large intestine. The *crypts* **C 21** are very deep (about 0.5 mm) and lie very close together. The epithelium of the crypts consists almost entirely, and that of the surface for the most part, of *goblet cells,* which produce lubricating mucus. The remaining epithelia have a *ciliated border* of microvilli, an expression of their great water-absorptive function. There are solitary lymphoid nodules. **C 22** Submucosa, **C 23** internal circular muscles, **C 24** tenia, **C 25** subserosa.

Movements of the large intestine. The intestinal contents are moved along in the colon as they thicken by slow peristalsis and antiperistalsis. By means of a few transport movements, the intestinal contents are moved into the distal part of the colon.

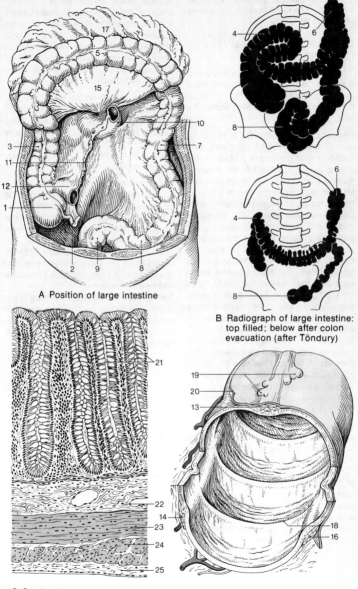

A Position of large intestine

B Radiograph of large intestine:
top filled; below after colon
evacuation (after Töndury)

C Section through intestinal wall

D Ascending colon, partly cut away

The cecum, appendix and rectum will be described in greater detail because of their medical importance.

Cecum and the Ileocecal Valve (Colic Valve)

The **cecum,** the pouchlike, widest part of the large intestine, is 6–8 cm long. It lies on the right upper part of the ilium (right iliac fossa) near the anterior abdominal wall.

Ileocecal (colic) valve BC1. The distal part of the ileum **AB2** protrudes as a round or oval papilliform projection (the colic or ileocecal valve) into the cecum. It pushes aside and forces apart the circular muscle layer of the cecum which, in turn, surrounds the lower end of the ileum like a muscular clamp. The ileum, together with its mucosa, forms an upper and a lower lip, which end in a posterior and an anterior frenulum in which the circular muscle fibers of the cecum are again united. The ileocecal valve lies in front of the mesocolic band (tenia mesocolica), the muscle fibers of which are inserted partly in the muscular valve of the circular muscle fibers of the cecum and which can open the end of the ileum. Some of them radiate into the longitudinal muscle layer of the lower ileum. Their contraction assists in dilating the end of the ileum which forms an ampulla immediately before the ileocecal valve. The ileocecal valve can actively change its shape like a sphincter.

Shortening of the longitudinal muscle layer of the invaginated end of the ileum and of the tenia results in shortening of the papilla and dilatation of its ostium. Contraction of the circular muscle layer of the ileum and cecum results in extension of the papilla and closure of the ostium. The sphincter opens periodically and allows the contents of the small intestine, i. e., of the ileum, to pass into the large intestine, the cecum, and hinders reflux. Furthermore, if the cecum is tightly filled, a mechanical valvular mechanism can come into operation to prevent reflux, as the lips of the valve become tightly pressed against each other. Contraction of the arched circular muscle fibers produces elevation and shortening of the cecum and so helps to empty the funnel-shaped outlet **BC3** of the appendix **A4.** The red line in **B** marks the line of the section in **C.**

Peritoneum. While the ileum lies intraperitoneally the cecum is often fused with the posterior abdominal wall: fixed cecum *(cecum fixum).* However, quite often there is a "mesocecum" which permits the cecum to become mobile, *cecum mobile.*

The vermiform appendix **A4** has a *mesoappendix* **A5** and this can account, at least in part, for the considerable variability of its position. Above and below the opening of the ileum into the cecum there is a peritoneal fold the *ileocecal fold* directed toward the appendix, behind which there are two peritoneal pockets, the *superior* **A6** and *inferior* **A7** *ileocecal recesses.* Frequently on the right side behind the cecum or the ascending colon there is a further peritoneal space, the *retrocecal recess* **A8.** The cecum already contains *semilunar folds* **B9. A10** Mesentery, **A11** tenia libera.

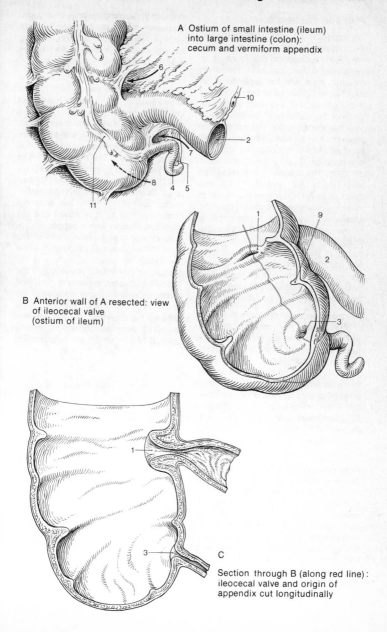

A Ostium of small intestine (ileum)
into large intestine (colon):
cecum and vermiform appendix

B Anterior wall of A resected: view
of ileocecal valve
(ostium of ileum)

C Section through B (along red line):
ileocecal valve and origin of
appendix cut longitudinally

Vermiform Appendix

The **appendix A 1,** part of the large intestine, arises like a funnel from the end of the cecum during fecal development, or sometimes in childhood, but only rarely in an adult. As a rule the appendix in an adult has a narrow exit on the medial wall of the cecum, but it is still possible for some of the contents of the intestine to pass into the appendix. The *mouth of the appendix* may be surrounded by a small, semilunar fold of mucosa. The appendix is about 8 cm (2–20 cm) long and 0.5–1 cm thick. The 3 tenia of the colon, the *tenia libera* **A 2,** *omentalis* **A 3** and *mesocolica* **A 4** run across the cecum to meet in a star-shape at the origin of the appendix. They form a closed longitudinal muscle layer in its wall. A surgeon can locate the appendix by following the course of the tenia. The *mesoappendix* **A 5,** a continuation of the mesentery of the small intestine, carries the appendicular artery and vein **A 6** to the appendix behind the end of the ileum **A 7.**

The **position** of the appendix in relation to the pelvis, the intestine and the peritoneum varies so greatly that it is hardly possible to indicate its "normal" position. In about 65% of cases the appendix is turned up behind the cecum, the *retrocecal position;* in 31% it hangs across the linea terminalis into the lesser pelvis, the *descending type;* in more than 2% it lies horizontally behind the cecum, the *paracolic position;* and in 1% in front of and in less than 1% behind the end of the ileum.

Projection of the appendix on the abdominal wall. It is important for the diagnosis of appendicitis to know the position of the appendix in relation to the abdominal wall (pain on pressure and muscular tension of the abdominal wall). In the normal position of the cecum the *origin* of the appendix is projected on the center of a line between the right anterior superior iliac spine and the umbilicus (*McBurney's point* **C 8**). In the *descending type* the *tip of the appendix* is projected roughly towards the border between the right and the middle third of a line connecting the two anterior superior iliac spines (*Lanz's point* **C 9**).

Fine structure. The mucous membrane of the appendix, like that of the rest of the large intestine, has crypts but no villi. But the appendix in man does not serve either for digestion or absorption of food; it is part of the *immunologic system* (see p. 105). It is sometimes called the "tonsil of the intestine". The mucosa is full of lymph follicles (*aggregated follicles* **D 10**) which also penetrate into the submucosa. As an organ of the defense mechanisms against infection, the appendix sometimes reacts in a violent and hyperactive manner with a danger of suppuration and of perforation of its wall into the abdominal cavity. In about 25% of the population the lumen is partly and 8% completely obliterated. Narrowing of the lumen is due to hypertrophy of the lymph follicles. Cell debris and residues of the intestinal contents are common in the lumen of the appendix. **D 11** Mesoappendix, **D 12** external longitudinal muscle layer, **D 13** internal circular muscle layer.

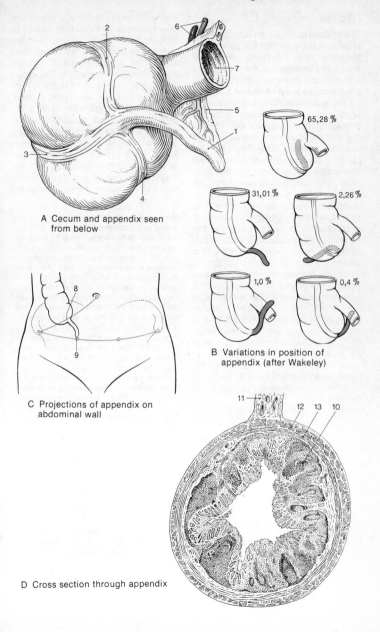

A Cecum and appendix seen from below

B Variations in position of appendix (after Wakeley)

65,28 %
31,01 %
2,26 %
1,0 %
0,4 %

C Projections of appendix on abdominal wall

D Cross section through appendix

Rectum

The **rectum** is 15–20 cm long and is S-shaped. It follows at first the convexity of the sacral bone, the *sacral flexure,* and then turns backward at the level of the coccyx and passes through the floor of the pelvis, the *perineal flexure.* Finally, it becomes the *anal canal* and ends at the anus. It also has a projection to the left formed by the Kohlrausch fold. The sacral flexure lies behind the peritoneum and is covered anteriorly by the peritoneum, see *rectouterine* and *rectovesicular excavations* (pouches) (p. 312). The perineal flexure runs extraperitoneally. Haustra (sacculations) and teniae (bands) are absent from the rectum, and the longitudinal muscle layer is continuous. In the upper third of the rectum is a portion of considerable extensibility, the *rectal ampulla.* If it is filled the feeling of the need to defecate occurs. Below the ampulla three constant transverse folds project like wings into the lumen of the intestine: 2 smaller ones on the left and between them a larger one on the right, *Kohlrausch's fold* **AB1**, at a distance of 5–8 cm from the anus. Through contraction of the circular muscle fibers **C2** the folds approximate to each other; and on contraction of the longitudinal muscle fibers **C3** they move apart from one another *("rectoanal pylorus").*

B On the right: the transverse folds and the anus as seen through a *rectoscope.*

Anal canal. In its lower ⅔ rds it is lined by a thin, lightly keratinized skin containing sensory innervation, which merges into the external skin. The latter reaches into the end of the anal canal. It has a cornifying, pigmented epidermis, hairs with sebaceous glands and sweat glands. The mucosa of the colon reaches the upper ⅓ of the anal canal. In this zone 6–10 roll-shaped longitudinal folds, the *anal columns* **B4** arch into the lumen. They are thrown up by knots of vessels and are covered by several layers of non-keratinized squamous epithelium. At their lower ends the anal columns are joined together by transverse folds. The grooves between the longitudinal folds end in shallow pockets **B5** at their anal ends. They are covered with a single layer of columnar epithelium. The region of the anal column, which is about 1 cm long, is called the *hemorrhoidal zone.* Branches of the *superior rectal artery* descend in the anal columns. They lie beneath the mucosa and form the bases of *internal hemorrhoids* (piles). The arteries are connected by nodular arteriovenous anastomoses **B6** with the anorectal venous plexus. The columns form a cavernous body which contributes to closure of the anus.

Closure of the anus. The anal canal is actively closed by smooth muscle fibers (the continuation of the circular muscle layer of the intestine), the *internal sphincter muscle of the anus* **AB7** and by striated muscle fibers, the *external sphincter muscle of the anus* **AB8**. The internal sphincter of the anus is about 2 cm high, and its hard lower edge can be palpated in the patient. The longitudinal muscle layer of the intestine radiates partly into the internal sphincter muscle and partly into the perianal skin which it draws into the anus. Above the external and the internal sphincter muscles lies the *puborectal muscle* **A9**. This is the most important muscle of the sphincter, it closes the anus and forms part of the *levator ani muscles* (see p. 306). It pulls the perineal flexure forward in a loop (closure). If the muscle relaxes the anus moves backward (opening). Damage to the puborectal muscle has a more serious effect on rectal incontinence than a lesion of any of the other sphincter muscles. Part of the *pubococcygeal muscle* also takes part in anal closure. The muscles are under permanent tension except during the act of defecation. **A10** Anococcygeal ligament.

Defecation. Defecation is preceded by transport of feces into the rectum. The increasing tension of the rectal wall acts as a stimulus to defecation. The involuntary sphincter muscles are relaxed by reflex action and the other intestinal muscles contract. The external sphincter and the levator ani muscles are relaxed voluntarily and pressure is applied by the abdominal muscles.

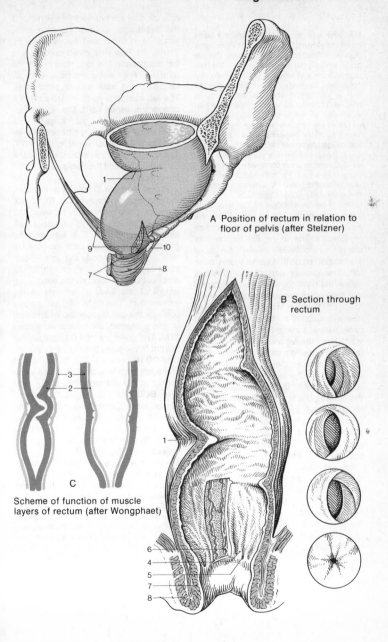

A Position of rectum in relation to floor of pelvis (after Stelzner)

B Section through rectum

C

Scheme of function of muscle layers of rectum (after Wongphaet)

Liver

The **liver** and the **pancreas** are the **large intestinal glands.** The *liver* acts as an *exocrine gland* in respect of production of bile. The bile acids emulsify fat in the intestine. The bile pigments are end products of the hemoglobin catabolism. Bile accumulates in the gall bladder and is discharged into the duodenum as needed. The most important action of the liver is its role as the largest organ involved in carbohydrate, protein and fat metabolism. These functions consume about 12% of the total oxygen content of the blood. The temperature of the blood in the liver veins reaches about 40°C. Compression or contusions can produce dangerous tears in the soft hepatic tissue. The liver is held together by a tense connective tissue capsule – *Glisson's* capsule.

Liver. The lower edge of the liver runs laterally along the costal arch. From the point where the medioclavicular line crosses the line of the 8th rib the liver margin runs obliquely through the upper abdominal region (the epigastrium) to the left. The greater part of the liver lies below the right dome of the diaphragm. We distinguish the convex *diaphragmatic surface* **D** which, in the upright position, has a horizontal surface and a curved anterolateral surface **B** that points downward from the *visceral surface* **C**. The latter ascends from the sharp lower edge obliquely backward and abuts posteriorly with a blunt edge on the diaphragmatic surface. Most of the liver is covered by the *peritoneum,* but posteriorly it is joined to the tendinous center of the diaphragm (the *area nuda* or bare area **BCD1**).

Diaphragmatic surface D. The reflect folds between the visceral peritoneum of the liver and the parietal peritoneum form band-like structures on the diaphragm. The *hepatic falciform ligament* **BCD2** divides the anterior surface of the liver superficially into right **BCD3** and left **BCD4** lobes of the liver. It is attached to the inner surface of the abdominal wall and its lower margin takes up the *ligamentum teres hepatis* and extends toward the umbilicus. The falciform ligament forms the *hepatic coronary ligament* on the superior surface of the liver beneath the diaphragm. This peritoneal

reflection which connects the liver and the diaphragm, delimits the *bare area,* which is not covered by peritoneum. There, the liver is in direct contact with the diaphragm. The fold, here called the *triangular ligament,* surrounds a triangular part of the bare area on both sides. The *left triangular ligament* **D5** runs in a connective tissue cord, the *fibrous appendix* **BCD6**. The *right triangular ligament* forms a blunt angle whose posterior fold forms the *hepatorenal ligament* **D7**. The *inferior vena cava* **CD8** runs to the diaphragm behind the peritoneum within the bare area.

Visceral surface. The **porta hepatis,** the entrance into the liver **C9, 10, 11–13,** forms a cross-connection between the sagittal grooves which together are shaped like an H. The **left sagittal groove** accommodates the remnants of fetal vessels: anteriorly the *ligamentum teres hepatis* in **BC2,** a remnant of the umbilical vein, posteriorly the *venous ligament* of the liver, a remnant of the venous duct. The **right sagittal groove** contains the *gall bladder* **BC14** anteriorly, and posteriorly the *inferior vena cava* **CD8**. The *quadrate lobe* **C15** bulges out in front of the porta hepatis, the *caudate lobe* **CD16** behind it. The lower surface of the left hepatic lobe **BCD4** bears the imprint of the stomach, that of the right lobe **BCD3** of the superior duodenal flexure, kidney, adrenal gland and of the right colic flexure. The *hepatoduodenal ligament* opens over the porta hepatis like a tent and affords space for the vessels passing through the porta to divide into two main branches (margin of the areas supplied: red line in **B**).

C9 Cystic duct, **C10** hepatic duct, **C11** choledochal duct, **C12** hepatic artery, **C13** portal vein, **D17** right triangular ligament.

Segmental structure. The liver is divided into variable segments which follow the divisions of the portal vein. Most of the segments reach the liver surface (broken line). The roots of the hepatic veins run within the segmental borders.

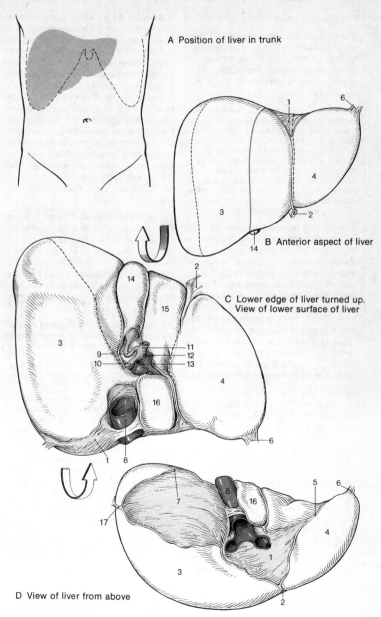

A Position of liver in trunk

B Anterior aspect of liver

C Lower edge of liver turned up.
View of lower surface of liver

D View of liver from above

Fine Structure of the Liver

The parenchyma and blood vessels form small *liver units*, which may be regarded as *liver lobules* (central vein lobules) or portal vein lobules depending upon which vessel is considered to be the center of the unit.

A delicate, sponge-like *connective tissue web* extends from the tough *capsule* into the liver. In it pass blood vessels and the epithelia of the liver run in its interstices. The millimeter-sized polygonal *liver lobules* **A1** are visible on the surface. The section shows blood-containing points 2–3 mm apart: **A2** the *central veins* of the lobule. Several lobules make up a *lobe* **A**. The *central veins* **A2** lead into intermediate veins **A3** which ultimately drain via sublobular and collecting veins into the hepatic veins.

The **hepatic lobule A1** is pear-shaped with a polygonal surface. Its diameter is 1–1.5 mm and its length about 2 mm. **Blood vessels**, branches of the *portal vein* **A4**, and of the *hepatic artery* **A5**, as wall as *interlobular veins* and *arteries*, reach the thicker end of the lobule to supply the surrounding vascular network. From there branches enter the lobule to carry blood from the portal vein *(vasa publica)* into the sinusoidal capillaries. The periphery of a lobule has a better O_2 supply than its center. The *central vein* **B2** emerges from the base.

Section. The triangular *periportal areas of connective tissue* **BD6** between the lobules each contain 3 vessels: branches of the *hepatic artery* **A5**, *hepatica propria* and of the *portal vein* **A4** and a *bile duct* **A7** *(Glisson's triad)*. Isolated veins **ABD3** are *intermediate, sublobular or collecting veins*. In the center is the *central vein* (intralobular vein) **ABD2**. The *epithelia* form cribose laminae, 1–2 cells thick, which radiate from the center of the lobule into the periphery. Between them run the 350–500 μm long *sinusoidal capillaries*.

Between the vessel wall **EF8** of the sinusoid capillaries **EF9** and the surface of the liver cell is a space, *Disse's space* **EF10**. The vessel wall consists of *en-*

dothelium **EF8** with gaps (about 100 mm) which is virtually devoid of a basal lamina. The surface of the hepatic epithelium **E11** contains microvilli **F11**. These come into direct contact with material which enters from the blood **EF9** through the capillary gaps into Disse's space **EF10**. There are also the *stellate Kupffer cells* **E12** within the endothelium. They are phagocytic and are probably involved in the breakdown of hemoglobin. The exchange of materials involved in metabolic processes takes place on the blood vessel side of the liver cell. The substances involved in the production of bile, however, pass right through the cells.

The **portal vein lobule** (triangle in **B** and **C**) is grouped around a Glisson's triad, i. e., with respect to a bile duct around the *excretory passage* of an "acinus", the "liver gland". The three branches of the portal vein from Glisson's triad form the axis of an **acinus of the liver C**. Many observations based on liver pathology regard the liver acinus as the liver unit.

The **bile canaliculi** *(capillaries)* **EF13** are tube-like spaces without walls of their own between the hepatic epithelial cells. Bile flows from the center toward the periphery of the lobule in the interlobar bile ducts **A7**, which have an epithelial lining. **F14** Cells nucleus, **F15** Golgi apparatus.

Division of labor. There is a division of labor in the *lobule* between the periphery, the mid-zone and the center over a 24-hr period cycle, *circadian* rhythm. The production of bile (secretion) starts in the morning in the peripheral zone and progresses towards the center, reaching its maximum in the evening. Glycogen storage starts in the center with its peak in the morning. Infiltration of fat from food usually commences during the afternoon in the periphery and reaches the center at midnight. Toxic fatty degeneration due to O_2 deficiency usually affects only the central region.

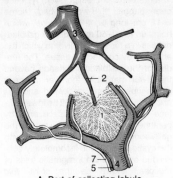

A Part of collecting lobule
(after Pfuhl)

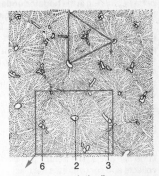

B Section through the liver seen
through a magnifying glass

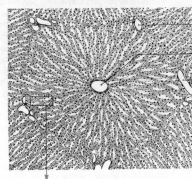

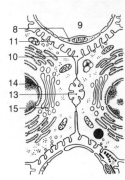

C Section through hepatic lobule
from B

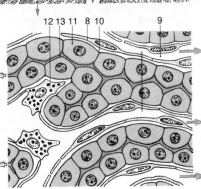

D Liver cells and capillaries, section from C

E Electron micrograph of hepatic
cells and capillaries

The *vessels* and *nerves* which enter the *porta hepatis* run in the **hepatoduodenal ligament**. The hepatic artery **A1** gives off branches to the gall bladder **A2**. Bile (hepatic) duct **A3** and in the depths the portal vein **A4**. **A5** Arrow in the epiploic foramen, **A6** common hepatic artery, **A7** celiac artery, **A8** left gastric artery, **A9** splenic artery, **A10** superior mesenteric vein, **A11** inferior mesenteric vein, **A12** splenic vein.

Bile Ducts and Gall Bladder

The bile reaches the duodenum via the *large bile ducts*. The nearby *gall bladder* collects and concentrates the bile.

The **large bile duct** has the diameter of a pencil. Proximal to the origin of the cystic duct to the gall bladder it is called the *common hepatic duct* and distal to it the *choledochal duct*. In the porta hepatis, the left and right hepatic ducts form the *common hepatic duct* **A3** which is 4–6 cm long. The *cystic duct* **A13** from the gall bladder which is 3–4 cm long, opens into it at an acute angle. The *choledochal duct* **A14** is 6–8 cm long. It passes behind the bulb of the duodenum to the posterior, medial side of the descending duodenum where, in 77% of cases, it penetrates the duodenal wall to open into the *major duodenal papilla* **AB16** with the *pancreatic duct* **A15**. In over 50% of cases, the ducts form a common *hepatopancreatic ampulla* **AB17**. Radiological studies have shown that it is more common for the ducts to open separately. Before its entry into the ampulla, each duct has a sphincter muscle **B18,19**. The ampulla may itself be closed by its own sphincter muscle, the *sphincter ampullae* (Oddi) **B20**. The mucosa of the ampulla has folds which restrict the reflux of bile and pancreatic secretion into the ducts. The thin wall of the bile ducts is made up of tall epithelial cells, a prominent elastic network, and a thin muscle layer. Mucus ducts open into the bile ducts.

The **gallbladder** (*vesica fellea*) **A21** is a pear-shaped, thin-walled bag, some 8–12 cm long × 4–5 cm wide, which holds up to 30–50 ml fluid. The gallbladder lies in a fossa in the liver to which it is attached by connective tissue. The fundus of the gallbladder extends beyond the lower margin of the liver. Its neck points upward and backward and lies above the duodenal bulb. The lower surface of the gallbladder is covered by peritoneum. The lumen of the neck of the gallbladder and of its connections with the cystic duct **A13** is incompletely subdivided by spiral diaphragmatic folds of mucosa, known as the *spiral fold (Heister's valve)*.

Histologic structure. The mucosa forms reserve folds that outline polygonal areas, which in histologic sections often appear as "mucosal bridges" **C22**. The simple columnar epithelium contains absorptive cells with microvilli as well as other cells which contain secretion granules. Mucus-producing goblet cells are also present and their number is increased in chronic inflammation, e. g., in patients with gallstones. The remainder of the wall consists of loose (alveolar) connective tissue and a thin muscle layer **C23**.

Bile flow. The course taken by the bile can be demonstrated on a radiograph by contrast media which are excreted by the liver after oral administration and absorption. The gallbladder becomes filled by reflux of the bile if the ampullary sphincter is closed. Tension in its wall is adjusted to the degree to which the gallbladder is filled. During a sudden rise of pressure (cough, abdominal straining) the diaphragms of the spiral valve are said to act as shutters. Opening of the ampullary sphincter and emptying of the gallbladder are controlled by hormones or by the autonomic nervous system. There is no reflux of intestinal contents into the ampulla. Overflow of bile into the pancreatic duct is pathologic and leads to activation of the pancreatic enzymes and to necrosis of the pancreas.

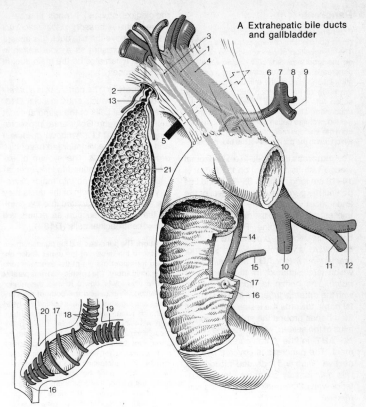

A Extrahepatic bile ducts and gallbladder

B Entry of common bile duct and pancreatic duct into duodenum

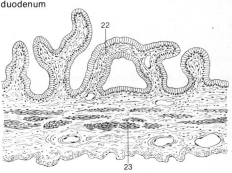

C Section through wall of gallbladder

Pancreas

The **pancreas** is the most important intestinal gland (for the islet cell apparatus see p. 168). The composition of pancreatic juice depends on the food ingested. Like gastric secretion, pancreatic secretion is activated by a nervous stimulation, then by the stimulus provided by the filling of the stomach and finally by the action of a hormone released from the duodenum. During the third phase partly digested proteins cause liberation of hormones from the mucous membrane of the duodenum, which reach the pancreas via the blood stream.

The **pancreas** is shaped like a horizontal wedge with its thin end on the left. It is 14–18 cm long, weighs 65–75 g and lies behind the peritoneum at the level of the 2nd lumbar vertebra. The *head* **AB1**, the thickest part, fits into the duodenal loop to the right of the spine. Posteriorly and below its *uncinate process* **AB2** hangs downward and embraces the *superior mesenteric artery* **AB3** and *vein* **AB4**, which are situated in the *pancreatic notch*. The horizontal *body* **B5** bulges with the *omental tuber* **A6** into the omental bursa towards the lesser omentum. It then bends around the spine toward the hilus of the spleen, which it reaches by its *tail* **AB7** in the phrenico-splenic ligament. The pancreas is covered by connective tissue and is divided into lobules. It is only loosely connected to the posterior wall of the trunk and it moves with respiration. The transverse mesocolon runs over the head of the pancreas along the anterior head of the pancreas. Thus, the anterior surface of the gland is divided into an upper part which lies in the posterior wall of the omental bursa, and an inferior part which faces the free abdominal cavity.

Its **excretory duct**, the *pancreatic duct* **A8**, is 2 mm in diameter and runs right through the length of the gland. It receives short, vertical tributaries from the lobules. In about 77% of cases (in anatomical preparations) the pancreatic duct ends together with the *common bile duct* on the *major duodenal papilla*, in the remainder (more common according to

radiological studies) it ends nearby. If present, the *accessory pancreatic duct* **A9** ends above the bile duct. It is absent in 3% and present as a side branch in 33%, but it can also be the main duct in 5–8% of cases.

Fine structure. The pancreas is a *purely serous gland* which ends in acini **C10**. The epithelial cells of the gland show at their apex (in the Golgi area) prosecretory granules **D11** (*zymogen granules*) and at their base widespread basophilic *ergastoplasm* **D12**. The system of excretory ducts is confined to long intercalated parts which lead into major excretory ducts. At the origin of the ducts they are said to be invaginated into the acini, so that in cross section an acinus will show "centroacinar cells" **D13**.

Position. The pancreas and the duodenum are situated in the middle of the upper abdomen. Along the upper margin of the pancreas runs the splenic artery. The splenic vein runs parallel to it a little lower down. It joins the inferior mesenteric vein behind the body of the pancreas and together they pass behind the head of the pancreas with the superior mesenteric vein **AB4** into the portal vein. The superior mesenteric artery **AB3** takes its origin from the aorta, runs behind the head of the pancreas and downward for several centimeters, then through the pancreatic notch on the uncinate process **AB2**, over the upper border of the horizontal part of the duodenum and into the root of the mesentery. The head of the pancreas lies in front of the inferior vena cava **B14** and the aorta **B15** and reaches upward to the celiac artery.

The root of the **transverse mesocolon** is fixed to the anterior edge of the pancreas and divides the abdominal cavity into the *upper* and *lower abdomen*. **A16** Duodenojejunal flexure.

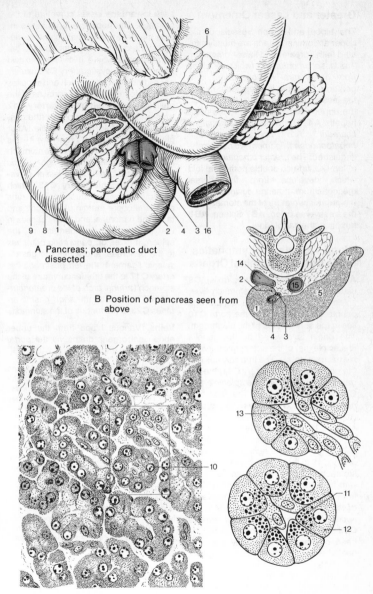

A Pancreas; pancreatic duct dissected

B Position of pancreas seen from above

C Section through pancreas

D Acini of pancreas, higher magnification (section from C)

Greater and Lesser Omentum

The blood and lymph vessels of the upper abdominal organs are mostly situated in the greater and lesser omentum. The *lesser omentum* **A 1** (which develops from the *ventral mesogastrium*) is a peritoneal fold that is stretched between the lesser curvature of the stomach **A 2**, the upper duodenum **A 3** and the porta hepatis **A 4**. The delicate *hepatogastric ligament* **A 1** and the tough *hepatoduodenal ligament* **A 5** can be distinguished. The *greater omentum* **AB 6** is a flap-like, fat-rich double peritoneal fold which develops from the dorsal mesogastrium. It hangs downward from the greater curvature of the stomach and the transverse colon. **AB 7** Spleen, **AB 8** fundus of the gall bladder.

Blood Vessels and Lymphatics of the Upper Abdominal Organs

Arteries. The **celiac trunk** (axis) **C 9** usually supplies the entire stomach, liver, spleen and part of the duodenum and pancreas. Its origin from the aorta **C 10** lies in the aortic hiatus of the diaphragm. The origin is covered by the ganglia (celiac plexus) of the autonomic nervous plexus (see Vol. 3). The trunk divides into the *common hepatic artery* **C 11**, the *left gastric artery* **C 12** and the *splenic artery* **C 13** (*Haller's tripod*). The *common hepatic artery* **C 11** divides into the *gastroduodenal artery* **C 14** and the *hepatic artery* **C 15**. The gastroduodenal artery **C 14** descends behind the duodenal bulb. It gives off the *right gastroepiploic artery* **C 16** to the greater curvature of the stomach, which forms an anastomosis with the *left gastric epiploic artery* **C 17** from the *splenic artery* **C 13** (*arterial arch of the greater curvature*). It ends in the *superior pancreatico-duodenal arteries* **C 18** which supply the duodenum and the head of the pancreas. They are connected with the *superior mesenteric artery* **C 19** via an anterior and posterior arterial loop. The *right gastric artery* **C 20**, a branch of the *hepatic artery* **C 15**, runs

back toward the lesser curvature of the stomach. The hepatic artery passes inside the hepatoduodenal ligament, medial to the common bile duct **B 21** to the porta hepatis where it divides into two branches. The *cystic artery* **C 22** usually arises from the right branch and supplies the anterior and posterior surfaces of the gallbladder. The **left gastric artery** **C 12** runs within a peritoneal pouch (the gastropancreatic fold) to the cardia. After giving off *esophageal branches*, it runs along the lesser curvature of the stomach and anastomoses with the *right gastric artery (arterial arch of the lesser curvature)*. The **splenic artery** **C 13**, which generally follows a tortuous course, runs along the upper margin of the body and tail of the pancreas, and finally passes in the phrenicosplenic ligament to the spleen. It gives off *pancreatic branches* to the pancreas, and within the gastrosplenic ligament, the *left gastroepiploic artery* **C 17** to the greater curvature of the stomach (*arterial arch of the greater curvature*), as well as the *short gastric arteries* **C 23** to the fundus of the stomach.

Veins. Venous blood from the upper abdominal organs drains via the *portal vein* **B 24** into the liver (see p. 248).

Lymphatics and lymph nodes. The lymph vessels and nodes follow the course of the arteries. The *right and left gastric lymph nodes* **D 25** lie alongside the lesser curvature; they drain into the *celiac lymph nodes* **D 26**. Lymph from parts of the gastric fundus and the upper portion of the body of the stomach in the neighborhood of the greater curvature also drains into the celiac lymph nodes via the *left gastroepiploic lymph nodes* **D 27**. Lymph from the lower portion of the corpus and the pyloric portion of the stomach drains via the *right gastroepiploic lymph nodes* **D 28** and the *pyloric lymph nodes* **D 29** partly to the *celiac lymph nodes* **D 26**, and they are partly connected with lymphatics of the liver via the *hepatic lymph nodes* **D 30**. This makes it possible for lymph to drain through the diaphragm into the *anterior mediastinal lymph nodes*.

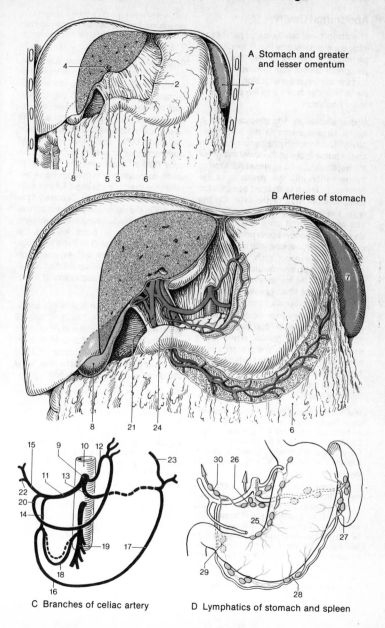

A Stomach and greater and lesser omentum

B Arteries of stomach

C Branches of celiac artery

D Lymphatics of stomach and spleen

Abdominal Cavity

The **abdominal cavity** is that part of the intra-abdominal space that is lined by the peritoneum. Behind it lies the *retroperitoneal space*. The transverse colon and the transverse mesocolon subdivide the abdominal cavity into the *upper* and *lower abdomen*.

Upper abdomen. The *omental bursa* **A** is the largest space in the abdominal cavity. The *foramen epiploicum (of Winslow)* – arrow in **B** – at the lower margin of the hepatoduodenal ligament **AB1** forms the entrance into the *vestibule* of the omental bursa. It is situated beneath the caudate lobe **AC2** of the liver. On the right, the inferior vena cava borders on the vestibule, and on the left lies the aorta. The *superior recess* points upward under the liver, between the inferior vena cava and the esophagus. The *inferior recess* extends downwards between the stomach and the transverse colon to the attachment of the greater omentum **ABC3**. The gastrocolic ligament **C4** connects the transverse colon with the greater curvature of the stomach. **A5** attachment of the greater omentum to the transverse colon. The greater omentum becomes adherent to the left abdominal wall across the left colonic flexure and forms the phrenicocolic ligament, the floor of the splenic fossa.

The *omental bursa* **C6** is a space in which the stomach can move. Its anterior wall is formed by the lesser omentum **C7** and the stomach **C8**. After their removal the posterior wall and the extent of the omental bursa are demonstrated in **A**. **A9** Cardia, **A10** spleen, **A11** duodenum.

Attachment (root) of the transverse mesocolon BC12 (at the level of the second lumbar vertebra) and of the **mesentery BC13** to the posterior abdominal wall. The *root of the mesentery* extends from the duodenojejunal flexure (left from the 2nd lumbar vertebra) to the opening of the ileum into the cecum. The mesentery is continuous with the mesoappendix of the appendix. **B14** Omental tuber at the posterior wall of the omental bursa, **BC15** duodenojejunal flexure, **B16** superior mesen-

teric artery and vein, **C17** course of the blood vessels in relation to the head of the pancreas **C18**, uncinate process **C19** and duodenum **C15**, **C20** transverse colon.

Drainage spaces in the *upper abdomen*, the *subphrenic recesses*, are found beneath the diaphragm along the anterior surface of the liver, on both sides of the falciform ligament **D21**. Between the ascending and descending colon and the lateral abdominal walls on either side lie the *left* **D22** and the *right* **D23** *parietocolic recesses* The *subhepatic recess* **D24** extends over the transverse colon, stomach and lesser omentum beneath the liver. It leads into a *left posterior subphrenic recess*. **D25** Coronary ligament. On the right side there is a narrow posterior passage below the diaphragm, through the wide base area to the *hepatorenal recess* **D26**. The drainage spaces (demarcation spaces) are important in the flow of peritoneal fluid and in the spread and encapsulation of infections.

The **lower abdomen** below the transverse mesocolon is divided into 2 peritoneal pouches: above the root of the mesentery the *right* **D27** and below it the *left* **D28** *mesenterocolic recesses*. They are connected to each other above the duodenojejunal flexure **BC15**.

The *recesses* in the lower abdomen are formed where the retro- and intraperitoneal portions of the intestine merge into each other (see p. 216). The lateral groove of the ascending colon and the medial and lateral grooves of the descending colon lead into the peritoneal recesses of the lesser pelvis: *rectovesical* and *rectouterine pouches*, see p. 312. Folds in the lateral grooves of the descending colon form the *paracolic sulci*.

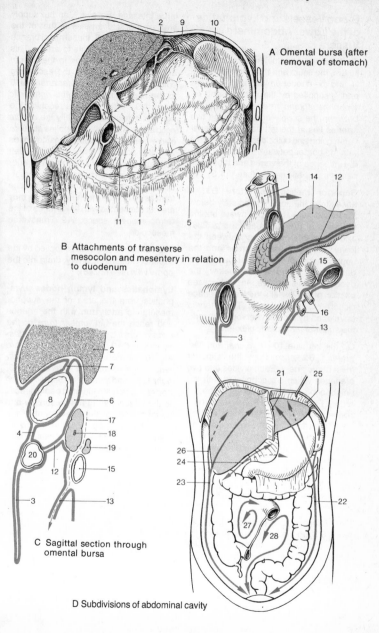

A Omental bursa (after removal of stomach)

B Attachments of transverse mesocolon and mesentery in relation to duodenum

C Sagittal section through omental bursa

D Subdivisions of abdominal cavity

Blood Vessels and Lymphatics of the Lower Abdominal Viscera

Of the lower abdominal viscera the jejunum, the ileum and the colon are wholly, and the duodenum and pancreas are partly, supplied by the *superior* and *inferior mesenteric arteries*. The border between the supply areas of these two arteries lies in the left half of the transverse colon and coincides approximately with the border between the areas under the influence of the vagus and the sacral parasympathetic nerves (see Vol. 3).

Superior mesenteric artery B1. It arises from the aorta immediately below the celiac trunk, runs downward behind the head of the pancreas (see p. 238) and enters the *mesentery* between the lower margin of the pancreas and the upper margin of the inferior part of the duodenum, about 3 cm below the duodenojejunal flexure. In rare cases traction of the vessel's trunk can produce closure of the duodenum. The artery is enveloped by a dense nervous network, the *superior mesenteric plexus*.

On the left side 10–16 *jejunal and ileal arteries* **B2** arise from the superior mesenteric artery. Each divides into two branches which are connected with the neighboring arteries (arcades of the 1st order). Further rows of transverse cross-connections follow (arcades of the 2nd–4th orders), producing an ever narrower vascular meshwork. The *formation of arcades* is more marked in the lower part of the small intestine than in its upper part. The parallel blood vessels which run from the outer arcades to the intestine are end arteries and their blockage will result in a local lesion of the intestine.

The *inferior pancreaticoduodenal arteries* **B3** are the first to arise behind the head of the pancreas from the right side of the mesenteric artery. They are connected with the arch formed by the *superior duodenal arteries* **B4** (see p. 240). They then give off 3 arteries to the

large intestine. Following on the supply area of the ileal arteries is that of the *ileocolic artery* **AB5**, the terminal branch of the superior mesenteric artery. This artery gives off branches to the cecum and to the lower part of the ascending colon, and the *appendicular artery* **AB6** to the appendix. Next is the *right colic artery* **AB7** with branches reaching the right colic flexure, and finally the *middle colic artery* **AB8**, which supplies about ⅔ of the transverse colon. It anastomoses with the adjacent branches of the left colic artery, which arises from the inferior mesenteric artery (see p. 246). The arcades of the arteries of the large intestine form a row of wide meshes. As a rule they are not connected with the arteries of the retroperitoneal space. **A9** Transverse mesocolon.

The **veins** from the area, supplied by the superior mesenteric artery, drain into the *portal vein* (see p. 248).

Lymphatics and lymph nodes. Lymphatics from the area of the superior mesenteric artery run with the arteries and reach the left *lumbar trunk* or the *cisterna chyli* **C11**, usually via one *intestinal trunk* **C10**. Along their course there are 100 or more small *superior mesenteric lymph nodes* **C12** – *ileocolic* **C13**, *right* **C14** and *middle* **C15** *colic lymph nodes* – which are situated partly directly at the attachment of the mesentery and partly directly at the attachment of the mesentery and partly further along toward the root of the mesentery. **C16** *Pancreaticosplenic lymph nodes*.

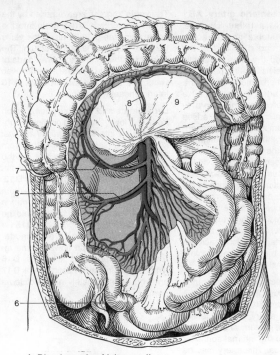

A Blood supply of jejunum, ileum, ascending and transverse colon: superior mesenteric artery and vein

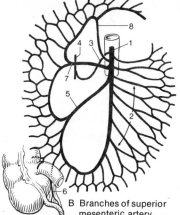

B Branches of superior mesenteric artery

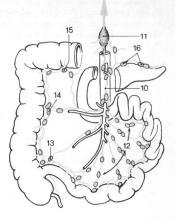

C Lymphatics in area supplied by superior mesenteric artery

The **inferior mesenteric artery AB1** arises ventrally at a rather low level (3rd–4th lumbar vertebra) from the aorta, turns to the left and runs across the psoas muscle and the linea terminalis into the lesser pelvis. Its first part often lies below the horizontal part of the duodenum. This artery is covered by a dense network of autonomic nerve fibers. The *left colic artery* **AB2** arises either directly from the inferior mesenteric artery or, together with the *sigmoid artery* **AB3**, from the short common stem of the inferior mesenteric artery. The left colic artery divides into an *ascending* branch, which continues the supply to the intestine from the middle colic artery (a branch of the superior mesenteric artery), and into a *descending* branch. This supply area is often followed by the *sigmoid artery* or arteries which reach the sigmoid colon in the pelvic mesocolon. **A4** Duodenojejunal flexure.

Arteries of the rectum. The last branch of the inferior mesenteric artery is the *superior rectal artery* **BC5**. It runs across the internal iliac artery into the lesser pelvis behind the rectum and supplies the latter down to the internal sphincter muscle of the anus. This artery may be divided into two branches. Branches of the superior rectal artery are joined above the pelvic floor **C6** on either side by the branches of the *middle rectal artery* **C7**, which comes from the internal iliac artery. Below the pelvic floor, which is pierced by the branches of the superior rectal artery, these are joined on each side by branches of the *inferior rectal artery* **C8** from the internal, pudendal artery. However, the anastomoses of the middle and lower rectal arteries with the superior rectal artery are inadequate as a substitute for the superior rectal artery. As the superior rectal artery has only a single communication with the sigmoid artery, the *sigmoidea ima artery* the former vessel **C5** must not be ligated below this junction (circle in **B**). **C9** Edge of the cut peritoneum.

The **veins** from the area supplied by the inferior mesenteric artery drain into the *portal vein* (see p. 248), as do the veins from the upper parts of the rectum. The veins from the middle and lower parts of the rectum drain via the internal iliac veins into the *inferior vena cava*. **A10** Inferior mesenteric vein.

Lymphatics and lymph nodes. Lymph from the lower anal region flows subcutaneously into the *superficial inguinal lymph nodes* (see p. 78), and that from the upper anal region reaches via the lymph nodes of the ischiorectal fossa **D11** the *internal iliac lymph nodes* **D12**. Lymph from the pelvic part of the rectum drains into the *sacral* **D13** and *common iliac* **D14** *lymph nodes*, as well as into the *inferior mesenteric lymph nodes* **D15** via the mesosigmoid fold. From the descending colon lymph flows via the *inferior mesenteric* and *left colic* **D16** *lymph nodes* into the intestinal trunk **D17** and from there ultimately into the *cisterna chyli*.

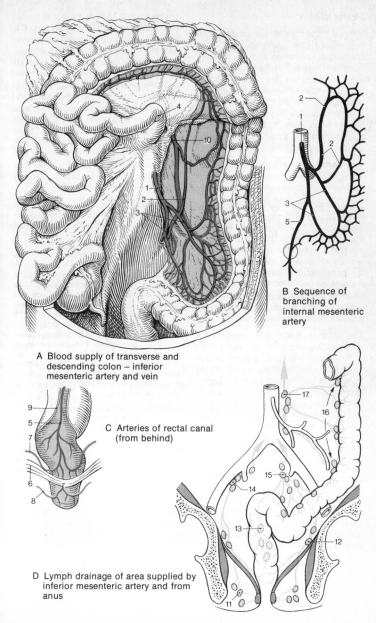

A Blood supply of transverse and descending colon – inferior mesenteric artery and vein

B Sequence of branching of internal mesenteric artery

C Arteries of rectal canal (from behind)

D Lymph drainage of area supplied by inferior mesenteric artery and from anus

Portal Vein

Venous blood from the unpaired abdominal viscera (the gastrointestinal tract, gallbladder, pancreas and the spleen), which are supplied by the three unpaired abdominal arteries (the celiac trunk and superior and inferior mesenteric arteries) reaches the liver via the portal vein and runs through the hepatic veins to the inferior vena cava. Nutrients absorbed from the intestine thus reach the central metabolic organ by the shortest route.

The **portal vein** is formed from 3 root veins, the splenic, inferior mesenteric and superior mesenteric veins.

The **splenic vein ABC1**, like the splenic artery, runs along the upper margin of the pancreas **A2**. In its course it receives the short gastric veins **BC3**, the left gastroepiploic vein **BC4** and the *pancreatic* and *duodenal* veins. Behind the body of the pancreas, it unites with the *inferior mesenteric vein* **ABC5**, and behind the head of the pancreas **A2** it drains together with the *superior mesenteric vein* **ABC6** into the **portal vein ABC7**.

The **inferior mesenteric vein ABC5** contains blood from the descending and sigmoid colon and from the upper rectum (*superior rectal vein*). Its course differs from that of the inferior mesenteric artery. Both vessels run together until the departure of the left colic artery. From there the vein follows that particular branch and then runs in a peritoneal recess, the *superior duodenojejunal fold*, across the duodenojejunal flexure behind the pancreas.

The **superior mesenteric vein** carries blood from the small intestine, the cecum and the ascending and transverse colon, accompanied by the superior mesenteric artery behind the head of the pancreas **A2**, and it receives the *duodenal, pancreaticoduodenal* and *right gastroepiploic veins* during its course.

The following veins end directly in the trunk of the portal vein: the *right* and *left*

gastric veins **BC8** from the lesser curvature of the stomach, the *cystic vein* **C9** from the gallbladder, the *prepyloric vein* from the anterior surface of the pylorus, and lastly the *paraumbilical veins* **C10** which accompany the ligamentum teres in the falciform ligament of the liver. They form anastomoses between the subcutaneous veins of the abdominal wall and the portal vein and open into the left main branch of the portal vein. Blue arrows in **C** represent connections between the afferents in the portal vein and the areas that drain into the superior and inferior vena cava: e. g., **D** congestion of the portocaval anastomoses.

Portocaval anastomoses. The drainage area of the portal vein borders on that of the superior and inferior vena cava at the following sites: *esophagus* (blood from the esophagal veins **ABCD11** flows through the azygos vein **D12** and the hemiazygos vein to the superior vena cava **D13**), *rectum* (blood from the middle and lower rectal veins **CD14** drains through the internal iliac vein **D15** to the inferior vena cava **D16**), *abdominal wall* (connected with the superior vena cava **D13** by the thoracoepigastric veins **D17** and to the inferior vena cava **D16** via the superficial epigastric veins **D18**. It is in those regions that portocaval anastomoses occur, i. e., *connections between the areas drained by the portal vein* **D7** *and the superior vena cava* **D13** *or the inferior vena cava* **D16**. Additional portocaval anastomoses may be present between the mesenteric veins and the retroperitoneal veins.

Clinical tips: In a case of portal vein congestion, e. g., in cirrhosis of the liver, part of the portal venous blood flow will bypass the liver via these portocaval anastomoses, and they become prominent as varicose veins, namely *esophageal varices* **D11**, *external hemorrhoids (piles)* **D14** and congestion of the subcutaneous abdominal veins **D10** ("*Caput Medusae*").

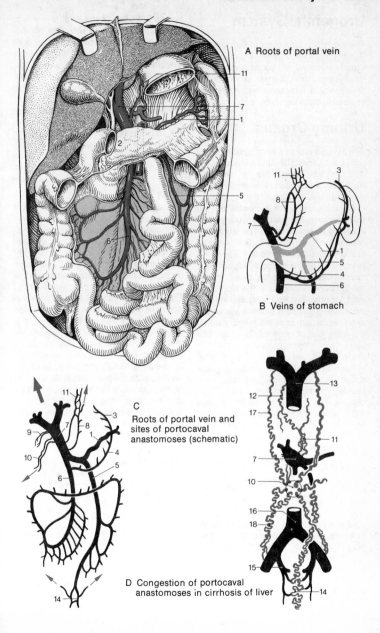

A Roots of portal vein

B Veins of stomach

C Roots of portal vein and sites of portocaval anastomoses (schematic)

D Congestion of portocaval anastomoses in cirrhosis of liver

Urogenital System

The excretory ducts of the urinary system and of the genital organs – the urogenital organs – are closely connected by their embryologic development as well as by their function. It is convenient, therefore, to discuss them together.

Urinary Organs

The *kidneys produce urine* by eliminating water and various harmful metabolic products, most of which arise in other organs. Thus the internal environment of the tissues is regulated, the equilibrium of minerals and water is maintained and the hydrogen ion concentration (pH) of the body is kept constant. The process of elimination takes place in 2 phases: at first an ultrafiltrate of blood plasma, the *primary urine*, is produced. This contains substances dissolved in blood in the same concentration as in the blood, except for proteins. Following this various materials, particularly glucose and water, are reabsorbed from the primary urine, the daily output of which is about 150 liters. During this process the amount of urine is reduced to 1% of its original volume to produce the greatly concentrated *secondary urine*. The urine then leaves the body via the *urinary tract organs*, i. e., the renal pelvis, ureters **AB1**, urinary bladder **A2** and the urethra. In addition, the kidneys have endocrine effects on blood pressure and hemopoiesis.

Anatomy. The **kidneys AB3** are bean-shaped and their longitudinal axes correspond approximately to that of the body. They converge upward toward the back. The kidneys are situated in a connective tissue space behind the peritoneal cavity in the lumbar region on either side of the spine. Their *upper poles* extend about as far as the upper margin of the 12th thoracic vertebra and in the adult their *lower poles* reach down to the 3rd lumbar vertebra. The *hilum* (renal porta,

see p. 252) lies at the level of the first lumbar vertebra **A4**. The 12th rib **A5** crosses the kidney obliquely at the border between its upper and middle thirds. Together with the 12th rib, part of the lumbar portion of the diaphragm **A6** and the costodiaphragmatic recess **A7** of the pleura lie across the upper third of the kidney. In 65% of cases the right kidney lies about half a segment lower than the left. In deep inspiration and in the upright position the kidneys descend by about 3 cm. Rotation and a dipping movement are also possible. The kidneys are kept in position by a *fascial sheath* and a *fatty capsule*.

The suprarenal (adrenal) gland **B8** which is embedded in the fat capsule, sits upon the upper pole of each kidney like a cap. The inferior vena cava **B9** and the descending part of the duodenum lie near the hilum of the *right kidney*. The anterior surface of the right kidney touches the liver **B10** and the right colic flexure **B11**. The abdominal aorta **B12** runs near the hilum of the *left kidney*. The anterior surface of the left kidney touches the stomach, pancreas and left colic flexure **B13**, and with its lateral border it touches the spleen **B14**. Posteriorly both kidneys border on the diaphragm above, medially on the psoas major muscle **B15**, laterally on the quadratus lumborum muscle **B16** and the transverse abdominal muscle. The kidney is crossed posteriorly by the subcostal, iliohypogastric and ilioinguinal nerves parallel to the course of the 12th rib.

The kidneys in *children* are lower than in adults, and the lower pole may reach the iliac crest. The kidneys of children are relatively larger than those of adults.

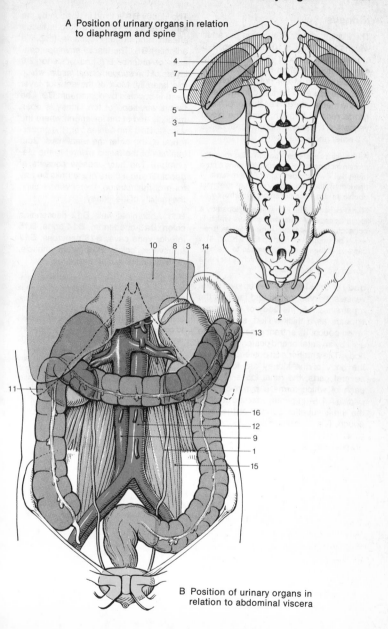

A Position of urinary organs in relation
to diaphragm and spine

B Position of urinary organs in
relation to abdominal viscera

Kidneys

The kidney (*ren or nephros*) in the adult weighs 120–300 g, is 10–12 cm long, 5–6 cm broad and about 4 cm thick. The convex lateral margin has a more pronounced curve at the poles, which appear, therefore, to be rolled in medially. On the medial margin lies the *renal hilum (porta renalis)* through which the blood vessels **A1**, nerves and the renal pelvis **A2** enter and leave.

As the kidney excretes urea, a waste product of protein metabolism, it enlarges following a pure meat diet – work hypertrophy. After removal of one kidney, the other kidney enlarges to almost double its size – compensatory hypertrophy.

Renal capsule. The kidney is enveloped by a tough *capsule of collagenous fibers* which is connected to the kidney by areolar tissue. It can easily be stripped off right up to the hilum. The capsule at the hilum is connected to the connective tissue of the vessels which enter into a central recess, the renal sinus.

Body of the kidney. After removal of the vessels, nerves, pelvis and fat from the *renal sinus* **A3**, the body of the kidney appears as a thick-walled, slightly flattened pouch. Its entrance is narrowed to a slit by an anterior and posterior lip of the body. The anterior and posterior walls of the body of the kidney are formed of several parts, the renal lobes (*renculi*) each of which forms a pyramid, *renal papilla* **A4**, 5–11 pyramids stand out from the inner superior surface of the renal pouch. The subdivision of the kidney into renal lobes does not correspond with its division into *vascular segments*, as each renal lobe is supplied by several branches of the renal artery, and each branch supplies more than one lobe.

There are usually five vascular segments to the kidney, a superior, anterosuperior, antero-inferior, inferior and posterior segment.

The kidney of the *newborn*, like that of certain animals, shows the clear demarcation of the original renal lobes (renculi), but in the course of the next 4–6 years this *renculi-marking* disappears.

The kidney **B6** is fixed in position by the renal *fascial envelope*, a fibrous tissue sheath, and a *fatty capsule* (capsula adiposa) **B5**. The fascial envelope consists of anterior **B7** and posterior **B8** layers of the *subperitoneal fascia*, which join laterally. Most of the anterior layer (leaf) is covered by peritoneum **B9**. The fascial envelope of the kidney is open medially and at the lower end where the vessels **B10** and nerves from the prevertebral space enter the renal hilus. Both laminae of the fascia extend to the diaphragm. The fatty capsule consists of depot fat and it therefore shrinks on any form of malnutrition, thereby increasing the motility of the kidney.

B11 Abdominal wall, **B12** descending colon, **B13** duodenum, **B14** aorta, **B15** inferior vena cava, **B16** pancreas, **B17** transverse colon, **B18** psoas muscle, **B19** musculus quadratus lumborum.

Clinical tips: Of all the variants and malformations with which 2% of the entire population is born, 30% affect the urogenital system. A *cystic kidney* results when one part of the embryonic cortex (a segment of the embryonic plate) fails to connect with the renal pelvis. *Supernumerary* kidneys are accounted for by early splitting of the ureter. A more common abnormality is *fused kidneys*; the most frequent type is the *horseshoe kidney* which lies in the midline in front of the spine. Congenital *displacement* of the kidney also occurs.

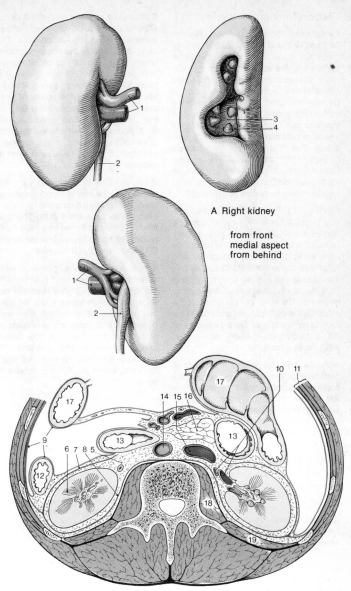

A Right kidney

from front
medial aspect
from behind

B Horizontal section through trunk at level of 3rd lumbar vertebra

Section Through the Kidney

Functioning of the kidney presupposes and ample blood supply. About 20% of the volume of blood which the heart pumps into the aorta flows into the kidneys. The *arborization of the renal blood vessels* has a peculiar pattern, which can be demonstrated by injecting suitable (colored) material into the vessels. The vascular tree, together with the tubules of the renal parenchyma (see p. 258), produces the characteristic macroscopic pattern on the cut surface of the kidney. Both a longitudinal (frontal) and a transverse section through the kidney show its division into a *cortex* and a *medulla*.

Cortex AB1. The brownish colored *cortex*, about 1 cm wide, lies beneath the fibrous renal capsule. Stripes (striae), radiating from the medulla, the *medullary rays* **AB2**, spread into the cortex and subdivide it into cortical lobules. Within the lobules there is an accumulation of dark red spots, the *renal corpuscles* or glomeruli. Between the pyramids of the medulla, the cortical substance lies in the form of *renal colums* ("Bertini's columns" **A3**) extending down to the renal pelvis. **AB4** Fibrous capsule.

Medulla A5. The medualla consists of several large *pyramids*. Their apices (the papillae) **B6** point toward the renal pelvis **AB7**. The pyramids show striations which converge towards the apex. It is also possible to distinguish in them a reddish *external* and a pale *internal zone*.

Blood Vessels of the Kidney

The *renal artery* **A8** and *vein* **A9** in the hilus enter the loose, fatty connective tissue **AB10** that lies between the renal parenchyma and the renal pelvis, except for the papillae. The artery divides into the following branches: the *interlobar arteries* **AB11** enter the medulla between the pyramids. They divide and run as *arcuate arteries* **AB12** like arches between the cortex and the medulla both of which they supply.

Cortex. The *interlobar arteries* **B13** radiate into the cortex and give off *afferent vessels* **B14** at regular intervals. They carry the blood to the renal corpuscles where they form tufts of blood vessels, the *glomeruli* **B15**. From there blood flows via the *efferent vessels* **B16** into the capillary net of the cortex. It finally drains via the *interlobar veins* **B17** into the *arcuate* **B18** and *interlobar veins* **B19**. Some interlobar arteries send *capsular branches* and some interlobar veins send *stellate venules* **B20** to the capsule.

Medulla. The *straight arterioles* **B21** radiate into the medulla from the efferent vessels of glomeruli near the medulla, or from the *arcuate artery* **B12**. The blood flows via the capillaries into the *venulae rectae* **B22**, and thence into the *arcuate* **B18** and *interlobar* **B19** veins.

In the *cortex* are the *glomeruli* which serve for ultrafiltration of fluid from the blood (see p. 256). The **medulla** contains the straight vessels, the *arteriolae* and *venulae rectae*, which are mainly concerned with *reabsorption* of fluids and of dissolved matter from the primary urine excreted in the glomeruli. The *tubules* of the renal parenchyma (see p. 256) play a dominant part in these functions.

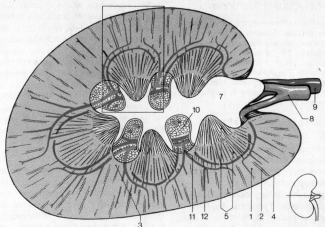

A Cross section through kidney

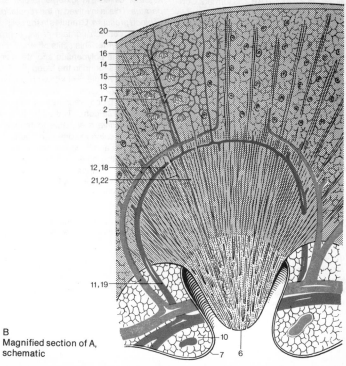

B
Magnified section of A,
schematic

Fine Structure of the Kidney

Besides the blood vessels, the kidney possesses a complicated system of tubules, the *nephrons* and the *collecting ducts*. The kidney of an adult contains 0.9–1.6 million nephrons. Each *nephron* is a unit which consist of a *glomerulus* and its *tubule*. The *glomerulus* contains a *filter* through which the *primary urine* is filtered from the blood. *Secondary urine* is produced by reabsorption into the tubules.

Structures Producing the Primary Urine

The glomerular *renal corpuscle* or *Malpighian corpuscle* **AB**, is 200–300 μm in diameter and is visible to the naked eye as a red spot. It contains the glomerulus with about 30 *capillary loops* **AB1**, which invaginate into the blind sac, the head of the tubular system at the *vascular pole*. In this way a double-walled capsule, *Bowman's capsule* **AB2**, is formed. Connective tissue cells at the vascular pole form the *mesangium* **AB3**. The space between the two walls of Bowman's capsule receives the primary urine and allows it to pass through its *open end* **AB4** into the tubular system. **AB5** *Vas afferens*, **AB6** *vas efferens*, **AB7** *macula densa*.

The **urinary filter** is 0.8 m² and comprises three structures contiguous with one another, endothelium, basal lamina and podocytes. *Endothelium* **CD8**: the endothelial wall, which is covered by the basal lamina, is thin and has regular, open pores 70–90 mm in diameter. The endothelial pores do not permit the passage of blood cells. The *basal lamina* **CD9** is 0.1–0.15 μm thick. It consists of a network of filaments enbedded in a mass of glycoprotein. The basal lamina prevents the passage of large molecules (e. g. ferritin, molecular weight 400,000), which are trapped in the part of the basal lamina near the vessels. The cell bodies of the *podocytes* **ABCD10** protrude into the lumen of Bowman's capsule. Their primary processes are given off into the superficial surface of the basal lamina. They give rise to numerous, short, secondary processes, which alternate with neighboring podocyte processes to produce slit pores of 40–50 nm. A band-like diaphragm (arrow in **D**) which has rectangular pores of 4×14 nm lies over the slit pores. The slit pore diaphragms form the densest filtering structure. They prevent the passage of substances with a molecular weight greater than 70,000. An adequate blood pressure (more than 50 mm Hg) is required for the urine filtration mechanism to work (filtration pressure hypothesis).

Juxtaglomerular apparatus. The *macula densa* **AB7,** the *pole cushion* **AB5** (= epitheloid cells of the vas efferens) and a group of *extraglomerular mesangial cells* **AB3**, which lie between the macula densa and the angle formed by the vas afferens and vas efferens are together called the *juxtaglomerular apparatus*. This is involved in the *regulation of blood pressure*. Granules in the cells of the pole cushion contain *renin*, a proteolytic enzyme. This splits off *angiotensin I* from the polypeptide angiotensinogen, which is found in the blood plasma. Further enzymatic breakdown produces the highly active vasoconstrictor (elevator of blood pressure) *angiotensin II*. The macula densa is thought to be a *chemosensitive area*, able to measure the sodium chloride content of the urine in the tubule and to cause the release of renin from the pole cushion cells. The juxtaglomerular apparatus regulates perfusion of the glomeruli by blood. The cells of the juxtaglomerular apparatus recieve *adrenergic* innervation.

The site of formation of the renal factor erythropoietin (stimulation of red cell production) is not definitely known.

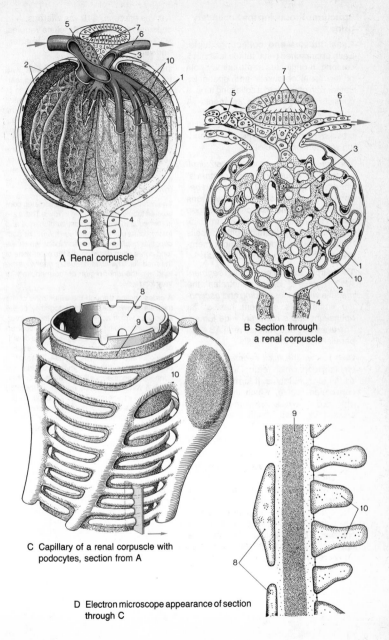

A Renal corpuscle

B Section through a renal corpuscle

C Capillary of a renal corpuscle with podocytes, section from A

D Electron microscope appearance of section through C

Structures Producing the Secondary Urine

Renal tubules and collecting ducts. Each unbranched renal tubule is about 5 cm long. It originates at the urinary pole of the renal corpuscle and opens together with others into a collecting tube. It consists of the following segments (**A** shows on the left the electron microscope appearance of cells in consecutive cross-sections through the corresponding segment of a tubule).

The *main section* **A1** is the *proximal convoluted tubule (tubulus contortus)* near the glomerulus **A2**, which then becomes straight *(tubulus rectus)* and runs as the descending limb of Henle's loop toward the medulla. It is 40–60 μm thick and is lined by tall, dense, epithelial cells (high microvilli, folded basement membranes and numerous mitochondria).

The hairpin-shaped *connecting segment* **AB3** (Henle's loop) penetrates the medulla with its *descending* and *ascending* limbs. This segment has a flat epithelium and a relatively wide lumen. The arteriolae and venulae recti **AB4** run parallel to it into the medulla.

Next follows the *thick segment* **AB5** of Henle's loop, which returns to the cortex as the straight tubule. It forms the distal convoluted tubule, which includes the macula densa, near the glomerulus from which it originated. The thick segment of the ascending limb of Henle's loop is about 45 μm and is lined by tall, clear columnar cells.

The *junctional tubule* **A6** leads into the *collecting duct* **AB7**, which merges with others into larger *papillary ducts* to end as the apex of the renal papillae **A8**. There are about 20–80 papillary ducts, which form a sieve-like plate, the area cribrosa. The collecting ducts and the papillary ducts are 200–300 μm thick. They are lined by tall, pale epithelial cells with relatively few mitochondria.

A9 Vas efferens, **A10** vas afferens, **A11** capsular branches, **A12** stellate venules, **A13** interlobar vein, **A14** arcuate artery and vein.

The architecture of blood vessels and tubules determines the macroscopic appearance of the cut surface of the kidney (see p. 254). **Cortex.** The glomeruli and the convoluted tubules of the main and thick parts of the nephron lie in the *cortical zone*. The *medullary rays* in the cortex each contain 4–8 collecting ducts and segments of the loops of Henle. The collecting ducts, the junctional tubules and the segments of Henle's loops extend into the *outer zone* of the **medulla**, but only 14% of them and the papillary ducts enter the *inner zone*, the pyramids.

The *renal corpuscles* and their associated convoluted tubules are arranged in *rows*. There are *juxtamedullary* and *subcapsular* glomeruli and *intermediate* glomeruli between them. The *subcapsular nephrons* have *short loops* which extend only a short distance into the outer zone of the medulla. The *juxtamedullary nephrons* extend into the inner region of the medulla with *very long loops*.

A cross-section through the outer zone of the medulla shows the bunching together of the structures which act according to the *counter-current principle* for the purpose of concentrating urine (see p. 260). At the center of a cluster of ducts there lies a group of blood vessels, the arteriolae and venulae rectae **AB4**. Toward the border of each group of these vessels, the connecting segments **AB3** increase in number. Often a venula medullaris recta is surrounded by several segments of Henle's loops. At the periphery of a bundle of ducts, there are mainly groups of thick and thin segments of Henle's loops **AB5**, of collecting ducts **AB7** and capillaries.

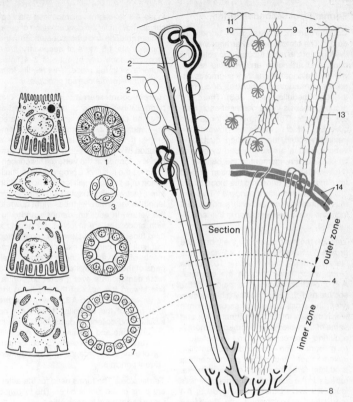

A Renal tubules and blood vessels in cortex and medulla (schematic)

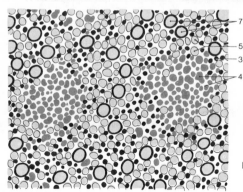

B Cross-section through renal medulla (after v. Möllendorff)

Formation of Secondary Urine

Countercurrent exchange – bundle of ducts. The almost parallel arrangement of the loops of the renal tubules, the collecting ducts and arteriolae and venulae rectae, with their elongated capillary segments, together form numerous *bundles of ducts* in the renal medulla (see p. 258). There is a group of vessels in the center of each bundle of ducts. The number of loops and collecting ducts increases toward the periphery; often the collecting ducts and venulae rectae are surrounded by several loops. This arrangement in the bundles of ducts is the basis of reabsorption of fluid from the primary urine into the blood by the *"hairpin-countercurrent principle"*. This permits counterflow of fluid with a concurrent increase in salt concentration in the remaining fluid.

Through the *main section* of the tubule (the proximal convoluted tubule) **A1** about 150 liters of primary urine flows (= 10% of the blood which flows through the glomeruli **A2**). From this, glucose, the majority of the sodium and chloride ions and other electrolytes, the small amount of serum albumin which passes through the urinary filter and some of the water is reabsorbed by energy-requiring processes; some substances (e. g., para-amino – hippuric acid, penicillin, sulfonamides and other pharmaceutical products) are excreted in the proximal tubules. In the *connecting segment* (Henle's loop) **A3** and in the *collecting ducts* **A4** more fluid is reabsorbed into the blood vessels. In the *thick segment of Henle's*

loops **A5** reabsorptive processes also occur, so that finally the concentration of urea, sodium chloride and other substances in the approximate 1.5 liters of *secondary urine* produced each day by man is 3–4 times higher than in the blood. This results from the *hairpin-countercurrent principle*.

Hairpin-countercurrent principle. Two parallel tubes S_1 and S_2 are connected at one end like a hairpin and are separated from each other by a semipermeable membrane. If this system is perfused with a solution of a mineral salt, a high concentration K is formed at the vertex. This concentration is the result of a progressive multiplication of a single, small local concentration effect (active salt transport) along the entire length of the 2 parallel tubes, right down to the vertex. In addition, osmosis and diffusion processes are involved. By adding a third tube R to the system, a concentrated salt solution can be excreted. Measurements of the various concentrations in all parts of the tubular system and of the vasa recta demonstrate ever greater concentration toward the apex of the papilla at the vertex of Henle's loop **A3**, as well as in the final part of the collecting tube **A4**. The tubules and collecting ducts together with the arteriolae and venulae rectae **A6**, fulfil the structural requirements for concentration of urine in accordance with the countercurrent principle.

Terminology: The terms used for the different parts of the tubule differ. The following terms correspond to each other.

Anatomical name	Term used in physiology		Term used in microscopic anatomy
Tubulus contortus	Proximal convoluted tubule		Main section (proximaltubule)
Tubuli recti	Henle's loop	First part / Mid part / Final part	Connecting segment (Henle's loop) (thin segment of Henle's loop)
Tubulus contortus	Distal convoluted tubule		Middle section (distal tubule) Thick segment of Henle's loop

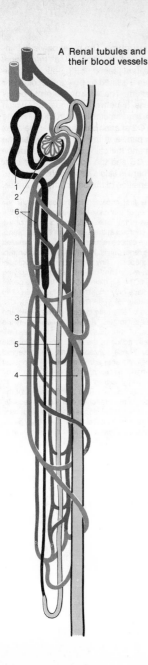

A Renal tubules and
their blood vessels

1
2
6
3
5
4

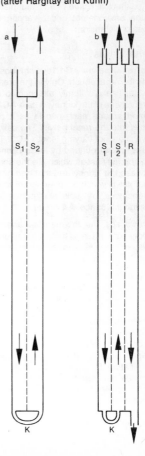

B Countercurrent model for
production of concentrated
fluids
Left without, Right with removal
of concentrated fluid from system
(after Hargitay and Kuhn)

a

b

S_1 S_2

S_1 S_2 R

K

K

Organs of the Urinary Tract

The *renal pelvis* collects the urine which comes from the apices of the papillae. It tapers into the *ureter* which passes the urine in small portions into the *urinary bladder*. The urine is discharged from the bladder through the *urethra*. The wall of the urinary tract organs has a muscle layer which is capable of producing peristaltic movements and alterations of tone. The mucosa consists mainly of transitional epithelium. Glands are only present in a few places. A loose adventitial connective tissue permits adaptation to the environment under varying degrees of distension.

Renal Pelvis

The **renal pelvis AC1** (*pyelon*) loosely lines the renal pouch. Only in the area of the papillae is it firmly attached to the parenchyma of the kidney. A *calix* **AC2** is formed around each papilla as it protrudes into the renal pelvis. The calices have a stalk of varying length **AC3**. The stalks are surrounded by loose connective and fatty tissue which fills the space between the kidney and that part of the renal pelvis which is not adherent to the parenchyma. Vessels and nerves run through this space. **A4** Ureter.

Clinical tips: As the papillae and the calices are often affected together in diseases, clinicians speak of the *"pyelorenal border region"*.

The **shape** of the renal pelvis shows individual and functional differences. It is usually possible to distinguish between an upper and a lower main branch. The pelvis can be uniform in shape, an ampulla-like bag (*pelvis renalis ampullaris*) **A**, or it can consist of several funnel shaped tubes **B5** (*pelvis renalis ramificatus*) **B**. Transitional forms are common. The formation of a particularly large ampulla-shaped pelvis may be due either to developmental processes or to disease (hydronephrosis). The average capacity of the renal pelvis is 3–8 ml.

Fine structure. The renal pelvis is thin-walled and is lined by a *transitional epithelium* (see p. 268), but over the papillae it is lined by polygonal columnar epithelium. The wall of the renal pelvis contains a network of muscles, which forms a sphincter-like structure in the fornix **C2** of each calix and near the apex of the papilla at the border between the pelvis and the calix: the *sphincter of the fornix* **C6** and the *sphincter of the calix* **C7**. There is also a *pelvic sphincter* **C8** at the transition of the pelvis into the ureter.

Urine is first collected in the calix while the sphincter of the calix is contracted. Then the muscles in the calix wall, the sphincter of the fornix, contract at the same time as the sphincter of the calices relaxes. Urine is then expelled into the renal pelvis. Not all the calices contract simultaneously. Any which are pathologically dilated are rigid and play no part in the emptying process. The expulsion of urine from the pelvis takes place only after the latter has been filled to a certain degree. The urine is quickly discharged by alternate closure of the pelvic and caliceal sphincters. An elastic network in the meshes of the muscle fibers facilitates antagonistic action of the corresponding muscle groups. These functions can be observed radiologically after administration of a contrast medium which is excreted by the kidneys, or by retrograde pyelography.

Vessels and nerves. The blood vessels of the renal pelvis are branches of the renal vessels, but the pelvic circulation is essentially independent of the renal circulation. The renal pelvis has sensory innervation and its distension is painful.

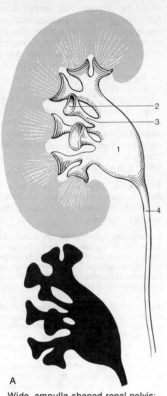

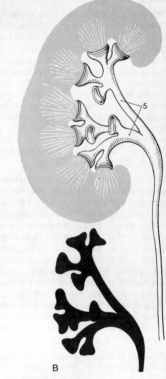

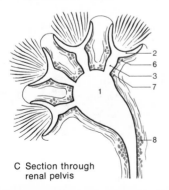

A Wide, ampulla-shaped renal pelvis:
cast and radiograph

B Tubular (branched) renal pelvis:
cast and radiograph

C Section through
renal pelvis

D Radiograph of renal pelvis:
left – collecting phase;
right – discharge phase
(schematic, after Narath)

Ureter

The **ureter A1** is shaped like a slightly flattened tube 4–7 mm diameter. Its length varies, being about 30 cm in the male and about 1 cm shorter in the female. The slitlike lumen of the ureter has a stellate appearance because its mucosa is folded longitudinally. The two ureters penetrate the wall of the urinary bladder at its fundus, 4–5 cm apart from each other, obliquely from behind and above toward the middle. They run for 2 cm in the bladder wall and end in a slit, the *ostium uretris* **A2**. For opening and closure of the ostium see p. 268.

Fine structure. The *mucosa* has a transitional epithelium **B3** (see p. 268) and its connective tissue permits limited movement of the mucosa. The *muscle layer* is interlaced with connective tissue fibers. Cross sections of the ureter show in its upper part a sparse inner longitudinal and a thick outer circular muscle layer. **B4** joined in the lower part by an outer longitudinal muscle layer. The different layers are formed by variation in the angle at which the spiral muscle bundles are arranged. The *loose areolar tissue* in which the ureter is situated allows some movement of it within its surroundings.

Transport of urine. *Peristaltic waves* propel urine through the ureter. After administration of a dye (e. g., methylene blue) that is excreted by the kidneys, observation through a cystoscope shows that the colored urine is squirted in jets about 1 to 4 times per minute through the ostium of the ureter into the bladder. Impacted calculi in the ureter cause increased peristalsis. If the flow of urine is impeded the ureter muscles above the obstacle quickly hypertrophy. During pregnancy the ureter dilates and elongates.

Position of the ureters. The *abdominal portion* of the ureter passes almost vertically downward along the fascia of the psoas muscle (see p. 250) and is covered by the peritoneum. The ureter crosses underneath the testicular or ovarian blood vessels. At the entrance into the lesser pelvis it crosses over the point where the common iliac vessels divide **A5**. The left ureter passes beneath the pelvic colon. The *pelvic portion*: The ureter crosses beneath the ductus deferens in the male and beneath the uterine artery in the female. It is palpable through the anterior wall of the vagina (of importance in ureteric calculi). Physiological *constrictions of the ureter* are found where it leaves the renal pelvis **A6**, where it crosses the iliac vessels **A7** and where it enters the bladder **A2** (danger of impaction of ureteric calculi). In the newborn and in babies the ureter is often tortuous. **A8** Urethra.

Nerves and vessels. The blood vessels and lymphatics of the ureter arise from those in its neighborhood. The ureter has sensory innervation.

Variants: Duplication of the ureter occurs in about 2% of the population, partial duplication being 3 times more frequent than total duplication. Total duplication, i. e., *ureter duplex (double ureter)*, is the result of a primary duplication of the primitive precursor, and partial duplication, *split ureter*, is due to early division of the ureteric anlage. In a case of double ureter, the ureter that arises at the higher level in the pelvis enters the bladder at a lower level (dystrophic opening) than one that arises lower in the pelvis (*Mayer-Weigert* rule). The following displacements of the ureteric orifices may be found: in the male – in the urethra, seminal vesicles, ejaculatory duct, ductus deferens and in the prostatic utricle; in the female – in the urethra, vestibule of the vagina, vagina, uterus or in Gartner's duct. In a case of *"megaureter"* the entire ureter is usually dilated, thick walled and tortuous.

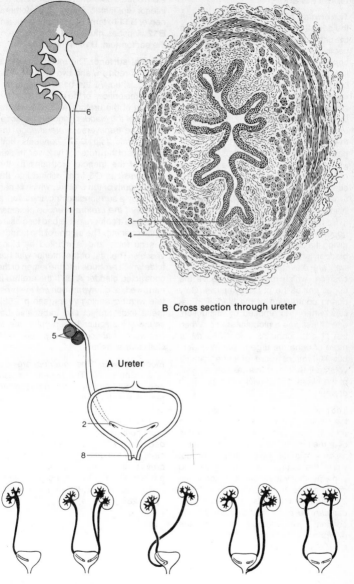

B Cross section through ureter

A Ureter

C Variations in course of ureters (after Langreder)

Urinary Bladder

The **urinary bladder** (*vesica urinaria*) in the adult lies in the lesser pelvis beneath the peritoneum, in the subperitoneal connective tissue space behind the pubic bones **A1**. In the newborn the bladder lies higher than the pubic bones. The *body (corpus)* of the bladder tapers upward toward the front into its vertex, the *apex* of the bladder; the *fundus (base)* of the bladder lies posteriorly. The *median umbilical ligament* **ABC2** ascends from the vertex to the umbilicus; it is the remnant of the embryonic urachus; the *lateral umbilical ligaments* **B3**; the remnants of the obliterated umbilical arteries, ascend from the body of the bladder to the umbilicus. Both ureters **BC4** enter the bladder at its base and the urethra leaves it from there **C5**. The vertex and the upper wall of an empty bladder collapse giving it a cuplike appearance. When the bladder is full the vertex and the upper wall are elevated and form a flat oval cushion, which may extend above the upper rim of the pubic sympysis. Only during contraction of the bladder muscles, when it is emptied by micturition, does it assume a globular shape. When the bladder contains about 350 ml or more of urine the urge to urinate is felt, but 700 ml or more of urine can be voluntarily retained. In a case of paralysis of the bladder its capacity may be much greater.

The **peritoneum** covers the bladder from the vertex to about the line where the ureters enter the bladder. It is movable over the bladder and forms a transverse reserve fold **B6** which straightens out when the bladder is filled. If filled to capacity the vertex of the bladder ascends between the abdominal wall and the peritoneum. In this position the bladder can be punctured through the abdominal wall without injuring the peritoneum. In front and laterally the bladder is surrounded by loose connective tissue, the *paracystium* (see p. 313) which contains vessels and nerves.

AB7 Ductus deferens, **AC8** prostate, **B9** inferior epigastric artery, **B10** femoral nerve, **B11** external iliac artery and vein, **B12** iliopsoas muscle, **B13** cut edge of the peritoneum, **B14** seminal vesicle.

Internal surface. The mucosa in life is soft and reddish, and two areas can be distinguished. At the base (fundus) between the orifices of the ureters **BC4** and the exit of the urethra **A15, C5** lies the *trigone of the bladder* (trigonum vesicae) **C16**. The transverse *ureteral ridge*, the interureteric fold **C17**, connects both openings of the ureters. The lower, thicker angle of the triangle protrudes as the *uvula vesicae* **C5** from behind into the inner mouth of the urethra, which is narrowed by a surrounding ring, the *internal ostium of the urethra*. Venous plexuses here form a compressible pad for closure of the opening. The trigone of the bladder has no folds and is marked by blood vessels. The rest of the interior wall has folds which protrude into the lumen of the contracted bladder **A18**. If the urethra is narrowed, e. g., hypertrophy of the prostate with advancing years (see p. 282), these folds project like trabeculae because of muscular hypertrophy due to overwork (*trabeculated bladder*). **C19** Colliculus seminalis.

Blood vessels. The *arteries* are described on p. 70. The superior vesical artery originates from the umbilical artery, the distal part of which becomes obliterated after birth and forms the lateral umbilical ligament. The *veins* form a prominent network **A20** under the bladder. The *lymphatics* run to lymph nodes alongside the umbilical artery and in the prevesical connective tissue. The nerves are described in Vol. 3. For the *female urethra* see p. 304, for the *male urethra* see p. 286.

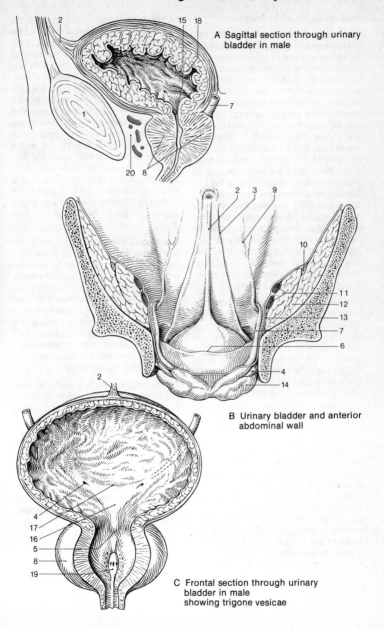

A Sagittal section through urinary
bladder in male

B Urinary bladder and anterior
abdominal wall

C Frontal section through urinary
bladder in male
showing trigone vesicae

Fine Structure of the Bladder

The **muscles of the bladder wall** with the exception of the trigone have a reticular structure, outer bundles run ventrally from the neck of the bladder and from the pubovesical muscle over the posterosuperior wall to the vertex of the bladder. From there muscle bundles pass into the median umbilical ligament and the prostate, or to the anterior wall of the vagina and the rectum. The outer muscle bundles **A1** radiate into the middle layer of more circular muscle fibers. From there arise the inner longitudinal muscle bundles, which produce the corrugated inner surface of the bladder. As the *bladder fills*, muscles that originally had little tone become stretched; if the bladder muscles are paralysed, hernialike bulges may develop in its wall. During *micturition* muscle tone increases and the bladder becomes rounded in shape. **A2** Ductus deferens, **AB3** ureter, **A4** seminal vesicle, **A5** prostate.

Muscles of the internal urethral meatus, the *internal urethral opening*.

Closure: The anterior part of the circumference of the internal urethral meatus is encircled by muscle loops **B6**, which arise from the longitudinal muscles **A1** of the bladder. They pull the anterior wall of the urethra backward and, together with other circular muscle fibers, they form the *involuntary sphincter muscle* of the bladder outlet. Muscle fibers from the pubovesical muscle **B7**, which run around the posterior circumference of the opening of the urethra, are also involved in the closure. Muscles **B8** that are situated in the uvula (see p. 266) and the veins of the mucosa also contribute to closure of the urethra. *Opening*: The following muscle bundles participate in the active, involuntary opening of the urethra: muscle bundles **C9** from the loops of the closure mechanism of the urethral orifices radiate into the involuntary sphincter of the bladder. When they contract they form a groove at the floor of the bladder, thus contributing to opening

of the outflow tract. Contraction of the muscle bundle in the uvula **B8** can draw it backward. In addition, fascicles of the pubovesical muscle, which are inserted in the anterior wall of the urethra as well as the fascicles of the retrovesical muscle **B10**, which radiates into the posterior wall, also participate in opening the outlet of the bladder. The *voluntary sphincter muscle* consists of muscle fibers that have separated from the *deep transverse perineal muscle* **B11**, and as the *muscle of the urethral sphincter* they surround the urethra with ascending, spiralling slings. The levator ani muscle **B12** may also take part in closure of the urethra. **B13** Pubic bone, **B14** pubococcygeal muscle, **B15** rectum, **B16** coccyx.

Muscles of the trigone and the opening of the ureter.

Along its lower third, the ureter is accompanied by an outer muscle layer **C17**, the ureteric sheath, which forms a loop surrounding the mouth of the ureter. If the muscle bundles are shortened e. g., by a stretch stimulus due to accumulated urine, the orifice is elevated and opened – *"opening loop"*. Muscular slings **C18** run between the 2 ureteric orifices and pull them downward to close them – *closure loop*.

The mucosa is lined by pseudo-stratified *transitional epithelium*, which can rapidly be transformed from high cyclindrical cells into low cells by movement of cells and folding of cell membranes during a change in the volume of the bladder. The uppermost layer of cells has a special outer layer as a protection against urine. There are mucous glands at the inner opening of the urethra. The mucosa over the trigone is firmly adherent to the muscle layer but elsewhere it is only loosely attached to the muscles.

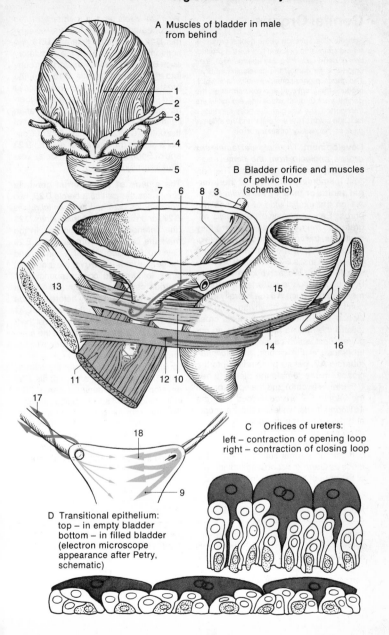

A Muscles of bladder in male from behind

B Bladder orifice and muscles of pelvic floor (schematic)

C Orifices of ureters:
left – contraction of opening loop
right – contraction of closing loop

D Transitional epithelium:
top – in empty bladder
bottom – in filled bladder
(electron microscope appearance after Petry, schematic)

Genital Organs

Specific differentiation of the genital organs is an indispensable precondition for the production of germ cells (ova and spermatoza), their union and for care of the developing embryo. The *genital organs* comprise the *gonads* which produce the *germ cells* and sex hormones, the *genital tract* through which the sex cells are transported, the *sex glands* whose secretions facilitate union of the sex cells, and the *external genitalia* necessary for sexual union.

Development. *The male and female sex organs originate from the same undifferentiated embryonal anlage*. The earliest (primordial) precursor of the **internal genital organs** are the *genital ridges* **A1** on the medial side of the mesonephros. Early embryonic *germ cells* migrate into the genital ridges, where they multiply in the cortical zone in the case of the ovary **C2**, or the central area in the case of the testis **B3**. Two pairs of excretory ducts, the *Wolffian duct* (mesonephric duct) **A4** and the *Müllerian duct* (paramesonephric duct) **A5** run laterally in the genital ridges and downward into the cloaca **A6**. The *Müllerian ducts* cross toward the midline and fuse into an unpaired duct which causes a dorsal protuberance in the cloaca, the *Müllerian tubercle* **A7**, before penetrating into the cloaca. In the *female* the uterine tubes **C8**, uterus (womb) and the upper part of the vagina **C9** are developed from the *Müllerian ducts*, while the *Wolffian ducts* and the rest of the mesonephric anlage atrophy and form the epoophoron **C10**. Parts of the mesonephron (Wolffian) duct can remain permanently for some distance as *Gartner's duct* (longitudinal duct of the epoophoron) **C11**. In the *male* the

Wolffian ducts and a remnant of the mesonephros develop into the epididymis **B12**, the ductus deferens **B13**, the seminal vesicles **B14** and the ejaculatory duct **B15**. The *Müllerian ducts* atrophy into the appendix testis **B16** and the prostatic utricle **B17**. The *gubernacular cord* **A18** develops into the gubernaculum testis **B19**, or into the ovarian ligament **C20** and the round ligament of the uterus **C21**. The anlage **A22** of *Cowper's glands* (bulbourethral glands) **B23** and of *Bartholin's glands* (glandulae vestibulares major) **C24**.

The **anlage of the external genitalia** comprises the *genital tubercle* **D25**, two *genital folds* **D26** and two *genital ridges* **D27**, and also the *urogenital sinus* **D28** (the anterior part of the cloaca). In the *male* the corpora cavernosa penis **E29** develop from the genital tubercle, and the genital folds close over the urogenital sinus to form the corpus spongiosum (corpus cavernosum urethrae) and the glans penis **E30**. Shortly before birth the testicles descend into the scrotum **E31** which has been formed by fusion of the 2 genital swellings. The orifice of the combined *Müllerian ducts* is hidden in the urethra (rectangle around in **B**; see colliculus seminalis p. 282). In the *female*, the clitoris and the glans clitoridis **F32** develop from the genital tubercle. The labia minora **F33** and the bulbus vestibuli develop from the genital folds which remain separated. The labia majora **F34** develop from the genital ridges. The aperture of the combined *Müllerian ducts* lies in the vagina (rectangle at the bottom of **C**).

Male	Undifferented Anlage	Female
testis	anlage of the gonads	ovary
seminal ducts	mesonephric (Wolffian) duct	(Gartner's duct)
prostatic utricle	Müllerian duct	uterine tubes, vagina, uterus
corpus cavernosum penis	genital tubercle	clitoris, glans clitoridis
scrotum	genital ridges	labia majora

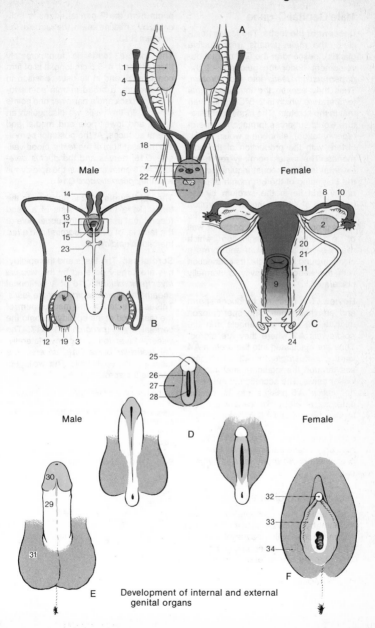

Male

Female

Male

Female

Development of internal and external
genital organs

Male Genital Organs

Descent of the testis. The *testes* (testicles), the male gonads (reproductive glands), descend at the end of fetal development under the guidance of the *gubernaculum testis* into the *scrotum*. Thus they escape the intra-abdominal temperature which is 2–5°C higher than that in the scrotum. The higher temperature would suppress formation of sperm (spermatozoa), although it would not interfere with the production of the hormones. The testes normally reach the external (subcutaneous) inguinal ring at the beginning of the 8th month of pregnancy and lie in the scrotum by the beginning of the 9th month (maturation times of the newborn). *Descent of the testis* proceeds along the posterior wall of a peritoneal evagination which reaches into the *scrotum* and forms a cavity around the testicle. Its connection with the abdominal cavity is normally obliterated.

Review. The *testis* **A1** produces sperm and sex hormones. The spermatozoa migrate via narrow channels into the *epididymis* **A2** where they are stored. They are ejaculated into the *urethra* **A4** via the *ductus deferens* **A3**, which passes through the inguinal canal into the lesser pelvis. The secretion of the *seminal vesicle* **A5** passes into the ductus deferens, an *ejaculatory canal* which perforates the *prostate* **A6**. The ducts of the prostate and of the *glands of Cowper* **A7** open directly into the urethra. **A8** *Ureter*, **A9** *corpus cavernosum penis*, **A10** *corpus cavernosum urethrae*, **A11** *urinary bladder*.

Testis and Epididymis

Both testes are suspended in the scrotum by a fibrovascular stalk, the spermatic cord, which leaves the inguinal canal through the external inguinal ring **B12**. The left testicle often hangs lower than the right. The scrotum serves to regulate the temperature of the testes. Its origin from the 2 genital ridges is indicated by a midline seam, the *raphe of the scrotum* **B13**.

The mature **testis** is approximately plum-shaped, 4–5.5 cm long. It is of firm consistency and in its usual position in the scrotum, a broad margin points toward the back and a narrower one points forward. In the testis we distinguish an upper and lower pole and medial and lateral surfaces. At the posterior border, the *mediastinum* of the *testis*, blood vessels **C15**, nerves and the *ductus deferens* **AC3** enter and leave the epididymis with the *spermatic cord* **C14**.

The **epididymis AC2** is attached to the testis like a tail. It consists of a head (caput) that lies above the upper pole of the testicle, of a *body (corpus)* and a *tail*, the *cauda epididymis*.

Scrotal sac. The testis and the epididymis are largely covered by the *visceral layer (epiorchium)* **C16** of a peritoneal sheath, the *tunica vaginalis of the testis*. This is reflected at the mediastinum of the testis and at the epididymis onto the parietal layer (*periorchium*) **C17**. The visceral layer forms a groove laterally, the *testicular bursa* **C18**, between the testis and the epididymis. The two layers enclose the scrotal sac.

At the head of the epididymis and at the upper pole of the testis there are often 2 small vesicles, the *hydatids*. The hydatid of the testis, called the *appendix of the testis* **C19**, is a remnant of the Müllerian duct; the hydatid of the epididymis, the *appendix of the epididymis* **C20**, is a remnant of the mesonephros. **C21** Section cut through the coats of the testis.

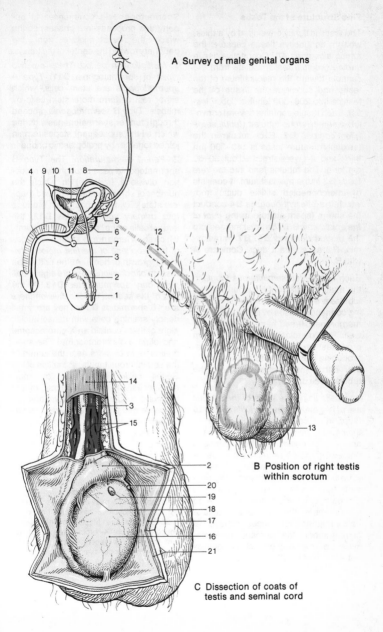

A Survey of male genital organs

B Position of right testis within scrotum

C Dissection of coats of testis and seminal cord

Fine Structure of the Testis

The testis is tightly enveloped by a thick, whitish, connective tissue capsule, the *tunica albuginea* **A1**. Connective tissue septae (*septula testis*) radiate from the capsule toward the mediastinum of the testis and subdivide the tissue of the testicle into 200–300 lobules (*lobuli testis*). Each lobule contains several convoluted seminiferous tubules (*tubuli seminiferi contorti*) **B2**. Each tubule in the (sexually) mature testis is 140–300 μm thick, and, if it were stretched out, 30–60 cm long. The tubules lead into the *rete testis* **B3** in the mediastinum. It consists of interconnected slitlike ducts from which the *efferent ductules* **B4** conduct the sperm (spermatozoa) to the *duct of the epididymis* **B5**. The latter merges into the *ductus deferens* **B6**. **B7** Paradidymis, **B8** an aberrant duct. Compare **A** with **B**.

Each *convoluted seminiferous tubule* in the mature testis contains a lumen. The tubules are separated from the surrounding connective tissue by a (hyaline) basal membrane **C9** and have a multilayered wall. The "germinal epithelium" in them produces *spermatozoa*. The hormone-producing interstitial cells are situated in the *connective tissue* between the tubules (see p. 276).

The *cells lining the walls of the tubules* are of two main types. *Sertoli's cells* **D10** (supporting cells) are recognizable by their nuclei, which contain little chromatin and a large nucleolus. Their interlaced processes form a protective and "feeding" lattice. Over half their height they are joined by occluding intercellular contacts, tight junctions, which separate two compartments. From the basal blood compartment, the spermatogonia take up the materials for mitoses. In the apical compartment, the primary spermatocytes develop antigenic characteristics but are protected from the blood stream 6= blood-testis-barrier) within the meshes of which the *spermatozoa* mature.

Spermatogenesis commences at puberty and proceeds in 3 phases, during which the cells migrate from the periphery toward the center of the tubule.

1. *Multiplication period:* There are two types of spermatogonia **D11**. *Type A spermatogonia* are stem cells, which either rest or form more stem cells by mitosis. *Type B spermatogonia* proceed through further spermatogenesis during which all the subsequent stages remain joined together by protoplasmic bridges.

2. *Period of maturation:* The Type B spermatogonia pass through two reduction divisions (*meiosis*) in which the number of chromosomes is halved (haploid state). Two spermatogonia produce four *primary spermatocytes* **D12**, the largest cells of the "germinal epithelium", which show large nuclei in various stages of the *prophase of the first meiotic division* ("growth period"). At the end of the *first meiotic division*, there are eight small *secondary spermatocytes* **D13**. At the end of the *second meiotic division there are 16 spermatids* **D14**. They are small deeply staining cells with dense nuclei; eight of them contain an X-chromosome and eight a Y-chromosome. The spermatids lie in clusters near the lumen of the seminiferous tubule, at the tips of the Sertoli cells. The steps of spermatogenesis follow the course of the testicular tubules, with a temporal arrangement so that stages of equal maturity are arranged in a spinal fashion around the walls of the tubules. **D15** Sperm.

The genetic sex is determined at the moment of fertilization by the chromosome combination produced by the sperm-ovum combination; the heterosomes XX characterize a female, the heterosomes XY a male cell nucleus. After the set of chromosomes has been halved by meiosis the "mature" (haploid) ovum is always bound to have one X-chromosome. It is, therefore, the sperm cell (spermatozoon) which determines the gender (sex) of the combined gametes at fertilization. In other words, it decides whether the future zygote will be female (XX) or male (XY).

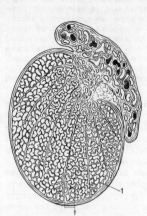

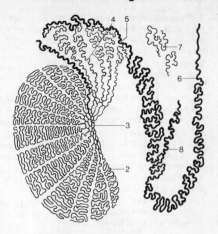

A Section through testis and epididymis

B System of tubules of testis and epididymis (after Rauber-Kopsch)

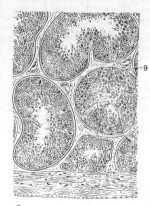

C
Section from A showing seminiferous tubules

D
Higher magnification of section from C; sector of seminiferous tubule

3. Period of differentiation ("spermiohistogenesis") (multiplication and maturation period, see p. 274). The spermatids lose their intercellular cytoplasmic bridges and are transformed into *spermatozoa* (sperm), which are a transportable form of sperm cells able to reach the ovum and to penetrate it. The most important structural changes are the *formation of the acrosome, condensation of the cell nucleus* and *formation of the tail*. The sperm tails stand out in the lumen of the seminiferous tubule. The spermatozoa that are released finally pass through the rete testis, the first part of the efferent pathway in the testis, into the epididymis. Residual bodies **B1**, the cytoplasm which the spermatozoa no longer require, remain in the Sertoli cells.

A sexually mature testis contains about 1000 million spermatogonia and can produce about 200 million spermatozoa per day.

Sperm (spermatozoon). The spermatozoon is about 60 μm long and consists of the following parts. 1. The *head* **BC2** contains the haploid cell nucleus. It is 3–5 μm long and appears oval in a frontal view, and pear-shaped from the side. Its pointed part carries a cap, the acrosome **C3**. 2. The *neck* **BC4** is a short connection between the head and the middle section and the following parts are joined flexibly to the head. The fibrils of the tail originate in the neck. 3. The relatively thick *middle section* **BC5** is about 6 μm long and contains the cilium around which the "spiral thread" is wound. 4. The following *main section* **BC6**, the longest part of the tail, is thinner than the middle section. 5. The *terminal part* is the thinnest part of the tail.

The entire spermatozoon is surrounded by a plasmalemma. The *acrosome* is formed from the lamellae of the Golgi apparatus and contains, amongst other things, hyaluronidase with which to penetrate the membrane of the ovum. The *neck* is connected to the cell nucleus by a "basal plate" and contains the *proximal centriole*, which is ready to form the division spindle after penetration of the sperm into the ovum. The axis of the *middle section* is formed from a

"9×2+2" structure of tubules like a cilium, which arises from the *distal centriole*. Externally, the tubules are surrounded by 9 thicker and wider, so-called "external fibrils" around which the *mitochondria* are densely packed ("spiral fibrils"). In the *main section*, the cilium is surrounded by a "fiber sheath" – two longitudinal "bands", which are bound together by "circular fibers". The *terminal part* only contains cilia.

Interstitial cells (*Leydig's interstitial cells*). In the loose areolar tissue between the seminiferous tubules, arranged around the blood vessels **D7**, are groups of large, polyhedral, *interstitial cells*, the *Leydig cells*, **D8**. They produce the male sex hormone, the androgen (testerone), and probably a small quantity of the female hormone, estrogen (steroid hormone producers). The *androgens* facilitate spermatogenesis and influence growth of the genitalia and development of the secondary sex characteristics. The interstitial cells are stimulated by the Interstitial Cell Stimulating Hormone (ICSH-LH) of the adenohypophysis (see p. 170).

Age differences. The testis grows throughout childhood and it only reaches full development between the age of 20 and 30 years. It shrinks again with advancing years. The testis of the newborn is under the influence of placental hormones for a short time and contains strongly developed interstitial cells. However, they disappear quickly and do not reappear until puberty. During childhood the "seminiferous tubules" of the testis are represented by cords of epithelial cells without a lumen. They contain Sertoli cells and stem cells of spermatogonia. Spermatogenesis commences at puberty and usually continues into old age. Both the seminiferous epithelium and the interstitial cells can be reduced by malnutrition, disease and advancing age.

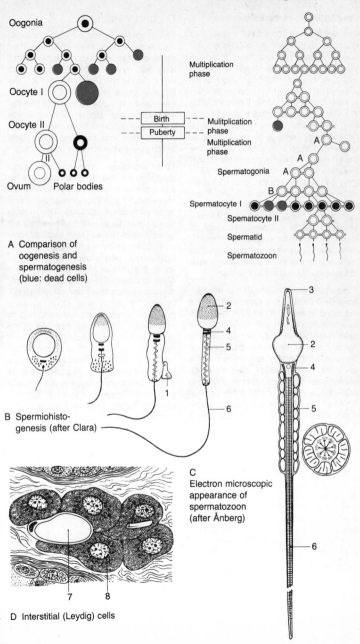

Oogonia

Oocyte I

Oocyte II

Ovum   Polar bodies

Multiplication phase

Birth — Multiplication phase

Puberty — Multiplication phase

Spermatogonia

A

A

A

B

Spermatocyte I

Spematocyte II

Spermatid

Spermatozoon

A Comparison of oogenesis and spermatogenesis (blue: dead cells)

B Spermiohisto-genesis (after Clara)

C Electron microscopic appearance of spermatozoon (after Ånberg)

D Interstitial (Leydig) cells

Fine Structure of the Epididymis

The **epididymis** consists of a system of ductules surrounded by connective tissue. The spermatozoa arrive there via the *rete testis* **A b**, which still lies in the testis, in 10–20 *efferent ductules (ductuli efferentes)* **A c**, which make up the greater part of the head of the epididymis **A 1**. Each efferent ductule is about 20 cm long and is convoluted to form a small conical body some 2 cm high. Its apex begins at the rete testis and at its base the efferent ductule opens into the *duct of the epididymis* **A d**. This duct is about 5 m long and is also very tortuous. Its lumen widens from 150 μm to 400 μm. It extends from the *head* to the tail of the *epididymis (cauda epididymis)* **A 2** where spermatozoa are stored. The cauda epididymis continues as the *ductus deferens* **A e**. Spermatozoa are immobile while still in the testicle and they reach the epididymis in a stream of fluid.

Fine structure. The position of the sections **B–D** through the system of ducts is indicated in **A** by the red lines **b–d**. The spermatozoa reach full maturity in the epididymis where they are surrounded by a colloidal secretion that protects them against an acid environment. The pH in the epididymis is 6.48–6.61 which immobilizes the spermatozoa. The *rete testis* **B** is a system of slits in the mediastinum of the testis and is lined by squamous epithelium. The *efferent ductules* **C** are thin-walled. Their epithelium consists of alternating patches of stratified ciliated columnar epithelium and of simple squamous epithelium. The lumen is stellate in cross section. The high ciliated columnar epithelia cells produce fluid and the low squamous epithelia can absorb fluid. The wall of the *ductus epididymidis* **D** contains only sparse muscle fibers and is lined by a double-layered columnar epithelium with "stereocilia", hair-tuftlike processes which are a sign of secretion.

Ductus Deferens

The **ductus deferens** is a continuation of the ductus epididymidis and serves for transport of spermatozoa. It is 50–60 cm long and runs with blood vessels and nerves in the spermatic cord through the inguinal canal (see p. 280). It dilates towards its spindle-shaped end into the *ampulla of the ductus deferens* which takes in the ostium of the seminal vesicle. It continues into the ejaculatory duct which pierces the prostate (see p. 282).

Fine structure. Although the outer diameter of the ductus deferens is 3.0–3.5 mm, its actual lumen is only 0.5 mm wide. It has a *strong muscle wall* and feels as hard as cartilage. The muscle fibers run in spirals which follow either left or righthand turns. Their angle of gradient changes but is more longitudinal in the external muscle layer **E 4**, more circular in the middle layer **E 5**, and again more longitudinal in the internal layer **E 6**. A cross section, therefore, shows a triple layered muscle wall. Suction and pressure mechanisms are assumed to play a part in the quick passage of the spermatozoa. The mucosa has 3–4 longitudinal folds. It is covered by a high columnar, double-layered epithelium **E 7** which still carries "stereocilia" near its origin. The muscle layer in the ampulla becomes thinner, the mucosa forms a reticular pattern of small folds and little niches appear.

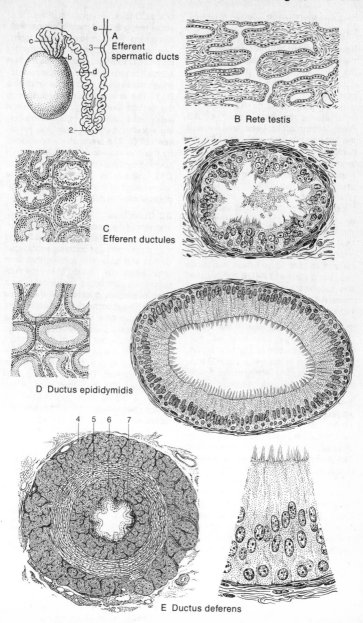

A Efferent spermatic ducts

B Rete testis

C Efferent ductules

D Ductus epididymidis

E Ductus deferens

Spermatic Cord, Scrotum and the Coverings of the Testis

Spermatic cord, *vessels and nerves of the testis*. As the testis descends through the inguinal canal into the scrotum it draws all the structures of the spermatic cord behind it. The *testicular arteries* **A1** originate from the aorta **A2** below the renal arteries; the *right testicular vein* **A3** drains into the inferior vena cava **A4**, and the left into the *left renal vein* **A5**. The testicular veins form a very marked, extended venous plexus, the *pampiniform plexus*. Congestion of the plexus may result in varicose dilatations. The *lymphatics* of the testis drain into lumbar lymph nodes on the inferior vena cava and the abdominal aorta. Vegetative *nerves* come from the celiac plexus (see Vol. 3). The bundle containing all these structures, including the cremaster muscle and its connective tissue, surrounds the *ductus deferens* **A–D6** and is called the *spermatic trunk*. **A7** Inguinal ligament, **A8** internal inguinal ring, **A9** external inguinal ring.

Scrotum and coverings of the testes. The layers of the abdominal wall take part in the formation of the *scrotum* and the coverings of the testis. *Each testicular (scrotal) layer is related* to and merges into *a layer of the abdominal wall*.

Inner layer (*peritoneum*). The *tunica vaginalis testis* **C10** (epiorchium, periorchium, see p. 272) is derived from the peritoneum **B11** and surrounds the serous *cavity of the scrotum* **B12**. A band-like remnant **B13** sometimes indicates its original connection with the peritoneal cavity. If, exceptionally, this connection, the *vaginal process of the peritoneum*, remains open, loops of the small intestine may enter the scrotum = *congenital inguinal hernia* (indirect). The cavity of the scrotum may be filled by a pathologic serous effusion causing marked enlargement of the scrotum – *hydrocele*.

Middle layer (*muscles and fascia of the abdominal wall*). The *internal spermatic fascia* **CD15**, an evagination of the transverse fascia **B14** of the abdominal wall, lies on top of the tunica vaginalis testis. The fascia is covered by the *cremaster muscle* **CD16** a thin, striated muscle, that arises from the internal oblique and oblique transverse abdominal muscles **B17**. The cremaster, which suspends the testicle like a sling, can elevate it either voluntarily or by reflex action during ejaculation (for the cremaster reflex, see Vol. 3). The cremaster is enveloped by the *cremasteric fascia* **CD18** and is covered by the *external spermatic fascia* **C19**, which is connected to the aponeurosis of the external, oblique abdominal muscle **B20**.

BC Outer layer (*skin of the abdomen*). The *skin of the scrotum* **C21** is a continuation of the skin of the abdomen **B22**. It is thin, pigmented and contains sebaceous glands and hairs with follicles that form small elevations. Instead of fat, the subcutaneous connective tissue contains a layer of smooth muscle cells, the *tunica dartos*. This is connected by elastic fibers to the adventitia of the subcutaneous blood vessels. During reflex contraction of the dartos, the skin of the scrotum becomes wrinkled, its surface area diminishes and the blood vessels get closer together. Loss of heat is thus diminished. On the other hand, if the dartos is relaxed more heat is given off. The scrotal skin and the dartos form the scrotum. It is divided by a fibrous partition, the *scrotal septum* **C23**, into 2 compartments, one for each testis.

B24 Site of the internal inguinal ring, **B25** inferior epigastric artery, **BC26** epididymis, **BC27** testis.

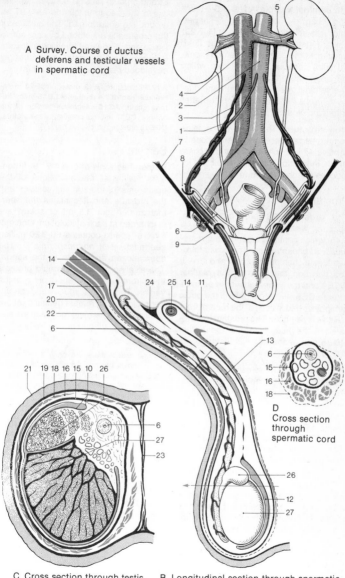

A Survey. Course of ductus deferens and testicular vessels in spermatic cord

D Cross section through spermatic cord

C Cross section through testis, epididymis and their coats

B Longitudinal section through spermatic cord and its coats (after Töndury)

Seminal Vesicles

The **seminal vesicle A1** is a 5–10 cm long, baglike, convoluted S-shaped gland. Its alkaline secretion (pH 7.29) which, together with the prostatic fluid, constitutes the bulk of the semen (seminal fluid) contains fructose from which the spermatozoa derive their energy. The seminal vesicle opens into the ductus deferens **AE2** shortly before the latter enters the prostate **A3**.

Fine structure. The thin wall contains some muscle fibers, and the mucosa is subdivided into chambers and niches, which in section produce a picture of mucous membrane bridges.

Prostate

The **prostate A3** produces a thin, opaque, weakly acidic secretion (pH 6.45) which contains among other materials acid phosphatase. The prostate resembles a chestnut in shape and size. It lies between the base of the bladder **ACE4** and the deep transverse perineal muscle **CE5**, 1–1.5 cm behind the pubic symphysis and in front of the rectum from which it is palpable. The prostate is perforated by the *prostatic part of the urethra* **CDE6** and by the two *ejaculatory ducts* **ACD7**.

Fine structure. The prostate consists of about 40 discrete tubuloalveolar glands **F8** which end, partly jointly, in about 15 little openings into the urethra around the *seminal colliculus.* The *dorsal part* of the prostate **E9** comprises a *right and a left lobe* which are stimulated by the male sex hormones. Between them is a middle lobe which responds to female sex hormones. In a case of hypertrophy of the middle lobe with advancing years the urethra is narrowed (*prostate hypertrophy*). The *anterior part* of the prostate **DE10**, in front of the urethra, contains only a few glands. The glands have a lumen of varying width and cells of varying height. In the lumen one commonly finds 1 mm size "prostatic" calculi **F11**,

i. e., inspissated glandular secretion. Strong smooth muscle bundles run between the separate glands as well as around the entire organ. The surface of the prostate is enveloped by connective tissue and is covered by visceral fascia. Between the two coats lies a strongly developed *periprostatic venous plexus.*

A12 Peritoneum, **A13** ureter, **ACE14** membranous part of the urethra, **ACE15** bulbourethral gland, **AC16** cavernous portion of the urethra, **CE17** internal opening of the urethra, **CDE18** prostatic utricle.

Semen

Semen has a pH of about 7.19. It consists mainly of the secretions of the epididymis, of the seminal vesicles and the prostate and it contains the spermatozoa. About 3.5 ml of sperm are discharged at each ejaculation. One milliliter of sperm contains 60–120 million spermatozoa (*"normospermia"*). The spermatozoa are motile in the sperm. The alkaline milieu of the fluid protects them from the acidity (about pH 4) of the vagina. The cervical mucus and the upper region of the female genital tract are alkaline. The spermatozoa reach the ampulla of the uterine tubes within 1–3 hours.

Clinical tips: If there are only 30–60 million spermatozoa per ml, a moderate diminution, the condition is called *hypospermia*; 5 million or less per ml, a serious diminution, is called *oligozoospermia*, and *azoospermia* is the complete absence of spermatozoa. After repeated ejaculations the number of spermatozoa quickly drops. 10–20% of the spermatozoa are always underdeveloped, over-age or deformed.

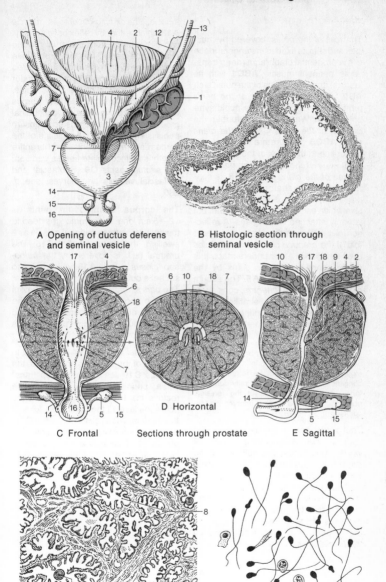

A Opening of ductus deferens
and seminal vesicle

B Histologic section through
seminal vesicle

C Frontal

Sections through prostate

D Horizontal

E Sagittal

F Microscopic section through prostate

G Microscopic appearance of
human spermatozoa

Penis

The *root of the penis*, covered by muscles and skin, is fixed to the undersurface of the urogenital diaphragm (deep transverse perineal muscle **ABC1** with its fascias) and to the sides of the pubic arch **AB2**. The movable *shaft of the penis* protrudes from beneath the pubic symphysis **C3**. We distinguish further the *dorsum of the penis* **C4** and the *glans penis* **ABC5**. The urethra ends in a slit-like orifice at the apex of the glans. The blunt edge at the base of the glans, the *corona of the glans penis* **AB6**, is separated from the shaft of the penis by a groove. The penis is covered by a thin skin which can slide across the shaft but is firmly attached to the glans. A reflection of the skin of the shaft (a "reserve fold"), the *prepuce or foreskin* **C7**, folds over the glans. At its undersurface it is tightly connected to the glans by the *frenulum of the prepuce* **A8**. The sebum of the prepuce, the *smegma*, consists of cells shed from the stratified, non-keratinized squamous epithelium of the glans and from the inner leaf of the foreskin.

Fusion between the foreskin and the glans is normal in the newborn, and in 20% of 2 year-olds the foreskin cannot be retracted. This conglutination disappears naturally during infancy. Abnormal tightness of the foreskin, phimosis, may need surgical intervention (circumcision).

The **penis** has 2 erectile bodies, the upper *corpus cavernosum penis* **ABC9** which serves only for erection, and the *corpus spongiosum penis* **ABC10** on the undersurface, through which the urethra **C11** runs and which ends in the *glans penis* **ABC5**. Each rectile body is enveloped by a tough, slightly elastic, 1–3 mm thick *tunica albuginea*, and together the corpora cavernosa are covered by the *fascia penis*.

The **corpus cavernosum penis** arises as two pointed *crura* **B12** from the descending rami of the pubic arch **AB2**. Each crus is covered by a thin striated muscle, the *ischio-cavernous muscle* **A13** which, either by reflex or voluntarily, can direct blood from the crus into the shaft of the penis. The two crura unite beneath the symphysis into one unpaired body. Its pointed end **B14** extends under the caplike corona of the glans penis. Along the undersurface of the corpus cavernosum penis runs a longitudinal groove in which the corpus spongiosum penis is embedded. In the midline of the corpus cavernosum penis along its entire length is an incomplete fibrous partition, the *septum penis* **C9**. Subfascial and epifascial vessels and nerves, see p. 280.

The **corpus spongiosum penis** is 12–15 cm long and commences deep to the deep transverse perineal muscle as a swelling, the *urethral bulb* **BC15**. The urethral bulb is covered by the *bulbospongiosus muscle* **A16**, which helps to squeeze out the contents of the urethra. The *pars spongiosa of the urethra* enters the bulb about 1 cm from its end. The urethra throughout its course is enclosed by the corpus spongiosum penis (corpus cavernosum urethrae).

The *seminal vesicle* **C17** is attached to the back of the urinary bladder **C18**, the *prostate* **C19** lies over the deep transverse perineal muscle **ABC1**. The rectovesical fossa **C20** usually extends below the level of the *transverse folds of the rectum (Kohlrausch's folds)* **C21**. Each *bulbourethral gland (Cowper's glands)* **ABC22** has a long excretory duct that opens into the urethra. **C23** Epididymis, **C24** testis.

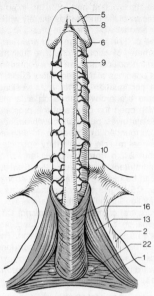

A Erectile bodies of penis and muscles at root of penis

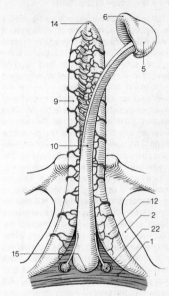

B Erectile bodies of penis with corpus spongiosum partly detached from corpus cavernosum penis

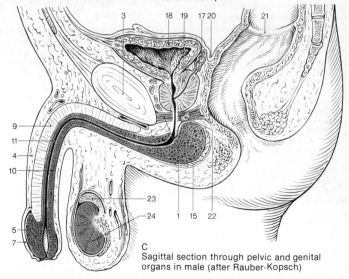

C Sagittal section through pelvic and genital organs in male (after Rauber-Kopsch)

Fine Structure of the Penis

The **corpus cavernosum penis ABC 1** is a porous framework of collagenous and elastic fibers and smooth muscle lined with endothelium. The *cavities* within it are slitlike when empty, they become several millimeters wide when engorged with blood and the *deep artery of the penis* **A 2** runs right through the center of the spongy substance at each side of the septum. Its arterial branches, the *helicine arteries*, are closed at the end by intima pads. The veins from the cavernous margins penetrate the *tunica albuginea* **A 3** to drain into sub- and epifascial veins. There are also arteriovenous anastomoses. The vascular spaces in the bulb of the **corpus spongiosum A 4** resemble those in the corpus cavernosum penis. In the shaft and in the glans of the penis are *venous networks* which, when engorged with blood, produce a swelling soft enough to let the semen pass through the urethra **AC 5**. **A 6** Fascia of the penis, **A 7** skin of the penis, **A 8** septum of the penis.

Vessels and nerves of the penis. In the middle of the narrow, dorsal longitudinal groove the *deep dorsal vein of the penis* **A 9** lies subfascially, the dorsal arteries of the penis on each side, and further laterally runs each *dorsal nerve of the penis* **A 10**. The superficial dorsal penile veins **A 11** lie above the fascia.

Erection of the penis. *Erection*, which is under neural control (see Vol. 3) occurs when the helicine arteries in the corpus cavernosum are opened and blood rushes in to fill the caverna and the tunica albuginea becomes tightened; the veins which penetrate through it are compressed. Simultaneously the trabecular muscles relax and the arteriovenous anastomoses are shut. Therefore, there is an *influx of blood while the outflow is restricted*, compare **E**. The venous network in the corpus spongiosum is also dilated. The penis becomes *flaccid* again when the helicine arteries contract.

Male Urethra

The **male urethra** is 20–25 cm long and has alternate *narrowings* and *dilatations*.

The narrow first part with the internal opening of the urethra lies entirely within the bladder wall. The wide *prostatic part* (see p. 284) is 3–5 cm long and has a diameter of about 1 cm. The fusiform seminal colliculus, 2 cm long, containing the openings of the ejaculatory ducts, is the continuation of the uvula vesicae. There are grooves on each side into which open the glands of the prostate. The narrowest part (*pars membranacea*) is surrounded by the deep, transverse perineal muscle (a voluntary sphincter) about 2 cm from the pubic angle. Its circumference is normally 1.2–1.5 cm but it can be passively dilated. The *spongy part* **C 5** begins below the muscle as a dilatation into which the pea-sized *bulbourethral gland (Cowper's gland)* **C 12** opens on both sides. This gland produces a stringy, alkaline secretion. The glans penis contains the wide, 2 cm long *fossa navicularis* **D 13**, which narrows toward the *external orifice of the urethra* **D 14**.

There is often a fold **B 15** in the roof of the fossa navicularis under which a catheter may become trapped. **BCD 16** Glans penis, **D 17** septum, **D 18** prepuce (foreskin) of the penis.

During *catheterisation* of the flaccid penis, its S-shaped curvature may be abolished. The curve beneath the symphysis disappears when the penis is lifted, and the curve below the urogenital diaphragm almost disappears when the penis is pushed backward, see **E**.

Fine structure. The mucosa has longitudinal folds (reserve folds) which help to close off the urethra. Recess of the wall (*lacunae urethrales*) (*Morgagni*) **B 19** open anteriorly. There are tubular *urethral glandules (of Littré)* in the whole of the pars spongiosa.

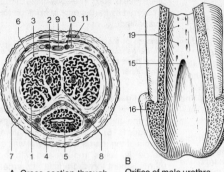

A Cross section through the shaft of the penis

B Orifice of male urethra opened from below

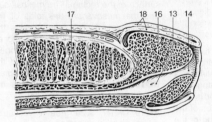

D Longitudinal section through tip of penis (after Feneis)

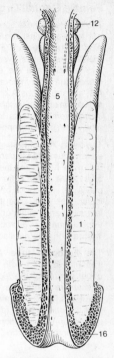

C Male urethra opened from above

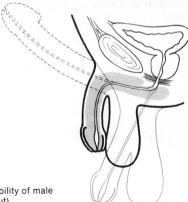

E Curvature and mobility of male urethra (after Testut)

Female Genital Organs

The female genitalia consist of the **internal** female genital organs: the *ovaries* **ABC1**, the *uterine (Fallopian) tubes* **ABC2**, the *uterus (womb)* **ABC3**, the *vagina* **AB4**, and the **external** genital organs: the *labia majora* **A5** and *minora* **A6**, and the *clitoris* **A7**, the *vestibule of the vagina* and the *vestibular glands*. The internal female genital organs are situated within the lesser pelvis.

The **ovary ABC1** is suspended from the *mesovarium*, the *suspensory ligament of the ovary* **ABC8** and the *ligament of the ovary* **AC9**. At the entrance to the lesser pelvis it is suspended obliquely. The Fallopian tube embraces the ovary superlaterally and anteriorly. It lies intraperitoneally in a groove between the internal and external iliac arteries.

The **Fallopian tube ABC2** runs intraperitoneally on both sides from the upper border of the lesser pelvis in a medial direction to the "tubal angle" of the uterus. The tube is mobile and is fixed over the *mesosalpinx* (duplication of the peritoneum). The abdominal end of the tube approaches the surface of the ovary.

The **uterus ABC3** is anchored between the urinary bladder **ABC10** and the rectum **BC11** by a muscular-connective tissue *retinaculum* (see p. 296), in the subperitoneal connective tissue space (see p. 312). This includes the *rectouterine muscle* which forms a peritoneal fold on each side, the *rectouterine fold* **C12**.

The uterus is also fixed by its *round ligaments* **ABC13** (*ligamentum teres uteri*) which keeps the uterus bent forward – *anteversion of the uterus*. The round ligament of the uterus originates at the tubal angle and runs beneath the peritoneum to the inner inguinal ring **C14**, through the inguinal canal into the labia majora and to the *mons pubis* **C15**, where it is anchored. The floor of the pelvis supports the uterus and keeps it in place (see p. 306). The body of the uterus is covered by peritoneum in front, on top and behind.

Because of the considerable mobility of the internal genital organs, it is better to speak of typical basic position and normal range of mobility, rather than a "normal position" of the organs.

Broad ligament of the uterus (ligamentum uteri) is the sheet of peritoneum lifted up by the uterus and the Fallopian tubes (cut edge **B16**). It passes from the lateral margin of the uterus to the lateral wall of the pelvis. It lies frontally and with the uterus is bent forward a little. It merges anteriorly into the peritoneal roof of the bladder and posteriorly into the anterior surface of the rectum. The *mesosalpinx* and the *mesovarium* are folds of the broad ligament. The latter divides the peritoneal space of the pelvis into an anterior and a posterior peritoneal pouch, the *vesicouterine* **B17** and the *rectouterine pouches* (of *Douglas*) **BC18**. The rectouterine pouch is the lowest point of the peritoneal cavity.

AB19 Ureter, **A20** greater vestibular glands (of *Bartholin*), **B21** pubic symphysis, **B22** deep transverse perineal muscle, **B23** external anal sphincter, **C24** inferior epigastric artery, **C25** medial umbilical ligament, **C26** median umbilical ligament.

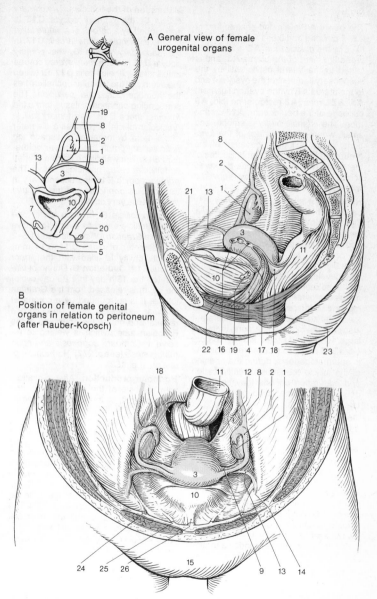

A General view of female
   urogenital organs

B
Position of female genital
organs in relation to peritoneum
(after Rauber-Kopsch)

C Position of internal female genital organs in lesser pelvis

Ovary

The **ovary A1** is almond-shaped and is 2.5–5 cm long and 0.5–1.0 cm thick. It is fixed by the *mesovarium* **A2**, a fold of the broad ligament of the uterus **A3**, and is joined to the lateral pelvic wall by the suspensory ligament of the ovary **A4** and to the uterus **A6** by the ovarian ligament **A5**. **A7** Ureter, **A8** rectouterine fold, **A9** cut edge of the peritoneum. **A10** uterine tube. The *infantile* ovary is smooth and small, whereas the surface of the *functional* ovary is lumpy, because of the protruding follicles **AB11** and fissures of the corpora lutea, as well as puckered scars. The small, *senile* ovary is covered with scars.

Histological structure. The ovary may be divided into a *cortex* and a *medulla*. The major blood vessels **B12** enter the center of the ovary from the mesovarium and run through the medulla. The surface is covered by a simple cuboidal epithelium **B13** The outer layer of the fully developed ovary are an epithelium and a subepithelial fibrous layer – the *tunica albuginea* **B14**, deep to which is a highly cellular cortical layer. There are the *Graafian follicles* in all their different stages of development. Among them are large *vesicular follicles* **AB11**, *corpora lutea* **B15** and their scarred remnants. As in spermatogenesis, during the ripening of the follicle (oogenesis) multiplication and maturation periods may be distinguished, but there is no period of differentiation.

Oogenesis. 1. The *period of multiplication* is completed during fetal development. At birth the *oogenia*, which develop from the primary germ cells, have reached their final number of about 1 million and have entered the prophase of the first maturation division as *primary oocytes*. 2. *Period of maturation*. After puberty the first maturation division has been completed and the second has begun (formation of the *secondary oocyte* and "polar body"). It is only after fertilisation that the second maturation division is completed (formation of the mature egg, *ovum* and a further "polar body").

Maturation of the follicles. *The primary follicle* **C**: the *primary oocyte* **C16** is surrounded by a corona of a *single layer* of *simple, follicular epithelial cells* **C17**. In the *secondary follicle* **D**, the *primary oocyte* **D16** lies in a *multilayered* corona of follicular epithelial cells **D17**. Between the ovum and the follicular epithelium lies the homogenous *zona pellucida*. The surrounding connective tissue forms the *follicular theca* **D18**. The *tertiary follicle* (vesicular follicle) **E** reaches a diameter of 0.5–1 cm, by formation of a space, the *follicular antrum* in the follicular epithelium **E17**. The wall of the antrum is lined by follicular cells, "granulosa cells". The *primary oocyte* **E16** lies eccentrically in the *cumulus oophorus* and the *follicular theca* **E18** is very obvious.

Graafian follicle **F**. In each cycle, one tertiary follicle enlarges over a few days to form a *Graafian follicle* (diameter 1.5–2 cm, diameter of ovum 0.11–0.14 mm) and its cavity is filled with the *liquor folliculi* **F19**. **Ovulation G**: During *ovulation* (on the 15th day in a 28 day cycle) the ovum is released from the Graafian follicle. It is surrounded by a corona of follicular epithelial cells, the *corona radiata*, and enters the open end of the Fallopian tube, which is waiting to receive it. *Compare oogenesis and spermatogenesis* (see p. 277). Hormonal regulation s. p. 153.

Hormone production. After ovulation, with the help of the theca interna, the follicular epithelium forms the *corpus luteum* **H**, a hormone producing gland with a thick, bag-like, folded wall **H15**, which produces *progesterone* under the influence of LH. **Follicular theca.** During maturation of the follicle, the connective tissue around the follicle develops into a multi-cellular layer, the *theca interna* **D–F18**. This is influenced by ICSH to produce female sex hormones, follicular hormones or *estrogens* (see p. 300). The theca externa surrounds the inner cell layer with connective tissue fibers.

Atresia of the follicle. Only about 400 follicles reach the stage of ovulation, the rest perish beforehand – atresia of the follicles. However a functional hormone-producing theca interna does occur around them for a short while.

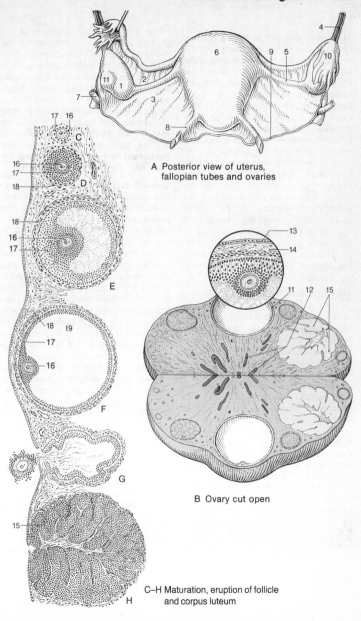

A Posterior view of uterus,
fallopian tubes and ovaries

B Ovary cut open

C–H Maturation, eruption of follicle
and corpus luteum

Fallopian Tube

The **Fallopian (uterine) tube** is 8–20 cm long. It lies intraperitoneally in the *mesosalpinx*, the upper extension of the broad ligament. Its abdominal end, the *infundibulum*, opens like a funnel into the abdominal cavity and ends in 1–2 cm long fringes, the *fimbriae* **AC1**. One of these, the *ovarian fimbria* lies on top of the ovary **A2**. Through the funnel the tube becomes dilated into an *ampulla* **C3** which is divided by longitudinal grooves. Toward the uterus the lumen of the tube narrows to become the *isthmus* in its medial third at the entrance into the uterus. The *uterine part of the tube* **C4** penetrates through the wall of the uterus and ends in the lumen of the uterus. The tube **C** shows *longitudinal folds* at its inner surface. They serve as "guide rails" for the ovum. The folds get flatter towards the isthmus and eventually disappear completely.

Fine structure. The tube consists of a mucosa, a muscle layer and a peritoneal covering. The **mucosa D5**: a cross section through the ampulla shows branching *folds* **D5**. The mucosa has a simple high columnar epithelium with *ciliated* **E6** and *glandular* **E7** *cells*. During the first half of the menstrual cycle ciliated cells predominate, and during the second half after ovulation glandular cells are in the majority. Toward the end of the cycle epithelial cells are also shed **E8**. The secretion of the tube consists of absorbed peritoneal fluid and secretion from the glandular cells. Among other purposes it is believed to provide nutriment for the fertilized ovum. The cilia produce a current of fluid which is mainly directed toward the uterus and which plays a part in the migration and distribution of spermatozoa. The **muscle layer D9** serves for movement of the egg and sperm, and the secretions of the tube. The muscle layers consist of: 1.) the dense, spiral *intrinsic muscle layer* near the mucosa, which can initiate peristaltic movement toward the uterus for the

transport of fluid, and an antiperistaltic wave for transport of spermatozoa, 2.) *muscle bundles – accompanying the blood vessels* **B10** contraction of which produces the constricting grooves of the ampulla and flexion of the entire tube; and 3.) a *subperitoneal muscle layer*, which forms lattice-like strands and a longitudinal band along the upper edge of the tube from the uterus to its funnel end. The muscle strands provide mobility for the tube and the fimbriae. The **peritoneal covering D11** enables the tube to shift against its environment.

Capture and transport of the ovum. Observations made during abdominal operations have revealed that during ovulation the fimbriae **A1** perform rhythmical movements. The ovary **A2** is moved upward and downward and is turned into its longitudinal axis by the muscles of the suspensory ligament **A12** and those of the ovarian ligament **A13**. During this process the ovarian fimbria reach the place where the follicle is ready to erupt, a process probably controlled by active locally produced substances (? enzymes). The ovum **A15** newly released from the follicle **A14**, reaches the funnel **A16** via the ovarian fimbria within 3–6 min. This movement is accomplished with the help of the current of fluid created by the cilia, and by suction exerted by rhythmical contraction of the tube. The ovum takes 4–5 days to reach the uterus but it can only be fertilized at the latest during the first 6–12 hours, while it is still within the ampulla of the fallopian tube. Pendulum movements of the tube help to mix the ovum with the spermatozoa and the secretions of the tube. Pushing and suction propel the egg in pendulum movements, step by step, through the chambers of the ampulla. The fertilized egg arrives at the uterus on the 4th–5th day. Its implantation in the mucosa of the uterus starts on the 6th day and is completed on the 12th day after eruption of the follicle.

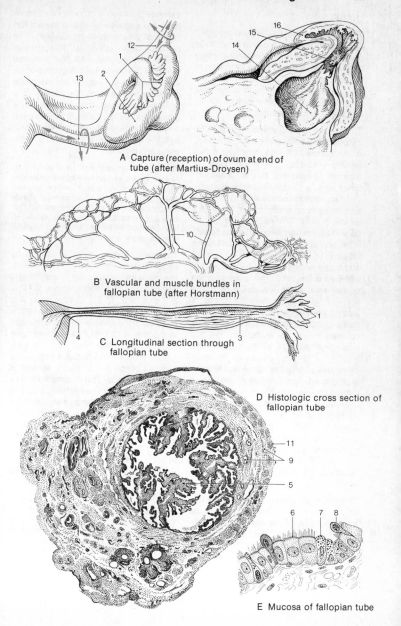

A Capture (reception) of ovum at end of tube (after Martius-Droysen)

B Vascular and muscle bundles in fallopian tube (after Horstmann)

C Longitudinal section through fallopian tube

D Histologic cross section of fallopian tube

E Mucosa of fallopian tube

Uterus

The *uterus* carries the *fetus* in pregnancy. The *mucosa of the uterus* is cyclically prepared for the implantation of the embryo and plays a part in the formation of the placenta. The *uterine muscles* become adapted to the growth of the fetus and provide the *motive power* for its expulsion during childbirth.

The **uterus** is pear-shaped and may be divided into a body and a neck. The *body of the uterus* **AC1**, the thick (and transversely enlarged) part of the "pear" is directed upward and has an antero-inferior and a postero-superior surface. The *fundus of the uterus* **AC2** in a mature woman rises above the "tubal angle" **A3** as the cupola. The *neck of the uterus, cervix* **C4**, the narrower part of the "pear" in the non-pregnant uterus comprises the lower ⅓ of its total length. It is rounded and is directed downward and posteriorly into the vaginal vault. The *vaginal portion of the neck of the womb* **C5**, which is known by gynecologists as the "portio" is that part of the cervix, about 1 cm long, that protrudes into the vagina **C6**, and is covered by vaginal epithelium. The *supravaginal part* **C7** of the neck of the womb is largely enclosed by subperitoneal connective tissue into which it is anchored. The *isthmus of the uterus* **C8** lies between the body and the neck of the uterus. The isthmus of the uterus, which loses its folds in pregnancy, is called the "lower segment" by the obstetrician. **C9** Urinary bladder. The cavity of the uterus **A10** within the fundus and uterine body forms a triangular slit with a downward pointing apex in the frontal plane.

The Fallopian tubes **A3** open into the two upper 'angles'. The lower 'angle' leads into the canal of the isthmus **A11**, the "internal mouth of the uterus", ("os uteri") of the gynecologist. The *cervical canal* **A12**, which follows downwards, is fusiform and dilated. It is filled by a *mucous plug*, which protects the uterus from

ascending infective agents, and facilitates the passage of sperm toward the lumen of the uterus by its alkalinity. The cervical canal opens at the *external orifice, ostium uteri* **A13**, the "external mouth of the uterus" of the gynecologist, at the level of the vaginal portion of the cervix in the vault of the vagina. In a nulliparous woman, this orifice is only a small, round depression **D14**. It has an *anterior* and a *posterior* lip. After childbirth the ostium uteri presents as a transverse fissure **D15**.

The vaginal part of the cervix is covered by several layers of non-keratinised squamous vaginal epithelium and therefore appears smooth and light pink at colposcopy. Purple sharply defined spots "pseudoerosions" (ectropia) are produced by islands of single-layered prismatic epithelium which has migrated from the cervix. **A16** Vaginal wall.

The **wall of the uterus** is formed by endometrium (mucous membrane), myometrium (muscular wall) and peritoneum. The *endometrium* **B17** is 2–8 mm thick. In the uterine cavity it is smooth and soft, and has broad, shallow plaques. In the cervical canal the mucosa is tougher **A12** and has palmate folds (*plicate palmatae*) and the *cervical glands* produce the mucous plug in the external orifice of the uterus. Menstrual cycle, see p. 300. The *myometrium* **B18** of the non-gravid uterus feels firm. It is about 2 cm thick and shows an ill-defined triple layer. The middle layer contains many blood vessels. Uterine muscle (see p. 298). The *perimetrium* (peritoneum) **B19** is firmly adherent to the myometrium. It merges into the broad *uterine ligament* **B20** at the lateral margin of the uterus. Most of the supravaginal part of the cervix **C7** is extraperitoneal.

Variations in the structure of the uterus **E** arise from defective symmetry or an asymmetrical junction of the 2 *Müllerian* ducts. All stages are possible from a double uterus with a double vagina (*uterus duplex separatus vagina duplex*) via a *bicornuate uterus*, to the *arcuate uterus* with an incomplete septum.

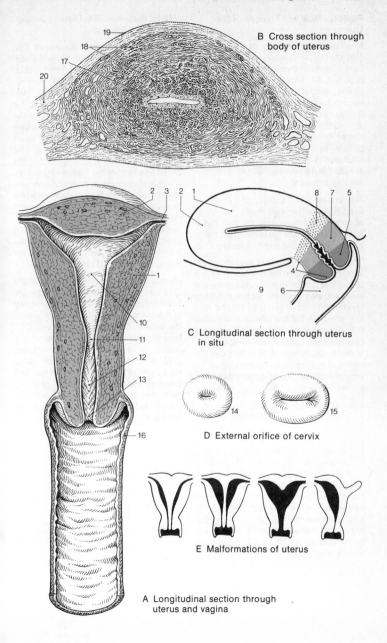

B Cross section through body of uterus

C Longitudinal section through uterus in situ

D External orifice of cervix

E Malformations of uterus

A Longitudinal section through uterus and vagina

Position, Size and Fixation of the Uterus

Position and mobility of the uterus.

The uterus is normally *anteflexed* and bent forward over the bladder (*anteverted*) **A1**. The longitudinal axis of the uterus and of the vagina is concave toward the front. It runs approximately in the direction of the guideline of the pelvis, an imaginary line that runs through the center of all the planes of the pelvis, parallel to the anterior concavity of the sacrum. *'Position'* refers to the relationship to the midline sagittal plane (right – left displacement).

The position of the body of the uterus over the bladder distributes the intra-abdominal pressure over the floor of the pelvis and protects the uterus from being pushed through the genital opening.

The fixation mechanism permits physiological mobility of the uterus. When the bladder is full the uterus becomes more erect **A2**; and when the rectum is full it is pushed forward. If the bladder and rectum are both full the uterus is elevated **A3**. *Retroflexion* and *retroversion* of the uterus **A4** are usually pathological conditions. A physiological condition can be restored by surgical intervention, e. g., by shortening both round ligaments of the uterus.

Size of the uterus.

Numerous different methods are used to determine the size and position of the uterus, but bimanual examination is commonly undertaken **B**. In the newborn, the uterus is 3.5 cm long and cylindrical in shape. In childhood the neck of the uterus remains longer than the body and the fundus does not protrude forward. After puberty, the uterus develops its typical shape, is 6–7.5 cm long and weighs 80–120 g. From the external orifice to the fundus, the uterine lumen is about 5.5 cm long. In old age the uterus atrophies, though the body remains relatively large whilst the cervix shrinks markedly. **B5** Urinary bladder, **B6** uterus, **B7** rectum.

Fixation mechanism of the uterus.

The uterus is mainly anchored in the cervical region **C8** by collagenous and elastic tissue strands, which also contain smooth muscle fibers (**retinacula**). They enable the position of the uterus to be adjusted actively and passively, and the tension of these strands on the vessel walls keeps the blood vessels open whatever the position of the uterus. The muscle strands extend upward into the broad ligament of the uterus where they are joined by muscle bundles from the myometrium. This fibrous apparatus (retinacula) keeps the uterus suspended in a position which is secured by the muscles and ligaments of the floor of the pelvis; it is situated in the parametrium, i. e., the subperitoneal periuterine connective tissue space (see p. 312).

Retinacula: The *pubovesical ligament* **C10** runs anteriorly past the bladder **C9** to the pubic bone. It counteracts prolapse of the bladder and of the anterior wall of the vagina. The *cardinal ligament* **C11** fans out toward the lateral wall of the lesser pelvis. The *rectouterine muscle* **C13**, which lies in the rectouterine fold, runs past the rectum **C12** to the sacrum. Between the two rectouterine folds the *rectouterine pouch* **C14**, a peritoneal pocket extends downward. The *round ligament of the uterus* see p. 288.

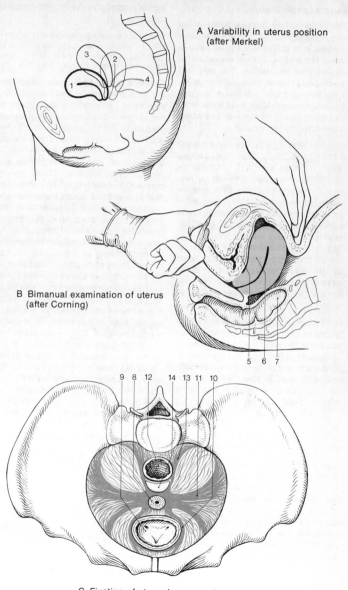

A Variability in uterus position
(after Merkel)

B Bimanual examination of uterus
(after Corning)

C Fixation of uterus by connective
tissue structures in lesser pelvis

Muscles of the Uterus

The **myometrium**, the *muscle wall* of the uterus consists of smooth muscle cells, vessels and connective tissue. Muscle tissue forms about 28% of the volume of the body of the uterus; it is less well developed in the isthmus and the cervix. In the *body* and *fundus* of the uterus there is a thicker *middle* layer and thinner *inner* and *outer* layers of myometrium which are not clearly demarcated.

The *middle* and by far the thickest *layer* **A1** is particularly well vascularised (*stratum vasculare* **A1**). Its muscle strands form a three dimensional network in the body, which mainly extends parallel to the surface of the uterus and is anchored onto the blood vessels. In the isthmus and cervix there are gradually ascending circular muscle bundles. In a Cesarean section through the isthmus region, therefore, the muscle bundles can be widely separated after a transverse incision. The middle layer provides the principle expulsive power during parturition.

The thin *inner layer* beneath the mucosa has many circular muscle bands (*stratum subvasculare* **A2**). It is said to assist contraction of the open blood vessels in the detachment of the placenta after parturition. The *outer layer*, which is also thin, consists of 4 lamellae containing muscle strands running alternately in a longitudinal and a circular direction (*stratum supravasculare* **A3**). Probably they help to stabilise the surface of the growing uterus during pregnancy.

In **pregnancy** the uterus grows rapidly by *hypertrophy* of individual smooth muscle cells, which increase to 7 to 10 times their original size. Fluid, which penetrates the connective tissue, permits displacement of the inner structures of the wall of the uterus. The isthmus is markedly elongated and becomes part of the "womb". It feels soft in contrast to the cervix on bimanual examination (*Hegar's sign of pregnancy*). **A4** Vaginal portion of the cervix.

Growth of the uterus during pregnancy is stimulated by estrogens and progesterone and by stretching of the wall. In the nongravid uterus the hormones prevent atrophy due to inactivity.

Cervix of the uterus. The structures of the neck of the uterus merit special consideration. The cervix remains closed for the duration of pregnancy, but during *parturition* it must quickly dilate to the width of the baby's head. There are active and passive mechanisms involved in this process: *passive enlargement* has already been prepared for by the increased fluid content of the cervix. The fluid is displaced during parturition and permeates into the connective tissue, the venous plexus dilates and the cervical glands become enlarged (see p. 312). An *active opening mechanism* is provided by restructuring of the muscle strands and the fibers of the connective tissue. In the non-pregnant uterus structures near the mucosa are arranged in a more circular pattern. By the 7th month of pregnancy these structures run a steeper course than in the cervix of the nonpregnant uterus. They merge partly into the longitudinal muscle bundles of the uterus (the "descending" strands to the cervix **B5**) and partly into those of the vagina ("ascending" strands **B6**). The pull from these longitudinal muscle bundles during parturition helps to produce opening of the internal and external orifices of the cervix. **B7** *Amniotic sac (opening phase of childbirth)*.

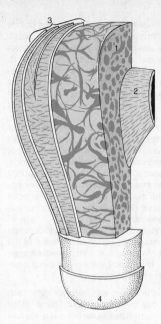

A
Model of arrangement of muscles in
body and isthmus of uterus
(after Wetzstein and Meinel)

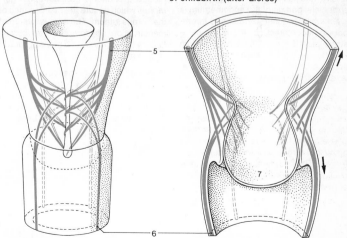

B Longitudinal muscle strands in
cervix of pregnant uterus

Left – closed cervix
Right – cervix in opening phase
of childbirth (after Lierse)

Endometrium

The *mucosa of the uterus* (**endometrium**) lies directly on the muscle **D1**. It has a simple, high columnar epithelium and some ciliated cells, and contains tubular glands, the *uterine glands*. The connective tissue of the mucosa has many cells but few fibers. A *basal layer* **D2** about 1 mm high, which is not shed during menstruation, can be distinguished from a *functional layer* **D3, D4**, up to 8 mm high.

During the period of sexual maturity, the ovarian hormones (see p. 170) cause *menstrual cycles* in the endometrium. They begin with the *menarche* from the 10th–15th years, with cycles that are at first still incomplete. They end with the *climacteric (menopause)* at about the 45th year. Higher hormonal regulation (see p. 153).

The **menstrual cycle** is divided into *phases*. In the following description the most common 28 day cycle has been taken as a basis. Counting starts with the first day of the period, compare **B**.

Phase of desquamation and regeneration **A**, the 1st to 4th day. This is mainly controlled by disappearance of progesterone and increase in estrogens. The superficial portion of the endometrium is shed while certain enzymes and thrombocytes counteract clotting of the menstrual blood. Eventually the epithelium and the connective tissue of the functional layer regenerate from the basal layer and the wound is closed.

Phase of proliferation **B C**. The 5th–15th day comprises the ovulation period. This phase is mainly controlled by estrogens and is therefore called the *"estrogen phase"*. The functional layer grows, the gland get larger and spiral arteries are formed. The body temperature rises after ovulation by 0.5–1 °C (*premenstrual hyperthermia*).

Phase of secretion **DE**, the 15th–28th day ("gestation phase"). This is mainly controlled by progesterone. The glands coil and produce a mucous secretion. The blood vessels increase in number and the spiral arteries grow longer **D**. Toward the 28th day the connective tissue cells near the surface change to large epitheloid *"pseudodecidua" cells* which resemble those of the mucosa of the pregnant uterus, the decidua. Because of this the superficial dense zone *(compacta)* of the endometrium **D3** can be distinguished form the loose, deep, *spongiosa layer* **D4** with its many glands. Toward the end of the cycle the spongy layer loses fluid (*premenstrual shrinkge of the mucosa*) **E** and the spiral arteries contract due to drying up of the progesterone supply. This results in lack of blood *(ischemia)* and tissue damage. Subsequently renewed dilatation of the blood vessels due to increased estrogen supply causes bleeding and *desquamation* of the spongy layer.

Even in cycles with intervals other than 28 day ovulation usually takes place 13–14 days before the first day of the new menstrual period *(constant length of the gestation phase)*.

In **pregnancy** the trophoblast (the nutritional shell of the fetus, see p. 316) produces *chorionic gonadotrophin hormones* which ensure the maintenance of the corpus luteum. The corpus luteum of menstruation becomes the *corpus luteum of pregnancy* and menstruation is suppressed.

Menopause. The menstrual cycles gradually disappear with the onset of the *climacteric (menopause)*. The mucosa and the muscle of the uterus atrophy and the vaginal portion of the cervix becomes smaller.

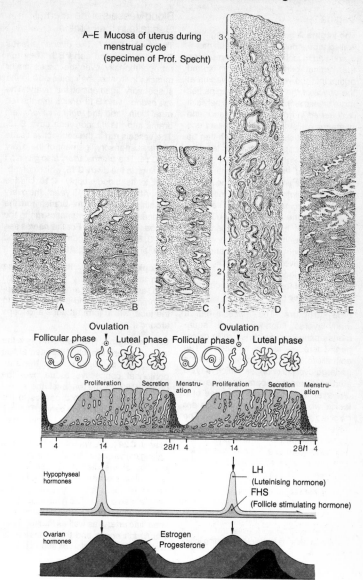

A–E Mucosa of uterus during menstrual cycle (specimen of Prof. Specht)

Ovulation
Follicular phase ! Luteal phase Follicular phase ! Luteal phase
Ovulation

Proliferation Secretion Menstru-ation Proliferation Secretion Menstru-ation

1 4 14 28/1 4 14 28/1 4

Hypophyseal hormones

LH (Luteinising hormone)
FHS (Follicle stimulating hormone)

Ovarian hormones

Estrogen
Progesterone

F Above: ovarian cycle (after Starck);
Middle: cyclical changes in endometrium
Bottom: blood level of hypophyseal hormones (yellow) and of ovarian hormones (gray and black) (after Taubert)

Vagina

The **vagina A1** is an 8–10 cm long, thin-walled tube, consisting of mucous membrane and muscle. At the top it embraces the vaginal portion of the cervix and opens into the vestibule of the vagina at the *orifice of the vagina*. The vagina lies more or less along the axis of the pelvis. It is flattened in front and the anterior and posterior walls are in loose contact and enclose an H-shaped slit, which can be opened without producing much tension. For the orifice of the vagina see p. 304. The *fornix of the vagina* forms a ring around the vaginal portion of the cervix, the shallow *anterior* **A2** and deep *posterior* **A3** *vaginal recesses*. The latter borders on the rectouterine pouch (of Douglas) **A4**. It is a site of danger of perforation during manipulation with instruments. Both the anterior and posterior walls have transverse folds, which protrude as longitudinal bulges in the central part of the wall. *Urethral carina of the vagina* (see p. 304).

Fine structure. The *mucosa* has a many-layered, non-keratinized squamous epithelium, which is rich in glycogen. There are almost no glands. The epithelium and the vaginal secretions undergo *cyclical changes*, which may be studied in smear preparations. The *muscle layer* is thin and is composed of a lattice work. The venous plexuses are involved in closure of the vagina. The adventitial connective tissue, the *paracolpium*, provides the connection with neighboring organs – a firm connection with the urinary bladder **A5** and a loose connection with the rectal wall **A6**.

The **vaginal secretion** is formed by the glands of the cervix and desquamate epithelial cells. It contains 0.5% lactic acid which is produced by lactic acid bacteria from the glycogen of desquamated cells. The acidic milieu (pH 4–4.5) of the vagina protects it against ascending infective agents.

Blood Vessels of the Internal Genitalia in the Female

The **ovary ABC7**. The *ovarian arteries* **B8** arise from the aorta **B9**. They run retroperitoneally over the psoas major muscle, in front of the ureter **B10** to the suspensory ligament of the ovary. The *left ovarian vein* **B11** drains into the left renal vein, and the *right ovarian vein* drains into the inferior vena cava **B13**. The vessels run in the connective tissue of the suspensory ligament of the ovary **ABC14**. The *uterine artery* also gives off a branch to the *ovary* **C15**, which contributes to its blood supply. **C16** Uterine vein. After the age of 45 years, the ovary is mainly supplied via the branch from the uterine artery. The *lymphatics* run to the *lumbar lymph nodes*. For the *nerves* see Vol. 3. **B17** Inguinal ligament, **B18** round ligament of the uterus, **AB19** uterus.

Fallopian tube A20.. Its arteries arise from the uterine artery (*tubular branch* **C21**) and the *ovarian artery* **BC14** (*tubular branch*). They send 6–9 branches to the tube, which form arcades and loops around it.

Uterus, vagina. The *uterine artery* is the largest visceral branch of the internal iliac artery, see p. 70. It runs in the subperitoneal connective tissue over the ureter **C22** to the base of the broad ligament, within which it passes to the uterus at the level of the cervix. There, it divides into the tortuous ascending *main branch* **C23** and the descending *vaginal branch* **C24** to the wall of the vagina. The coiling of the blood vessels allows for the mobility of the uterus. The *veins* form large *plexuses* around the body and the cervix of the uterus, and around the vagina, see p. 314. The *lymphatics* drain into the lymph nodes alongside the common iliac artery, as well as into the *inguinal lymph nodes* and the nodes on the walls of the pelvis. *Nerves* see Vol. 3.

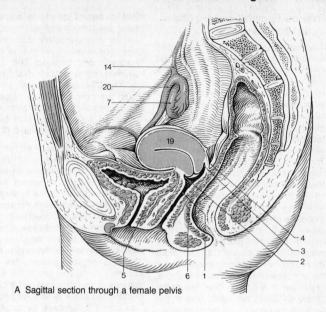

A Sagittal section through a female pelvis

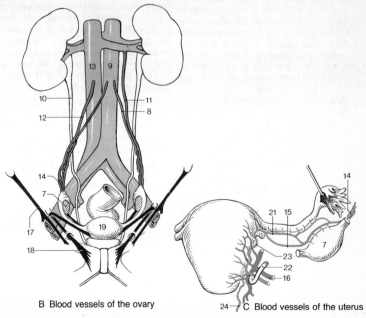

B Blood vessels of the ovary

C Blood vessels of the uterus

External Female Genitalia (the Vulva) and the Urethra

The **labia majora C1** are skin folds which form the borders of the pudendal cleft. Together with the *mons pubis* **C2** they form a triangle. Their hair cover merges at the mons pubis with the triangular *pubic hair*. The mons pubis and the labia majora contain fat, sebaceous glands, sweat and scent glands. The **labia minora C3** are thin skinfolds that surround the *vestibule of the vagina*. They are connected at the back by a band of skin, the *frenulum* **C4**, which disappears during the first childbirth. Anteriorly the labia minora divide into 2 folds on either side, the two inner folds forming a cordlike connection to the glans of the clitoris **AC5**, while the two outer ones meet in front of the glans to become the *prepuce of the clitoris* **C6**. The labia minora contain connective tissue without fat and sebaceous glands. The *clitoris* originates as two limbs, the *crura of the clitoris* **C7**, from the lower rami of the pubic bone. Beneath the pubic symphysis the crura form the 3–4 cm long shaft, the *body of the clitoris* **C8**, which turns backward and ends in the *glans clitoridis* **AC5**. The *body of the clitoris* is enveloped by a fascia and is fixed to the symphysis. Each crus of the clitoris is covered by one *ischiocavernous muscle* **C9**. The clitoris contains corpora cavernosa (compare with the corpora cavernosa of the penis, p. 286) and is amply supplied with sensory nerves. The **orifice of the vagina C10** may be partly closed by the vaginal flap, the *hymen*. The latter is torn at the first coitus and after childbirth its scarred remnants turn into verrucose *hymenal caruncles (carunculae hymenales)*. In rare cases *Gartner's ducts* may end on either side of the vaginal orifice (see p. 270). *Small vestibular glands* lie between the urethra and the orifice of the vagina. The duct of the *greater vestibular gland (of Bartholin)* **C11** ends on either side of the orifice of the vagina **C12**. The pea-sized gland itself lies behind the vaginal orifice, medial to the bulbospongiosus muscle behind the vestibular bulb and beneath the deep transverse perineal muscle. Its duct is about 1.5–2 cm long. The corpora cavernosa of the vestibule (*bulbi vestibuli vaginae*) **C3** consist of venous plexuses which are covered by the *bulbospongiosus muscles* **C14**. **C15** Centrum tendineum of the perineum, **C16** external sphincter muscle of the anus, **C17** anus.

The **female urethra** is 2.5–4 cm long and runs in a curve, concave toward the front, between the symphysis and the anterior wall of the vagina. It projects into the latter as the *carina of the urethra* **A18**. The urethra begins below the *trigone of the bladder (trigonum vesicae): internal urethral orifice* and ends as a slit or star shape 2–3 cm behind the glans of the clitoris: *external urethral orifice* **AC19**. The 1–2 cm long *paraurethral ducts (Skene's ducts)* also end in that area. It consists of an *intramural part* within the bladder wall and a *cavernous part* **B20**. The lumen is narrowed by longitudinal folds which contain venous plexuses, the *corpus spongiosum urethralis*, and they ensure closure of the urethra. The lumen can be widened to 7–8 mm. The *muscle wall*, a continuation of the bladder muscle, is enveloped by fibers from the *deep transverse perineal muscle* **B21**, the voluntary sphincter muscle.

Cross section of the urethra **B20**, vagina **B22** and rectum **B23** at the level of the floor of the pelvis, see p. 300. **B24** Lower ramus of the pubic bone, **B25** levator ani muscle, **B26** tip of the coccyx.

For the *vessels and nerves* see p. 302.

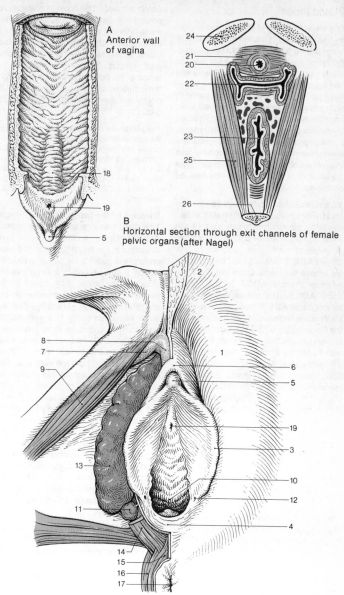

A
Anterior wall
of vagina

B
Horizontal section through exit channels of female
pelvic organs (after Nagel)

C External genitalia in female. Right side dissected

Pelvic Floor

The lower aperture of the pelvis is closed by the voluntary muscles and the fascia of the **pelvic floor**. There are openings in the latter for the urogenital tract and for the rectum. The muscles and fascia are arranged in three layers, partly in a wing-like fashion, and are much the same in both sexes. From above downward (i. e., from the deepest to the most superficial) the layers are: 1. the **pelvic diaphragm**, 2. the **urogenital diaphragm**, and 3. the **sphincter muscles of the rectum and the urogenital tract.** *Each diaphragm consists of a muscle layer covered on both sides by a fascia.*

Muscles of the pelvic diaphragm. The **levator ani muscle A1** originates as two wing-like plates from each side of the inner wall of the lesser pelvis. The tendon of origin of each plate, the *arcus tendineus* **ABC2**, extends from the symphysis across the fascia **C3** of the obturator internus muscle **ABC4** to the ischial spine **ABC5**. The entrance to the obturator canal **AB6** lies above the tendon. Each of the two muscle plates consists of the following part. The *puborectal muscle* **C7** originates from the pubic bone and with its partner from the other side forms the puborectal sling. Their medial portions, the *crura of the levator ani muscle*, surround the *genital hiatus* **C8**, which is about 3 cm wide at the pubic symphysis. Some muscle bundles **C9** cross in front of the rectum **C10**. Behind it the puborectal muscle follows the *iliococcygeal muscle* **C11**, it stretches as far as the coccyx and the *anococcygeal ligament* **AC12**. A third muscle, the *pubococcygeal muscle* **C13**, extends even further over and across the other two parts and runs from the pubic bone to the coccyx.

The **coccygeal muscle AC14** is attached posteriorly to the levator ani muscle. It arises from the ischial spine **ABC5** and fans out into the side of the lower sacrum and the coccyx. The muscle lies above the *sacrospinal ligament* **B15** with

which it fuses. The muscle may be replaced by connective tissue. Between the levator ani and coccygeal muscles is a triangle which contains no muscle because of the reduction of the iliococcygeal muscle. It is only closed by the fascia of the pelvic floor. Abscesses from the ischiorectal fossa can reach the subperitoneal space through this triangular gap.

The **piriformis muscle ABC16** together with the sacrum forms the *posterior wall* of the space above the pelvic floor. **AB17** Suprapiriform foramen, **AB18** infrapiriform foramen, **B19** lesser ischiadic (sciatic) foramen, **B20** sacrotuberal ligament, **C21** tendinous arch of the pelvic fascia, **C22** greater ischiadic (sciatic) foramen.

Muscles of the urogenital diaphragm. The deep transverse perineal muscle ABD23 forms a thin muscle layer stretched in the angle between the lower crura of the pubic bone, transverse to the genital hiatus, which to a large extent it closes from below. The fibers of the posterior portion of the muscle run in a roughly transverse direction. The anterior fibers run partly in a semicircular and partly in a circular direction around the urethra **D24**, and act as a voluntary sphincter. In the female most of these fibers go around both the vagina and the urethra. Spiral fibers of this muscle ascend within the wall of the vagina or the prostate. The angle immediately below the symphysis is closed only by connective tissue, i. e., by the *transverse perineal ligament* **D25**. The deep dorsal vein of the penis passes between the symphysis and the transverse perineal ligament and the dorsal artery and nerve of the penis pass between the ligament and the deep transverse perineal muscle through the urogenital diaphragm; compare with p. 310.

The **superficial transverse perineal muscle** may reinforce the posterior margin of the corresponding deep muscle with thin superficial transverse fibers that arise from the ischium (see p. 310).

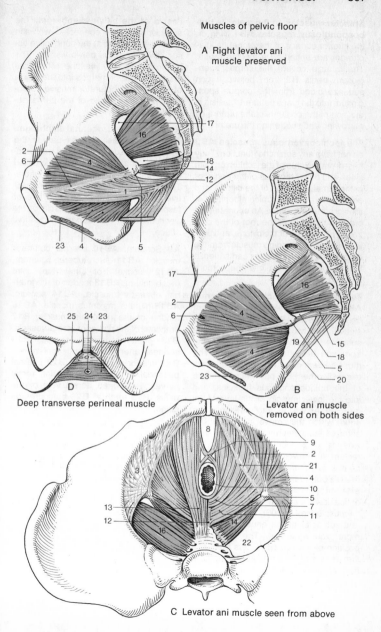

Muscles of pelvic floor

A Right levator ani
muscle preserved

Deep transverse perineal muscle

B Levator ani muscle
removed on both sides

C Levator ani muscle seen from above

Spincter muscles. In the male the **bulbospongiosus muscles AB1** lying on the inferior surface of the urogential diaphragm are united by a median raphe. The muscle covers the corpus spongiosum penis **B2** from below. It can squeeze blood from the corpus spongiosum into the glans penis and assists in ejection of the contents of the urethra by narrowing and shortening the urethra.

The **ischiocavernosus muscles AB3** do not have any sphincher function. They arise from the crus of the ischium on both sides and cover the crus of the penis, or in the female the crus of the clitoris, and are inserted into the tunica albuginea or the corpora cavernosa. An upper bundle crosses the back of the root of the crus penis, or the crus of the clitoris, and then joins the corresponding muscle from the other side. These muscles can squeeze blood from the crura into the corpora cavernosa.

The **external sphincter of the anus AB4** encircles the termination of the gut below the levator ani muscle **A5** like a sleeve. The two halves of the external sphincter muscle cross in the midline in front and behind the anal canal. The muscle fibers go partly into the anococcygeal ligament **B6**, partly into the raphe of the bulbospongiosus muscles, and into the lower margin of the levator ani muscles, as well as into the skin of the anus, which they move in unison. The external sphincter muscle of the anus extends for 3–4 cm upward along the rectum. As the voluntary sphincter of the anus it is permanently contracted and relaxes only during defecation. While the external sphincter of the anus closes the orifice down to a sagittal slit, the puborectal muscle (see p. 306), the most important sphincter of the anus, produces a more transverse slit. The criss-cross position of the two sphincters favours a firm closure of the anus. **AB7** Superficial transverse perineal muscle.

Fascia of the pelvic diaphragm. The *superior* fascia is connected with the pelvic visceral fascia, the connective tissue covering of the pelvic viscera. The *inferior* fascia parly lines the ischiorectal fossa. Laterally it merges into the fascia of the internal obturator muscle. Compare with the spaces of the pelvis described on p. 312.

A Fascia of the urogenital diaphragm. The *superior* fascia takes part in the connective tissue lining of the ischiorectal fossa. The *inferior* fascia (also called *membrana perinei*) faces the perineum. The root of the penis lies in the connective tissue space between the inferior fascia and the *superficial fascia of the perineum* **A8**, in the *superficial perineal space*.

A9 Scrotum, **AB10** internal obturator muscle, **AB11** sacrotuberal ligament, **AB12** socket of the hip joint (acetabulum), **AB13** tendon of the internal obturator muscle, **AB14** sacrum, **AB15** gluteus maximus muscle, **AB16** tendon of the piriformis muscle, **B17** branches of the posterior scrotal vessels, **B18** posterior scrotal nerves, **B19** perineal artery, **B20** pudendal nerve, **B21** inferior rectal artery, **B22** internal pudendal artery, **B23** inferior rectal nerves, **B24** deep dorsal vein of the penis, **B25** urethra, **B26** bulbourethral gland.

For the branches of the internal pudendal artery see the internal pudendal artery p. 71 and the inferior rectal artery p. 315. For branches of the internal pudendal vein see internal pudendal vein p. 71. For branches of the pudendal nerve, see p. 70 and Vol. 3.

Pelvic floor and perineum
in male, from below

A
Superficial layer:
left – fascia preserved
right – fascia removed

B
Fascia and first part
of corpus spongiosum
penis removed

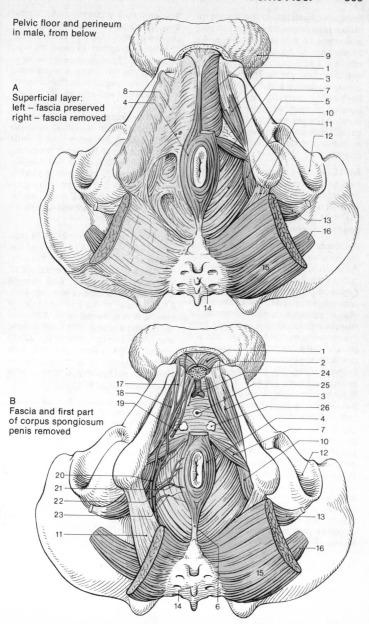

Sphincter muscles. In the female the **bulbospongiosus muscles A1** on each side cover the bulb of the vestibule **A2**. Fibers pass to the undersurface of the clitoris **A3** and into the mucosa of the vestibule of the vagina. The two muscles are connected by a *raphe* behind and in front of the vestibule. Isolated muscle bundles on each side pass beyond the dorsal raphe into the contralateral external sphincter ani **AB4** to form a figure-of-8. **Ischiocavernosus muscles A5** (see p. 308).

AB6 Levator ani muscle, **B7** coccygeal muscle from below shown by division of sacrospinal ligament **B8**; **B9** deep transverse perineal muscle partly removed on the right, **A10** superficial transverse perineal muscle.

Perineum. The prominent soft tissue bridge between the anus and the origin of the scrotum, or of the posterior commissure of the labia majora is called the **perineum**. In the center of the perineum is a tendinous layer, the *centrum tendineum perinei* **A11**, which is formed by muscular and tendinous radiations from the following muscles: both levator ani muscles **AB6**, the deep **B9** and superficial **A10** transverse perineal muscles, the bulbospongiosus muscle **A1** and the external sphincter ani **AB4**. Destruction of the tendinous center will interfere with the function of many muscles. On their external (inferior) surface the perineal muscles are covered by the *superficial perineal fascia*, which is attached at the posterior margin of the deep transverse perineal muscle. The fascia adheres laterally to the periosteum of the ischium and the pubic crus and radiates into the superficial abdominal fascia in front (see p. 308). Pathologic processes can spread from the perineum to the lower abdomen both beneath and above this fascia.

Loading on the Pelvic Floor

In quadrupeds the viscera are carried by the abdominal wall as in a hammock and the pelvic floor is drawn inward. In contrast, the viscera of man with his upright posture press on the floor of the pelvis. The following factors increase the load on the pelvic floor: raised intra-abdominal pressure on coughing or straining (defecation), relaxation and flabbiness of the abdominal wall and the diminishing suction effect of the lungs with advancing years. The closure of the lower aperture of the pelvis must withstand all these strains and at the same time it must permit opening of the pelvic apertures. There is, therefore, a risk that intra-abdominal pressure may push the pelvic viscera through their own openings (prolapse of certain organs, e. g., uterus, rectum) if the pelvic floor and its muscles are incompetent.

AB12 Sacrotuberal ligament, **A13** posterior labial nerves, **A14** artery of the vestibular bulb, **AB15** orifice of the vagina, **AB16** orifice of the urethra, **AB17** greater vestibular gland, **A18** perineal artery, **AB19** socket of the hip joint, **AB20** tendon of the internal obturator muscle, **A21** pudendal nerve, **A22** internal pudendal artery, **A23** inferior rectal artery, **A24** inferior rectal nerves, **AB25** tendon of the piriformis muscle, **A26** gluteus maximus muscle, **B27** transverse perineal ligament, **B28** anus, **B29** anococcygeal ligament.

For the branches of the internal pudendal artery compare with the internal pudendal artery p. 71 and the inferior rectal artery p. 239. For branches of the internal pudendal vein see p. 307. For branches of the pudendal nerve see p. 71 and Vol. 3.

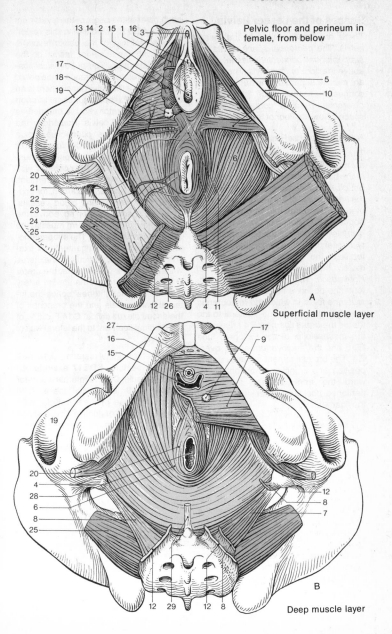

Pelvic floor and perineum in female, from below

A

Superficial muscle layer

B

Deep muscle layer

Spaces of the Lesser Pelvis

There are three spaces in the lesser pelvis. The *peritoneal space* meets the *subperitoneal connective tissue space* above the floor of the pelvis. The levator ani muscle separates the subperitoneal connective tissue space from the *ischiorectal fossa*, a connective tissue space below the floor of the pelvis.

Peritoneal Cavity in the Female

The peritoneum of the anterior abdominal wall **ACE1** stretches over the surface of the bladder. It forms the *vesicouterine pouch* **ADE2** between the bladder and the anterior wall of the uterus. Behind the uterus it descends into the *rectouterine pouch* **ADE3**, (*pouch of Douglas*), which is the lowest point of the peritoneal cavity. It extends as far down as the posterior fornix of the vagina to reach approximately the level of the *rectal fold of Kohlrausch*. The entrance into the rectouterine pouch is narrowed by the *rectouterine folds*, in which the rectouterine muscles run (see p. 296). Lateral to the uterus the *broad ligament* **E4** arises with the *mesosalpinx* of the Fallopian tube **ED5** and the *mesovarium* of the ovary **E6**. The broad ligament is transverse in position and is inclined horizontally toward the front, thereby forming a peritoneal pocket on either side in front of the ligament.

Subperitoneal Space in the Female

The *subperitoneal space* **CDE7** is part of the retroperitoneal space and extends from the symphysis, *retropubic space* **AD8**, back to the sacrum **DE9**. It is called the *paracystium* around the bladder, the *parametrium* around the uterus, the *paracolpinum* around the vagina and the *paraproctium* around the rectum. The space is lined by the *pelvic fascia* which covers the viscera in the pelvis. The fascia is fixed to the ventral, lateral and dorsal bony walls of the lesser pelvis. The superior and inferior fascia of the pelvic diaphragm are connected with each other at the origin of the levator ani muscle **ACD10**. Medially in this region the visceral leaf of the pelvic fascia is fused with the superior fascia of the pelvic diaphragm. The *pubovesical ligaments* are reinforced strands of the pelvic fascia. Above the pelvic fascia there is an opening into the obturator canal through which the obturator vessels and nerves leave the lesser pelvis. Posteriorly the pelvic fascia forms connective tissue arches around the anterior foramina of the sacrum, and covers the piriform muscle **DE11**. Between the posterior wall and the floor of the lesser pelvis remains a gap, the *infrapiriform foramen*, for passage of the nerves and blood vessels. There is a smaller opening above the upper margin of the piriform muscle: the *suprapiriform foramen*. The subperitoneal space contains the ligaments for fixation of the viscera, particularly the uterus within the lesser pelvis. Here, the ureter **D12** crosses beneath the uterine artery **D13**. In the subperitoneal space the internal iliac vessels and the branches of the sacral plexus divide. **C14** Fixation of the cardinal ligament to the lateral wall of the pelvis.

A15 Urethrovaginal septum, **A16** rectovaginal septum, **CDE17** external obturator muscle, **CDE18** internal obturator muscle, **C19** ischiorectal fossa, **C20** deep transverse perineal muscle, **C21** subcutaneous, connective tissue space beneath the deep transverse perineal muscle, **D22** ovarian ligament, **DE23** lower ramus of the pubic bone.

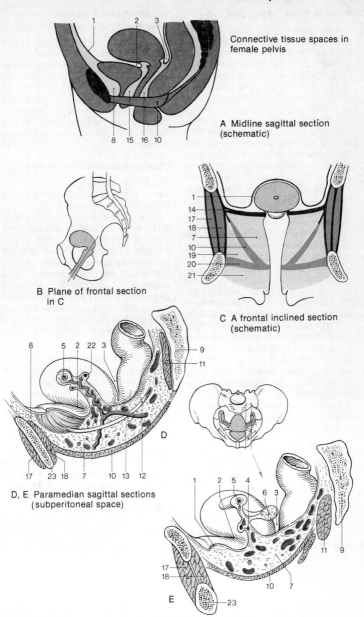

Connective tissue spaces in female pelvis

A Midline sagittal section (schematic)

B Plane of frontal section in C

C A frontal inclined section (schematic)

D, E Paramedian sagittal sections (subperitoneal space)

D

E

Ischiorectal Space in Female and Male

The connective tissue space below the pelvic floor is divided by the levator ani and external sphincter ani muscles into the right and left **ischiorectal fossa CD1**, which are filled with adipose tissue. The fat forms a movable pad that is shifted downward and sideways during defecation and parturition. The medial wall of each fossa is formed by the lower surface of the levator ani muscle **CD2** and by the external sphincter ani **D3**. These muscles are covered by the lower fascia of the pelvic diaphragm. The lateral wall of the ischiorectal fossa consists of the lower part of the internal obturator muscle covered by its fascia and surrounded by the bony frame of the ramus and the tuberosity of the ischium. The grooved roof of the fossa is formed by the origin of the levator ani muscle. Following the course of the groove anteriorly there is an angle **C1** which is bordered medially by the levator ani muscle **C2** and at its base by the urogenital diaphragm **C4**. The *pudendal canal (of Alcock)* **D5a** is situated within the fascia of the internal obturator muscle **CD5** near the ischium. The pudendal blood vessels and nerve follow a rather complicated course: after leaving the subperitoneal space through the infrapiriform foramen, they pass round the sciatic spine and the sacrospinal ligament. Then they re-enter the ischiorectal fossa through the lesser ischiadic (sciatic) foramen and finally pass through Alcock's canal toward the symphysis. For their further course see pp. 308 and 310. **C6** Subcutaneous connective tissue space, superficial perineal space, in the male beneath the deep transverse perineal muscle. **C7** Superficial perineal fascia.

Peritoneal Cavity in the Male

The peritoneum lines the *rectovesical pouch* **A8** behind the bladder, the lowest point in the peritoneal cavity in the male. The pouch extends to the fundus of the seminal vesicles at the level of the *transverse fold (of Kohlrausch)* of the rectum.

In exceptional cases the rectovesical pouch may extend down behind the seminal vesicles to the prostate. Lateral to the viscera in the lesser pelvis the peritoneum is slightly elevated by the ductus deferens which runs immediately underneath it. The entrance into the rectovesical pouch is narrowed by the two *rectovesical folds* over the rectovesical muscles.

Subperitoneal Connective Tissue Space in the Male

The subperitoneal space extends from the symphysis *(retropubic space* **A9**) to the retroperitoneal space **A10** in front of the sacrum. It includes the *paracystium* **C11** next to the bladder and the *paraproctium* **D12** next to the rectum. The prostate and seminal vesicles lie in the subperitoneal space. The ureter crosses underneath the vas deferens. The *puboprostatic ligaments*, reinforced strands of the pelvic fascia, firmly attach the prostate to the pubic bone. The rectovesical muscles run in the *rectovesical folds*. **A13** Urorectal septum, **C14** external obturator muscle.

Veins of the lesser pelvis. The subperitoneal connective tissue space in both sexes contains an extensive *venous plexus*. It is mainly situated around the rectum and internal female sexual organs or the prostate. The vessels drain mainly into the internal iliac vein **E15**, but are connected with the afferent portal venous system via the superior rectal vein **E16**. A further connection with the veins of the abdominal wall is via the inferior epigastric vein **E17**. **E18** External iliac vein, **E19** internal pudendal vein, **E20** levator ani muscle.

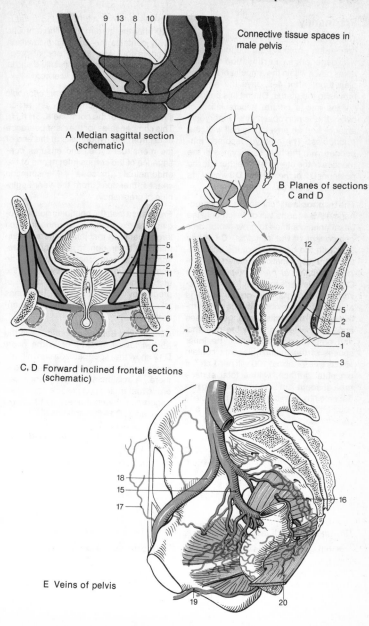

Connective tissue spaces in male pelvis

A Median sagittal section
(schematic)

B Planes of sections
C and D

C, D Forward inclined frontal sections
(schematic)

E Veins of pelvis

Pregnancy

The ovum must be fertilised within 6–12 hours after eruption from the follicle or it will degenerate. **Fertilisation** usually takes place within the ampulla of a Fallopian tube. After 4–5 days the embryo moves to the uterus. During this period it divides into a central accumulation of cells, the *embryoblast* (the embryonic anlage) and an enveloping cell wall, the *trophoblast*. The embryo becomes embedded with the help of enzymes in the mucosa of the uterus **AB1**, either into the posterior **D2** or the anterior **D3** wall of the cavity of the uterus.

The **trophoblast** ("chorion" of later pregnancy) **AB4** forms *villi* of which only the basal ones continue to grow (*embryonic component of the placenta*) **C4**. They fuse with the basal part of the endometrium; the combination is called the *"decidua" (maternal part of the placenta)* **C5**. The two parts together form the disc-shaped **placenta**, the organ which is attached to the embryo through the umbilical cord. It permits exchange of gases and metabolites between the maternal and embryonic circulations. The inner spaces of the mature placenta contain villous trees – about 100 villi per 1 cm^2 of placental surface, with a total surface area of about 7 m^2. After the embryo has become firmly embedded, it is possible to distinguish the *basal, capsular* and *parietal decidua*. **ABC6** Cavity of the uterus, **ABC7** uterine muscle.

In the **embryoblast** a cavity forms above and below the embryo – the yolk (vitelline) sac and the amniotic sac. The *vitelline sac* **C8** degenerates into a vesicle but the *amniotic sac* **BC9** grows together with the *embryo* **AB10**, called a *fetus* **C10** after the third month. The amniotic cavity contains the *amniotic fluid* which amounts to 1 liter at the end of pregnancy. The fetus floats in this fluid and is attached only by the umbilical cord.

Hormones. The embryo forms *chorionic gonadotropins* which have an action similar to that of the hormone ICSH (LH). This compensates for the disappearance of LH from the hypophysis at the end of the menstrual cycle, and ensures continuance of the corpus luteum and of the endometrial mucosa. Menstruation ceases, therefore, from the very beginning of pregnancy.

Pregnancy test: the chorionic gonadotropins are excreted in the urine in such large quantities that infantile female animals can be brought to maturity by injection of urine from pregnant women. The *corpus luteum of pregnancy* keeps the uterus quiescent until the 5th month of pregnancy when placental hormones take over and the corpus luteum degenerates.

Clinical tips: Ectopic pregnancies in the *abdominal cavity* **D11** or in the *ovary* **D12** show that spermatozoa can migrate into the abdominal cavity and fertilise ova there. A misimplanted pregnancy can erode maternal blood vessels and cause dangerous hemorrhages. **D13** *Tubal pregnancy*. **D14** Implantation of the fertilised egg in the isthmus leads to *placenta previa*, i. e., the placenta is lodged proximal to the birth canal.

Growth in length of the embryo or fetus:

1 month	0.8 cm vertex to coccyx
2 month	3 cm vertex to coccyx
3 month	7 cm vertex to heels
4 month (4×4) 16	cm vertex to heels
5 month (5×5) 25	cm vertex to heels
6 month	(6×5) 30 cm vertex to heels
7 month	(7×5) 35 cm vertex to heels
8 month	(8×5) 40 cm vertex to heels
9 month	(9×5) 45 cm vertex to heels
10 month	(10×5) 50 cm vertex to heels

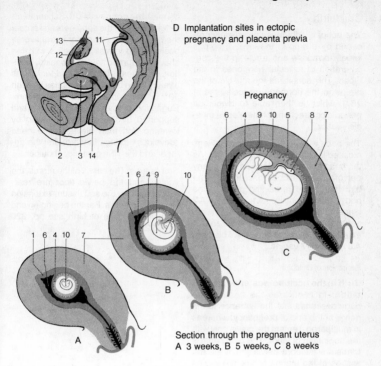

D Implantation sites in ectopic pregnancy and placenta previa

Pregnancy

Section through the pregnant uterus
A 3 weeks, B 5 weeks, C 8 weeks

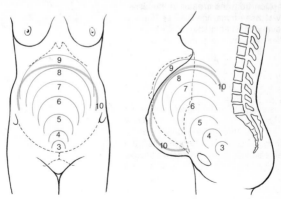

E Position of uterus during course of pregnancy (after Eufinger)
1st–10th lunar months

Childbirth

The act of childbirth is immediately preceded by a drop in the level of corpus luteal hormones and a rise in the concentration of follicular hormones in the blood. This sensitizes the muscle of the uterus to the hormone oxytocin (see p. 154), which is then able to trigger off periodic contractions of the uterine muscle – *labor*.

The baby with its body bent and its legs crossed is "parcelled" ready for birth. As the head has the largest diameter of any part of the baby's body, once it leaves the birth canal the remaining parts of the body follow easily. **A1** Uterus, **A2** placenta (the umbilical cord is not visible behind the baby), **A3** internal orifice, **A4** external orifice of the cervix, **A5** bladder, **A6** rectum **A7** vagina.

Mechanism of Childbirth

Birth in the occipito-posterior presentation: In primiparae the head of the baby descends into the entrance to the pelvis at the end of pregnancy, whereas in multiparae it enters the pelvis only with the beginning of labor. The *birth canal* is formed by the bony pelvis and by the soft tissues of the uterine cervix, the vagina and the floor of the pelvis. In the normal female pelvis the *pelvic inlet* (i. e., the transition from the greater to the lesser pelvis, the terminal line **B8**, see Vol. I) is an oval with a transverse axis. The *pelvic outlet* between the symphysis **C9**, the tuberosities of the ischium **B10** and the retracted coccyx **C11** (see Vol. 1) is an oval with a longitudinally directed axis. The position of the longest diameter of the baby's head, i. e., its sagittal diameter, is adjusted to the longest diameters of these ovals. Thus, the head of a baby on its way through the pelvis has to undergo a 90° screw turn. Further, the head also follows the anteriorly concave *guideline* of the pelvis and of its *soft tissue "extension tube"* **C12**. In this way the head is guided from its initial flexed position into extension before passing

beneath the symphysis **C9**. Subsequently the breadth of the shoulders first enters the pelvis with its diameter in transverse position and leaves the pelvis with the shoulders turned into the sagittal position. This means that the head, which has already been born, again rotates through 90° in the same direction as before. During this process the head is held and supported by the obstetrician, who by lowering and then elevating the head "develops" (brings forward) first the anterior and then the posterior shoulder.

Soft tissues: The cervix of the uterus, the vagina and the pelvic floor are rearranged to form the soft tissue "extension tube" of the pelvis. For the beginning and expulsion phases of birth see pp. 320 and 322.

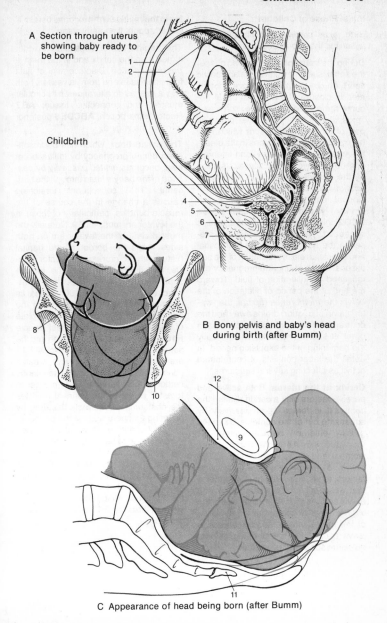

A Section through uterus
showing baby ready to
be born

1
2

Childbirth

3
4
5
6
7

8

10

B Bony pelvis and baby's head
during birth (after Bumm)

12

9

11

C Appearance of head being born (after Bumm)

Initial Phase of Childbirth

Birth takes place in two phases – the *initial* and the *expulsion phases.*

During the **initial phase** (opening phase) the soft tissues (cervix, vagina and pelvic floor) which up until then have cooperated in closure of the genital canal become remoulded into a wide, *soft tissue "extension tube".* This initial phase is part of the process of labor which now begins. Muscle contractions in the uterus follow each other at fairly long intervals and do not involve any conscious effort by the mother. By the labor activity an *amniotic sac* **C1** is formed. It is a protrusion of fetal membranes (chorion and amnion) lying in front of the baby's head **BCD2** into which the amniotic fluid is squeezed. The fixed amniotic sac precedes the baby through the birth canal and effects an elastic dilatation of the soft tissues, which during the pregnancy are loosened by infiltration of fluid. Toward the end of the period of dilatation of the cervix the *membranes rupture*, the "water" (amniotic liquor) drains away and the contractions of labor begin to follow each other at shorter intervals. This is really the beginning of the expulsion period. In detail the latter consists of coordinated activities affecting several structures.

Cervix of the uterus. Both active and passive factors play a part in dilating the cervix. It is forced open *passively* by squeezing out of the contents **C3** of the grossly enlarged cervical glands **B4** (compare with **A4** in the non-pregnant woman) and of the venous plexus. The *active* reshaping is due, among other factors, to the pull of the muscle bundles that descend from the uterus into the cervix and ascend to it from the wall of the vagina. Rearrangement of the structure of the more circular muscle fibers also plays a part. The parametrium affords sufficient space for the dilatation.

In the *primipara* the cervix is opened up step by step from the internal **CDE5** to the external **BCDE6** orifice of the cervix but in the *multipara* the external ostium is already open at an early stage.

The **vagina** has a considerably wider lumen than the cervix and in the main is *passively* dilated. Displacement of fluid from the tissue and from its vessels both play a role, as do alterations in its circular muscle, and connective tissue. **AB7** Rectouterine pouch, **ABCDE8** posterior fornix of the vagina.

The **pelvic floor**, which has been softened during pregnancy by infiltration of fluid, becomes *dilated passively by passage of the baby's head* through the birth canal. The consequent stretching causes a change in the course of the muscle bundles, particularly of those in the levator ani muscles **F9**. Normally the muscle shelf of the levator ani on both sides forms the border of the genital hiatus with the crura of the levator muscle, but during parturition the muscle shelf of the levator ani is pushed downward and turned by about 90°, so that its upper surface comes to lie against the birth canal. The transverse muscle plate of the deep transverse perineal muscle **F10** is altered in a similar fashion and the sagittal bulbospongiosus muscles **F11** are widened into a ring. This creates considerable tension on the tendinous center of the perineum **F12**. In order to forestall a tear in the perineal muscles the obstetrician conteracts this drag by pressing 2 fingers against the perineum *(support of the perineum)*. If the tension is considered to be very marked a tear in the perineum can be prevented by an incision in it to relieve the pull *(episiotomy)*. After parturition the structures of the pelvic floor are restored to their original shape and position. **F2** Head of the baby, **F13** external sphincter ani muscle.

Section through the cervix uteri

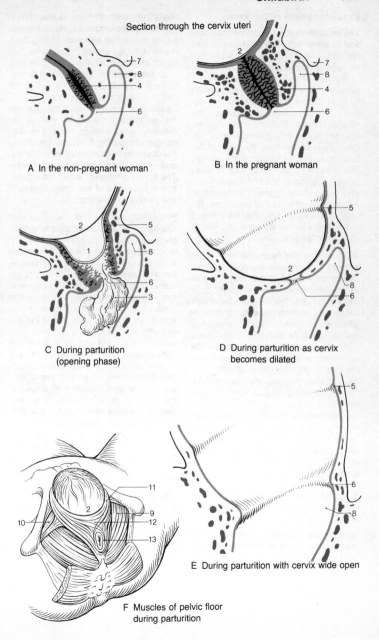

A In the non-pregnant woman

B In the pregnant woman

C During parturition
(opening phase)

D During parturition as cervix
becomes dilated

E During parturition with cervix wide open

F Muscles of pelvic floor
during parturition

Expulsion Phase of Childbirth

The **expulsion phase** commences after the external orifice of the cervix has opened. The *muscular contractions of labor* get stronger and follow each other at shorter intervals. The mother aids expulsion of the fetus by actively pushing with her abdominal muscles in the rhythm of labor *(expulsion pains)*. The muscle fibers of the uterus are considerably shortened during expulsion so that they are retracted over the head of the baby towards the fundus of the uterus. The muscle of the uterus has its fixed point in the anchorage of the cervix uteri and in the round ligament of the uterus **A1** on each side. **A2** Fallopian tube, **B3** urethra, **B4** vulva, **B5** anus, **B6** external orifice of the cervix, **B7** internal orifice of the cervix, **B8** placenta.

After *birth of the head*, delivery of the anterior and posterior shoulders and expulsion of the rest of the body follow in quick sequence. The *umbilical cord*, which connects the newborn baby with the as yet unborn placenta, is then ligated and severed. The CO_2 which has accumulated in the blood of the newborn child activates the respiratory center in its brain and the baby begins to breathe with the *first "cry"*. Simultaneously the fetal circulation switches over to the postfetal circulation (see p. 324). The obstetrician will check on the signs of *full maturity* of the newborn; body length not less than 48 cm, weight at least 2500 g, good turgor of the skin, scalp hair 2 cm long, nails covering the tips of the fingers, complete descent of the testes into the scrotum, the labia majora cover the labia minora, etc. It is also important to make sure that there are no malformations present that could endanger the life of the baby.

Birth of the placenta. After the birth of the child the uterus contracts to about 15 cm in length, the fundus being at the umbilical level. Moreover, the solid placenta detaches from the uterine wall and moves into the inferior uterine segment. The contracting uterus feels firm, it "swells". Muscle contractions and straining by the mother and manual help from the obstetrician deliver the placenta 1–2 hours after birth of the child. Bleeding from the placental site in the endometrium is controlled by contraction of the uterus. If hemostasis is incomplete, e. g., because a piece of placenta has remained stuck to the uterine wall, or because of general atony, the mother is in danger of bleeding to death. Therefore, the obstetrician must be satisfied that the maternal part of the placenta is complete. The fetal membranes which have been torn by bursting of the amnion remain attached to the placenta.

Involution. Involution of the uterus and restitution of its normal structure take place during the *puerperium* (the immediate post partum period). The perineal muscles retract into their original position in a matter of hours. The size of the uterus is quickly reduced by degenerative changes: after only 10 days the fundus has descended to the level of the symphysis, the placental site in the endometrium becomes covered by epithelium and the internal orifice of the cervix closes. Up to that moment there is a discharge from the endometrium, called the *lochia*, which is at first bloodstained but later is paler and serous in nature. The body mobilizes local and general defense mechanisms as a protection against ascending infections (e. g., *puerperal fever*). The blood vessels of the uterus are partly adapted to the reduced requirement for blood flow by thickening of their walls, and partly they disappear.

The size of the uterus. Red = immediately after parturition; blue = 5 days, black = 12 days after parturition.

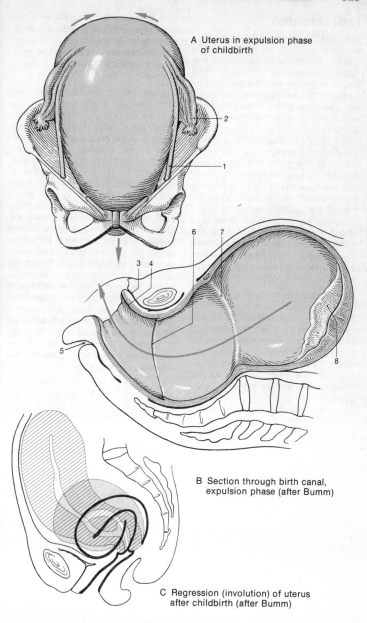

A Uterus in expulsion phase
of childbirth

B Section through birth canal,
expulsion phase (after Bumm)

C Regression (involution) of uterus
after childbirth (after Bumm)

Fetal Circulation

Placenta. The fetus receives oxygen, nutrients etc. from the maternal blood into which it returns CO_2 and waste products of metabolism. Proteins can also be exchanged between the fetus and the mother (e. g., immunoglobulins and insulin). The *placenta* **A1** is the connecting organ through which all materials are exchanged. The *villi of the placenta* (with a total surface of about 7 m^2) float in small pot-shaped chambers filled with maternal blood. The gases and solutes must pass right through the epithelial cell layer (*syncytiotrophoblast* and *cytotrophoblast*), through the connective tissue and through the capillary wall of the villi (= "placental barrier"). Usually there is no mixing of maternal and fetal blood. The mother and her baby may have different blood groups. In fact, loss of blood from the fetal placental vessels into the maternal circulation may occur.

Fetal circulation. Fetal blood which has taken up oxygen in the placenta flows through the *umbilical vein* **A2** beneath the liver. A small proportion flows through the liver **A3,** but the major part by-passes that organ and flows via a primary short circuit, the *ductus venosus (duct of Arantius)*, into the *inferior vena cava* **A4**. There it mixes with the markedly deoxygenated venous blood from the inferior vena cava **A5**. The blood in the right atrium is directed by the valve of the vena cava toward the septum of the atria and flows through a second short circuit, the open *foramen ovale* **A6**, into the left atrium. Blood from there either passes via the left ventricle and the aorta **A7** into the body of the fetus or via the two *umbilical arteries* **A8** back into the placenta. Deoxygenated (venous) blood which returns from the head and the arms via the *superior vena cava* **A9** into the right atrium crosses the bloodstream which comes from the inferior vena cava to reach the right ventricle. From there it flows through the *pulmonary trunk* **A10** and via a third short circuit, the *arterial duct (of Botallo)* **A11**, into the aorta and so into the peripheral arterial circulation. Very litte blood flows through the lungs. **AB12** Pulmonary arteries, **AB13** the pulmonary veins, **AB14** the superior vesical arteries.

Changeover of the circulation after birth (compare **A** with **B**). The change from the fetal to the postfetal circulation takes place after birth. The blood pressure in the aorta rises with the *closure of the umbilical blood vessels* **AB15**, which is brought about by contraction of the vessel wall and ligation of the umbilical cord. The partial pressure of CO_2 in the baby's blood rises and stimulates the respiratory center in the brain. With the onset of pulmonary breathing (the first "cry") the lung expands and the resistance in the pulmonary circulation drops. The pressure gradient causes a reversal of the blood flow in the arterial duct and the pulmonary circulation becomes filled with blood. The backflow of blood from the lungs increases this pressure in the left ventricle and effects *mechanical closure of the foramen ovale* by drawing its flap-shaped borders closely together. The *arterial* and *venous ducts close*, mainly by *contraction of the muscles in their walls*.

The umbilical vein **A2** becomes the *ligamentum teres hepatis* **B2**, the venous duct becomes the venous ligament of the liver **B3**, the ductus arteriosus **A11** becomes the ligamentum arteriosum **B11**, the umbilical arteries **A8** become obliterated to form the medial umbilical ligaments **B8**.

B4, 5 Inferior vena cava, **B7** aorta, **B9** superior vena cava, **B10** pulmonary trunk.

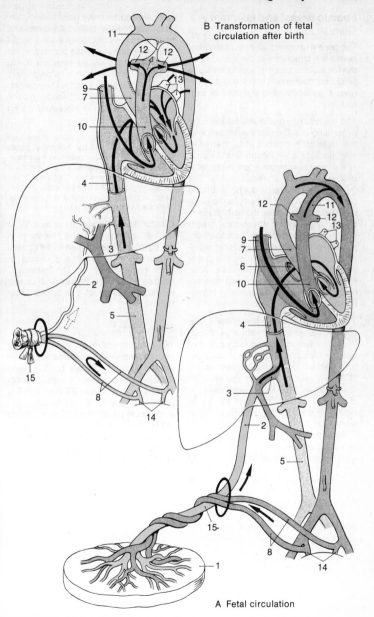

B Transformation of fetal circulation after birth

A Fetal circulation

Female Breast and Mammary Gland

The **breast** (*mamma*) and the **mammary gland** are structures of the skin, but as the female mammary gland has a close functional connection with the genital organs it is discussed in the present context.

The *mammary gland* of the newborn (of either sex) is still under the influence of the maternal hormones. It is, therefore, still relatively large for the first 3 weeks and it may discharge a few drops of a milklike secretion ("witch's milk"). The gland grows only slowly during childhood but with the onset of puberty its development accelerates. After the early stage of a *"budding" breast*, the typical breast shape develops later with increased deposition of fat. The gland grows considerably during pregnancy, toward the end of which it begins to secrete milk. After weaning the gland undergoes spontaneous regression with increasing formation of connective tissue.

The **mature breasts** are shaped like slightly deformed hemispheres (bowl, hemisphere or cone-shaped breasts, frequently race specific). The breasts extend from the level of the 3rd through 7th rib in the middle between the sternum and the axilla. They are *movable* over the chest wall against the pectoral fascia (see **D**) and they are approximately *symmetrical* in their position and formation. Frequently a process of the mamma extends across the margin of the pectoral muscle into the axilla (see **C**). The groove between the 2 breasts is called the **sinus mammarum**. The **nipple A** is 10–12 mm high and lies just below the center of the breast. It points a little upwards and sidewards and the milk ducts open on its surface on 12–20 orifices like pores. If the nipple is flat or retracted *(sunken nipple)* a baby may not be able to suck on it properly. The puckered skin of the papilla and its surrounding circular area, the *areola*, are usually somewhat more dark-ly pigmented than the neighboring skin. This is particularly marked in women who have borne a child. Only the tip of the nipples remains free from pigment. In the periphery of the areola there are 10–15 tubercles *(areolar glands, Montgomery glands)* arranged in a circle. They contain apocrine, eccrine and sebaceous glands and their secretion is increased during lactation.

Variants: the presence of more or less well developed supernumerary mammary glands *(accessory mammae)* is known as *hypermastia*. If only accessory papillae are present, this is known as *"hyperthelia"* or *"polythelia"*. The accessory glands also take part in the changes which accompany puberty and pregnancy. Their positions are shown in **E**.

Clinical tips: Disturbances of mobility or marked asymmetry of the breasts or of the position of the nipple may be caused by disease of the breast (e. g., cancer) or of the locomotor system. The figures in **C** indicate the *relative frequency of cancer* in each quadrant of the breast (after Bailey).

Male Breast

The basic structure of the male breast and mammary gland corresponds in principle to that of the female, but it remains relatively undeveloped. The gland is only about 1.5 cm wide and 0.5 cm thick. It is possible to develop the mammary gland in the male by administration of female sex hormones. During *puberty* the breast may temporarely show a more marked development – *gynecomastia*. Rudiments of supernumerary mammae can also occur in men.

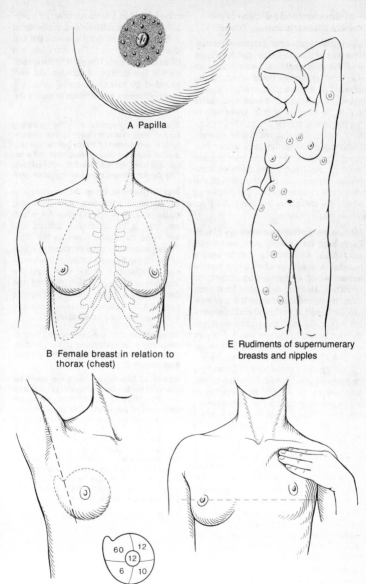

A Papilla

B Female breast in relation to thorax (chest)

C Extension of mammary gland into axilla and frequency of breast cancer (after Bailey)

D Mobility of breast

E Rudiments of supernumerary breasts and nipples

Fine Structure and Function of the Female Breast (Mammary Gland)

Glandular size and adipose tissue. The size of the *glandular part* **A1** varies less than the total size of the breast as the latter is largely determined by the amount of *adipose tissue* **A2** in it. The glandular part rests on the *pectoral fascia* **A3**. It is thickest at the lower lateral portion and is composed of 12–20 coniform lobes. Connective tissue strands (*retinacula*) divide the fatty tissue into small chambers that extend from the skin to the pectoral fascia. The *tautness* of the breast depends on the condition of the connective tissue and on the degree to which the chambers are filled. **A4** Pectoralis major muscle.

Mature, nonlactating mammary gland. Each lobe has a milk duct, *lactiferous duct* **AB5**, an epithelial tubule with a number of small branches and a narrow lumen, which is sometimes absent. Its branches **AB1** are separated from each other by connective tissue **B6** and they are thickened at the distal end to resemble a bud. Beneath the nipple each lactiferous tubule broadens into a 1–2 mm wide spindle-shaped ampulla, *lactiferous sinus* **A7**, which leads into the lactiferous duct at the base of the papilla. During the *menstrual cycle* after ovulation the breast enlarges by 15–45 cm^3 by sprouting of the lactiferous ducts. The enlargement reaches a climax before menstruation and returns to normal by the 7th day of the cycle.

Lactating mammary gland. During pregnancy the lactiferous ducts proliferate, there is less connective tissue and the breast undergoes a soft swelling. During the 5th month of pregnancy the terminal "buds" are transformed into *alveoli*. There is marked increase in the vascularization of the gland. In the 9th month yellowish *colostrum* begins to form which contains fat droplets and cell debris. About 3 day after childbirth the production of milk begins. The fat globules of the milk, produced by apo-crine secretion, are surrounded by a protein membrane and have a diameter of 2–5 μm. They expand the alveoli **B8** to a diameter of 1.2 mm. The ampullae are dilated to 5–8 mm. The walls of the alveoli and of the lactiferous tubules are surrounded by smooth muscle cells, the *myoepithelium*, which helps to empty the breast.

Spontaneous regression. The **weaning period** leads to accumulation of milk and distension and rupture of alveoli before milk production ceases. Phagocytes appear in the tissue and take part in removal of milk residues. The glandular tissue undergoes partial atrophy.

Nipple (papilla). Deep to the nipple and its areola is a system of circular and radial smooth muscle bundles **A9**, which are fixed to the lactiferous tubules and the veins by elastic fibers in the skin. These muscle bundles produce erection of the nipple by contraction of the areola and by dilatation of the veins and the lactiferous tubules. When *feeding*, the baby empties the ampullae by alternate pressure from the lips and the gums and subsequently they refill with milk.

Galactogram. This can be important in diagnosis and in demonstration of the glandular ducts on a radiograph after injection of a contrast medium **C**.

Effects of hormones. *Estrogens* affect the growth of the tubular system and prepare it for the *effect of progesterone*, which induces development of the alveoli. Estrogens and progesterone, also produced by the *placenta*, tend to suppress milk production. After reduction in the flow of these hormones at the end of pregnancy, *prolactin* stimulates milk production. Release of milk is controlled by *oxytocin* which causes contraction of the myoepithelia in the gland. The release of prolactin and of oxytocin is maintained by tactile stimulation of the nipples (*neurohormonal reflex*).

For the vessels, see pp. 53f., 77 and 81.

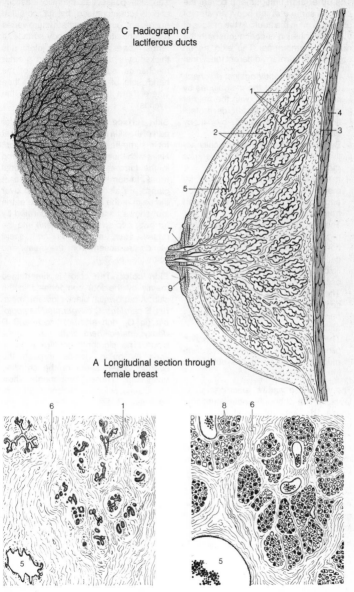

C Radiograph of lactiferous ducts

A Longitudinal section through female breast

B Histology of mammary glands;
left – at rest (inactive)
right – during lactation

The **outer skin** (integument) covers the external surface of the body, an area of about 1.6 m² in the adult. At the orifices of the body the skin is continuous with the mucous membranes. The skin as an organ performs many different functions:

- It *protects* the body against *mechanical, chemical* and *thermal injuries* by its epithelium which, with the help of glandular secretions, affords protection against invasion by many pathogenic microorganisms.
- By its content of immunologically active cells it takes part in the *defense mechanisms of the body*.
- It *maintains body temperature* by means of variable blood circulation and by the discharge of fluids (sweat) from its glands.
- The skin also plays a controlling role in *regulating water balance*, because it both protects the body against desiccation and provides a route for loss of fluid and mineral salts via the secretion of its glands.
- The skin, through its sensory innervation, acts also as *sense organ* for the *perception of pressure, temperature and pain*.
- The facilities of blushing, pallor, piloerection etc. make the skin a *communicating organ* of the autonomic nervous system.

The skin also shows electrical resistance, which changes under emotional stress, because of emotionally conditioned secretion by the skin glands. This phenomenon forms the basis of the *lie detector test*. The skin is of special medical interest because more than any other organ it is accessible to observation and because it contributes to the symptomatology of many systemic diseases, e. g., cyanosis in heart disease, circumscribed erythemata and rashes in infectious diseases etc.

Particularly when clothes made of artificial fibers are worn in a dry atmosphere, the skin may become electrostatically charged up to a potential of 1000 volts.

The skin possesses intrinsic elasticity and is characterized by its cornified epithelium, but its overall structure varies at different sites on the body surface. It forms *"spare folds"* over the joints, it is thicker over areas exposed to greater mechanical stress (e. g., palm of the hand, sole of the foot, the back) than where there is less stress (e. g., the eyelids).

Skin surface (epidermis). The greater part of the skin shows *polygonal areas;* at their summits the sweat glands open. Hairs with their sebaceous glands stand in the furrows between the polygonal areas. In some parts of the body scent glands are also present. The skin over the palm of the hand and the sole of the foot shows *parallel ridges* separated by furrows, and sweat glands lead into the ridges. Hair, sebaceous and scent glands are absent from the palms and soles (see p. 336).

Skin color. The color is determined mainly by the four components in the skin: **A** a brownish black pigment, *melanin;* **B** *carotene;* **C** *oxygenated hemoglobin* (in O_2 rich arterial blood) and **D** *deoxygenated hemoglobin* in venous blood. The pigments acquire a bluish tinge by being viewed through the opaque upper layers of the cornified epithelium. These components show characteristic local differences in concentration, which may be due to the environment, e. g., solar radiation or food rich in carotene.

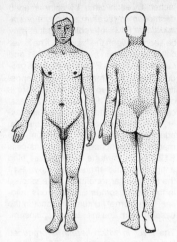

A Melanin

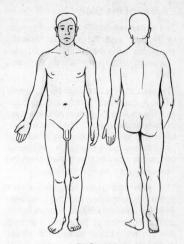

B Carotene

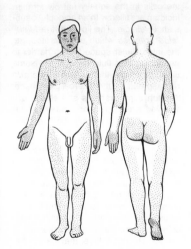

C Arterial blood

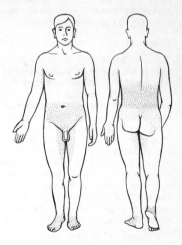

D Venous blood

Distribution of color components in living skin
(after Edwards and Duntley)

Layers of the Skin

The **skin covering** (integument) consist of the *skin* and the *subcutis*. Specific structures in the skin include the cutaneous appendages – *skin glands, hairs* and *nails.*

The **skin** (cutis) consists of the epidermis **AC1**, a stratified, cornified, squamous epithelium, and the corium **AFC2**, a layer of connective tissue. The corium contains the *papillary layer,* in which the papillae are interlocked with the epidermis, and a *reticular layer,* which gives the skin resistance against tearing.

The **subcutis** (*tela subcutanea*) **A3** forms the connection with the structures which lie beneath the skin, such as fascia and periosteum. It frequently contains fat. The major blood vessels and nerves of the skin lie in the subcutis (see p. 336).

Glabrous skin: A4 eccrine sweat glands, **A5** lamellar nerve endings (Paccini's corpuscles).

Hairy skin: B6 hair, **B7** sebaceous glands, **B8** muscle to erect the hair (arrector pili), **B9** apocrine scent glands.

Epidermis

The cells which originate in the basal layer of the **epidermis** undergo step by step transformation leading to their *cornification* and stratification in the epithelium. Ultimately the keratinised cells are shed from the outer surface of the epidermis. The migration of cells from the basement layer to the surface takes about 30 days. There is a layer of *regeneration,* of *formation of the keratin* and a *keratinised layer.* The first two of these layers also contain other layers of cells. The stratification of the epidermis is best developed in glabrous skin and is hardly developed at all in hairy skin.

The *regeneration layer (stratum germinativum)* consists of high basal cylindrical cells (stratum basale **F10**) and many successive layers of larger, rounder or polygonal cells which are attached to each other by spine-shaped *desmosomes (stratum spinosum or spinous layer* **F13**). Intercellular spaces may be seen between the "spines". The lower cell layers contain *melanocytes* **F11** and *Langerhans-cells* **F12.** Melanocytes produce the pigment melanin which is deposited in the basal epithelial cells. There is one melanocyte for each 4–12 basal cells. Melanin protects the basal cell layer, with its cells constantly in mitosis, against dangerous UV irradiation. The markedly branched *Langerhans-cells* **F12** of the immune system can bind antigens. In the *stratum spinosum* **F13** the intracellular tonofibrils are anchored in the *desmosomes* **E**. Numerous mitoses, with diurnal fluctuation, occur in the basal stratum and the stratum spinosum.

The *layer in which keratin is produced* includes the *granular layer* **F14** and the *stratum lucidum* **F15**. The epithelial cells of the narrow granular layer contain numerous lamellated granules and basophilic, *keratohyaline* granules, whilst the cells in the equally narrow stratum lucidum are filled with an acidophilic substance, *eleidin.* The lipids of the *lamellated granules* which are secreted into the intercellular space form a barrier to protect against fluid loss. At the same time lipids from the surface of the skin may spread in the intercellular space. Keratohyaline granules and eleidin are involved in keratinisation. Poorly keratinised hairy skin generally lacks a complete keratinising layer.

In the *keratinised layer* **F16**, the epithelial cells and the horny substance fuse into plates, which are finally desquamated as horny squames. They are acid-resistant but swell in alkalis (soap suds). The cornification is controlled by the level of vitamin A in the body; in vitamin A deficiency overcornification (hyperkeratosis) occurs. The horny layer may be 10–330 μm thick and is particularly well developed in glabrous skin.

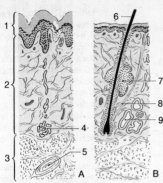

A Section through glabrous skin
B Section through hairy skin

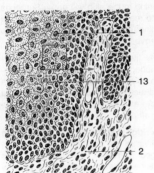

C Spinous layer of epidermis

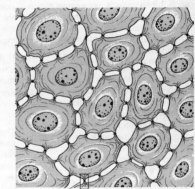

D Magnified section from C

E Electron microscopy of
section from D: desmosomes

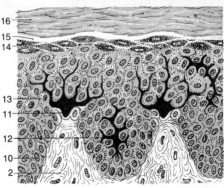

F Melanocytes in epidermis

Corium

The **corium** *(dermis)* is a dense network of collagen fibers, intermingled with elastic fibers. This enables the skin to withstand stretching and to return to its normal form after deformation. Leather is made from the corium of tanned animal skin ("leather skin"). The elasticity of the skin is mainly due to changes in the alignment and angles of the collagen fiber network of the corium. The elastic fiber network is responsible for rearrangement of the fiber layers after deformation of the skin. The following structures are located in the corium: hair roots, glands, blood vessels, connective tissue cells, free cells of the immunologic system and neural structures. On the basis of the fiber arrangement two layers may be distinguished in the corium, the *papillary layer* and the *reticular layer*.

The **papillary layer** *(stratum papillae)* borders directly on the epidermis and is interlocked with it by projections of collagen fibers, the *connective tissue papillae*. These penetrate into corresponding indentations in the epidermis and thus counteract shearing forces on the epidermis, see **B**. The height and number of the papillae are consistent with the mechanical strain on that part of the surface of the body; thus, there are very few in the eyelids, but they are well developed over the knees and elbows.

In *hairy skin*, the papillae are often poorly developed. They lie in groups connected to hair follicles and the ducts of the sweat glands, making structures such as cockade-like, epithelial ridges and rosette-like epithelial mounds.

Glabrous (hairless) skin has tall papillae. In a cross section of each ridge, 2 connective tissue papillae can be seen to project into the epithelium. The epithelial trough between the papillae carries the opening of the sweat gland duct.

The **reticular layer** is a continuation of the papillary layer with the subcutis. It contains interlacing bundles of the strong collagen fibers that are responsible for the resistance of the skin to mechanical stress. The direction of the collagen fiber network differs in different parts of the body. A pinprick in the skin produces a slit rather than a circular hole. The direction of the line of the split differs according to the different tensions in the skin. Systematic investigations have revealed a *system of preferred lines for incisions* (see **A**), which also indicates differences in tension in the skin.

Clinical tips: If an incision is made at right angles to the course of the lines, the opening in the skin will gape. Surgeons try to make incisions in the direction of the lines in order to hasten healing and to improve the cosmetic result.

If the living skin is considerably overstretched, e. g., the skin over the abdomen during pregnancy, ruptures occur in the structure of the corium which become visible as pale stripes, so-called *distension striae*.

The **ridges of the skin** are due to the papillae of the papillary layer being arranged in rows. This arrangement is genetically determined and is characteristic for each particular individual. This fact forms the basis of *finger printing* (dactylography) as a means of identification in criminal investigations. Four main but variable types of ridge patterns on the finger tips can be distinguished: *arch* **C I**, *loop* **C II**, *whorl* **C III** and *double loop* **C IV**.

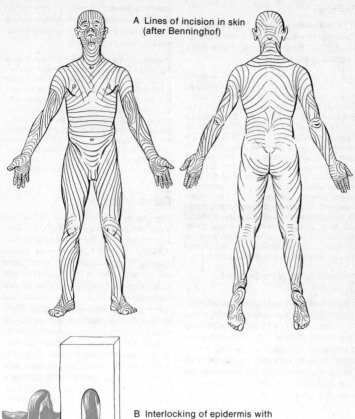

A Lines of incision in skin
(after Benninghof)

B Interlocking of epidermis with
papillae of corium
(after W. Schmidt)

C Papillary ridges on fingertip (after Wendt)

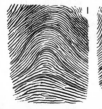

I II III IV

Subcutis

The **subcutis** provides the connection between the skin and the superficial fasciae of the body and is the basis of the mobility of the skin.

The subcutis is a fat store and an insulator. Fatty tissue is subdivided by dense connective tissue strands, like a quilt, into "shock absorbers" as an *integral part* of the structure e. g., in the sole of the foot. More often, the fat is just a *deposit*, e. g. as a fatty cushion, the panniculus adiposus under the skin of the trunk. The distribution of fat over the body surface is genetically determined and is under the control of several factors, including hormones (sex differences in local distribution of the panniculus adiposus). By means of connective tissue strands, *retinacula cutis*, the subcutis may connect the skin with the underlying tissue. In some places the subcutis is loose and free of fat, e. g., eyelids, external ears, lips, penis, scrotum etc. Over the face and scalp, the skin is fairly adherent to the underlying muscle and tendon; this is the anatomic basis of mimicry.

Blood Vessels of the Skin and Subcutis

The **arteries A1** form a network between the cutis and the subcutis from which branches descend to the hair roots and the sweat glands **A2**. Other branches ascend to the papillary bodies where they form a *subpapillary plexus* from which *capillary loops* **A3** penetrate into the papillae. The density of the loops varies from 20–60 loops per mm². The capillaries are kept open because the pressure within the tissue is lower than the blood pressure inside the capillaries.

Clinical tips: If the external pressure rises to 60–80 mmHg or more, as it may do between the skin and the bed during bedrest, then the main capillaries become occluded. If this is prolonged, then nutrition of the skin becomes so impaired that bedsores (decubitus) result.

The **veins A4** form networks below the papillae in the corium and between the cutis and the subcutis (*cutaneous venous plexus* **A5**). The rate of flow through them is influenced by **arteriovenous anastomoses.** At the ends of the fingertips, small arteriovenous glomus organs are present (see p. 42). The **lymph vessels** form three plexuses.

The blood vessels serve both for **nutrition of the skin** and **temperature control**. There is a temperature gradient between the blood which is mostly warmed in the liver and the muscles, and the peripheral blood. This is regulated by the arterioles and arteriovenous anastomoses of the skin. A higher blood flow results in a greater heat loss and in increase in temperature of the *"body shell"*. If blood flow falls, there is a reduction in heat loss and in the body shell temperature. The *"central core"* of head and trunk is always maintained at constant temperature.

Glands of the Skin

The *glands of the skin* **C** like the hair and nails, are appendages of the skin. They are situated in the corium (compare 'Glands' p. 150). The **eccrine sweat glands D**, *coiled glands*, of which there are about 2 million in the body are almost ubiquitous, but there are more on the forehead and sole. They are coiled tubules with a straight outer end. Their acid secretion impedes the growth of bacteria *(protective acid mantle)* and by evaporation aids thermoregulation. Sweat also provides a route of elimination from the body for various substances. The **apocrine sweat glands E** *(scent glands)* are present in certain circumscribed hairy areas, namely the axillae, mons pubis, scrotum, labia majora, around the anus, and at the mamma. The glands are branched and have wide, excreting acini. Their alkaline secretion contains aromatic substances. Because of the absence of the protective acid mantle at those sites infections *(sweat gland abscess)* are liable to occur. Secretion from the scent glands commences at puberty. The **holocrine sebaceous glands F** usually originate from hair rudiments and lead into the hair funnel of the hair follicle. They are berry shaped and contain a stratified epithelium. The *sebum* formed in them lubricates the hair and the adjacent skin.

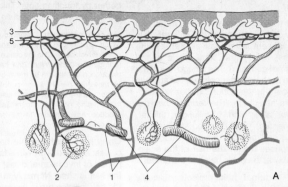

3
5

2   1   4   A

Blood vessels of skin (after Horstmann)

B   Isothermal zones in "body shell"
(after Aschoff and Wever)
left – low outer temperature
right – high outer temperature

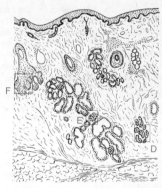

F

E

D

C   Skin glands of axilla

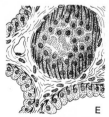

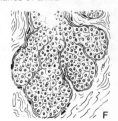

D   E   F

Hair

Hair (pili) serves the *sense of touch* and provides for *preservation of heat*. It is only absent from a few skin areas (the palms, soles, and some parts of the external genitalia). The *lanugo hair* of the fetus is short, thin, unpigmented and has its roots in the corium. It mostly disappears after birth. The *terminal hair* develops after birth. It is longer, thicker, pigmented, concentrated in specific areas and rooted in the upper subcutis. It comprises mainly the hair of the scalp, the axilla, the pubic hair and the beard.

The **terminal hair** projects obliquely above the surface (vortex) from its cylindrical root sheath into which a *sebaceous gland* **A1** opens. Above the gland is the hair funnel, below it is a smooth muscle, the *arrector pili muscle* **A2**, which extends to the undersurface of the epidermis. This muscle can erect the hair (hair raising, goose pimples) and can compress (squeeze) the sebaceous glands.

Fine structure. We distinguish the **hair root A3**, the hair *bulb* **A4**, which rests on the connective tissue of the hair *papilla* **A5**, and the **hair shaft A6**, which extends above the surface of the skin. The bulb, papilla and the surrounding connective tissue are called the *hair follicle*.

The hair is a modified type of keratin formation which arises from a deep circumscribed depression in the epidermis **A7**. The *hair* is the tip of the keratin, the *epithelial root sheath* **A8** is the epidermal funnel of the depression and the *connective tissue layer of the root sheath* **A9** (hair follicle) is the papillary body. The hair grows from the cells of the hair bulb and is nourished by the hair papilla. The papilla forms the epidermal matrix of the hair root and its destruction results in loss of the hair forming structure. The *cuticle of the root sheath* is interlocked with *that of the hair* and so ensures its anchorage. The hair shaft consists mainly of a *cortex* of elongated cornified cells stuck together by tonofilaments. Thick hair contains a less cornified *medulla*.

The **color of the hair** is mainly due to *melanin*, which is produced by the melanocytes in the matrix which deliver it to the cells of the hair bulb (compare p. 332). When the *hair turns gray* production of melanin by melanocytes ceases or they just perish. If air bubbles enter the center (central canal) of hair it also looks white. In the *albino*, however, the melanocytes do not produce any melanin at all because of a hereditary anomaly.

Growth of the hair. The hair grows in cycles: a period of growth is followed by a period of involution and rest after which the hair is shed. Normally about 80% of the hair follicles are in the growing stage and 15% are at rest. There is a loss of 50–100 hairs a day. The melanocytes retreat temporarily, the hair bulb **BC10** is detached from the papilla and pushed out, compare **BCD** (*"clubbed hair"* **D11**). From the cells which remain in the attenuated papilla **C12** a new bulb **D13** rises from which the new hair grows. Terminal hair grows about 1 cm per month and single follicles can remain for months and even years.

The **hair cover E** is under the control of *hormones*. Typical of the male is a rhomboid configuration of the pubic hair which ascends to the umbilicus (navel), hair covering the inside of the thigh, the chest and, of course, the beard. There are many individual variations. Typical of the *female* is the triangular shape of the pubic hair and the absence of terminal hair over the trunk. In patients with endocrine disturbances (e. g., of the adrenal cortex) a male-type hair cover can develop in a female (*hypertrichosis, hirsutism*).

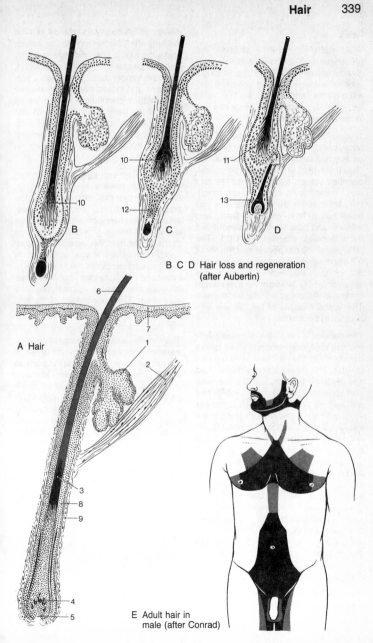

B C D Hair loss and regeneration
(after Aubertin)

A Hair

E Adult hair in
male (after Conrad)

Nails

The **nails** (*ungues*) protect the phalanges of the fingers and toes and give them a *sense of touch* by providing a counter support against pressure exerted on the tactile elevations. If a nail is lost, there is a diminished sense of touch in the affected phalanx.

The **nail** is about 0.5 mm thick. It is a horny plate of the epidermis, which rests on the *nail bed* and is anchored in the "hyponychium". It consists of polygonal, cornified scales, tightly packed like roof tiles. Three layers of criss-cross tonofibrils are fused in the structure. The *nail body* has a free margin in front, and laterally and posteriorly it is surrounded by a skinfold, the *nail wall* **BC1**. The latter forms the *nail pocket* which is about 0.5 cm deep in the region of the *nail root* **B2**. An epithelial cuticle, the *eponychium*, grows from the free margin of the nail wall onto the surface of the nail. The lateral margins of the nail are embedded in the *nailfold*.

The *nail bed* is the epithelial tissue below the nail root, from which the nail grows constantly at the rate of 0.14 to 0.4 mm per day. The proximal part of the nail bed lies deep in the nail pocket and the distal part appears light pink, near the proximal nail wall as the *lunula*, which is convex distally. The nail bed extends distal to the lunula **A3** as the dark red *"hyponychium"* **A4**, which is visible through the nail. It is an epithelial layer over which the nail is pushed distally. The connective tissue, which corresponds to the papillary body of the skin and lies beneath the epithelium of the nail bed and of the "hyponychium", is arranged in longitudinal channels. The blood in the capillaries of these connective tissue channels produces the pink color of the nails. **C5** Finger bone.

Skin as a Sense Organ

The skin is amply supplied with **nerves**. A small proportion of them are *autonomic* nerves which supply the glands, but the majority are *sensory nerves*. The nerves make the skin a vital and indispensable *sense organ* for the sensations of touch, pain and temperature. The sensory perceptions and indeed the sensory nerves vary in their distribution over different parts of the body. Distinct types of *end corpuscles* are concerned with particular sensations. Figure **D** gives a general idea of these distinctions *(details in Vol. 3)*.

Regeneration and aging. The skin has considerable powers of regeneration. After skin lesions the immune cells in the corium deal with local infection, the capillaries and the connective tissue regenerate, the epithelium grows from the edge of the lesion over the regenerating connective tissue and a *scar is formed*. At first the reddish color of the scar is due to increased capillarization, but later whitish collagen fibers are visible through the epithelium. The appendages of the skin (the glands and hair) are not formed again in scar tissue.

The *manifestations of old age in the skin* are mainly those of atrophy: the layers of the skin become thinner and the papillae flatten. There is an associated alteraion in the chemical composition of the basic connective tissue substance, resulting in paucity of fluid and loss of elasticity in the elastic fibers of the corium and the subcutis. As a result skin folds which have been lifted from the underlying tissue are slow to disappear. Irregular (random) spots of pigmentation also appear.

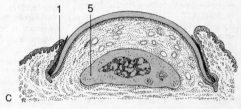

B Longitudinal section through nail bed (after Rauber-Kopsch)

A Fingernail

C Cross section through nail bed (after Rauber-Kopsch)

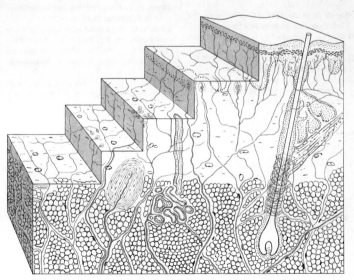

D Diagrammatic representation of innervation of skin (after Weddell)

References

Only a small selection of the literature (textbooks, handbooks, monographs and journals) referring to the themes of each chapter can be cited here. These publications also contain further references.

General Reading List

Arey, L. B.: Developmental Anatomy, 7th Ed. Saunders, Philadelphia 1974.

Bargmann, W.: Histologie und mikroskopische Anatomie des Menschen, 7th Ed. Thieme, Stuttgart 1977.

Bargmann, W.: Handbuch der mikroskopischen Anatomie des Menschen. Springer, Berlin 1929ff.

Benninghoff, A., K. Goerttler: Lehrbuch der Anatomie des Menschen, Vol. 1, 2. 13/14th Ed. by K. Fleischauer, Urban & Schwarzenberg, München 1985.

Bloom, W., D. W. Fawcett: A Textbook of Histology, 10th Ed. Saunders, Philadelphia 1975.

Braus, H., C. Elze: Anatomie des Menschen, 3. Ed. Vol. II. Springer, Berlin 1956.

Bucher, O.: Cytologie, Histologie und mikroskopische Anatomie des Menschen, 9th. Ed. Huber, Bern 1977.

Fawcett, D. W.: An Atlas of Fine Structure. The Cell. Saunders, Philadelphia 1966.

Feneis, H.: Anatomisches Bildwörterbuch, 5th Ed. Thieme, Stuttgart 1982.

Grant, J. C. B.: An Atlas of Anatomy, 6th Ed. Williams & Wilkins, Baltimore 1972.

Grashey, R., R. Birkner: Atlas typischer Röntgenbilder vom normalen Menschen, 10. Ed. Urban & Schwarzenberg, München 1964.

Ham, A. W.: Histology, 7th Ed. Lippincott, Philadelphia and Toronto 1974.

Hamilton, W. J.: Textbook of Human Anatomy, 2nd Ed. Macmillan, London and New York 1976.

Hamilton, W. J., H. W. Mossman: Human Embryology, 4th Ed. Williams & Wilkins, Baltimore 1972.

Hamilton, W. J., G. Simon, S. G. I. Hamilton: Surface and Radiological Anatomy, 5th Ed. Williams & Wilkins, Baltimore 1971.

v. Herrath, E.: Atlas der normalen Histologie und mikroskopischen Anatomie des Menschen, 4. Aufl. Thieme, Stuttgart 1977.

Hollinshead, W. H.: Textbook of Anatomy, 3rd Ed. Harper and Row, Maryland 1974.

International Anatomical Nomenclature Committee: Nomina anatomica, 4th Ed. Excerpta Medica Foundation, Amsterdam 1977.

Langman, J.: Medical Embryology, 7th. Ed. Williams & Wilkins, Baltimore 1985.

Leonhardt, H.: Histologie und Zytologie des Menschen, 7th Ed. Thieme, Stuttgart 1985.

Leonhardt, H.: Human Histology, Cytology and Microanatomy. Thieme, Stuttgart 1976.

Pernkopf, E.: Atlas der topographischen und angewandten Anatomie des Menschen, Vol. 1, 2 Ed. by H. Ferner, Urban & Schwarzenberg, München 1964.

Rauber, A., F. Kopsch: Lehrbuch und Atlas der Anatomie des Menschen, Vol. I, 20th Ed. neubearb. von G. Töndury. Thieme, Stuttgart 1968.

Snell, R. S.: Clinical Anatomy for Medical Students. Little Brown, Boston 1973.

Starck, D.: Embryologie, 3. Aufl. Thieme, Stuttgart 1975.

Starck, D., H. Frick: Repetitorium anatomicum, 12. Aufl. Thieme, Stuttgart 1972.

Töndury, G.: Angewandte und topographische Anatomie, 5th. Ed. Thieme, Stuttgart 1981.

Williams, P. L., R. Warwick: Gray's Anatomy 35th Ed. (British) Saunders, Philadelphia 1973.

Williams, P. L., C. P. Wendell Smith: Basic Human Embryology, 2nd Ed. Pitman, London 1969.

Angiology

Allen, L.: Lymphatics and Lymphoid Tissues. Rev. Physiol. 29: 197–224, 1967.

Bargmann, W., W. Doerr: Das Herz des Menschen, Vol. I. Thieme, Stuttgart 1963.

Barthelheimer H., H. Küchmeister: Kapillaren und Interstitium. Thieme, Stuttgart 1955.

Clemens, H. J.: Die Venensysteme der menschlichen Wirbelsäule. De Gruyter, Berlin 1961.

Decker, K., H. Backmund: Angiographie des Hirnkreislaufs. Thieme, Stuttgart 1968.

Düx, A.: Koronarographie. Thieme, Stuttgart 1967.

Fishman, A. P., H. H. Hecht: The Pulmonary Circulation and Interstitial Space. University of Chicago Press, Chicago 1969.

Hammersen, F.: Anatomie der terminalen Strombahn. Urban & Schwarzenberg, München 1971.

Guasp, F. T.: The Electrical Circulation. Guasp. Denia 1970.

Kaindl, F., E. Mannheimer, L. Pfleger-Schwarz, B. Thurner: Lymphangiographie und Lymphadenographie der Extremitäten. Thieme, Stuttgart 1960.

Kaplan, H. A., D. H. Ford: The Brain Vascular System. Elsevier, Amsterdam 1966.

Kappert, A.: Lehrbuch und Atlas der Angiologie, 8th. Ed. Huber, Bern 1969.

Kinmonth, J. B.: Some general aspects of the investigation and surgery of the lymphatic system. J. cardiovasc. surg. 5: 680–682, 1964.

Kinmonth, J. B., G. W. Taylor: The Lymphatics. Disease, Lymphography and Surgery. Arnold, London 1972.

v. Lanz, T., W. Wachsmuth: Praktische Anatomie, 2rd. Ed. Vol. I/3, 4, Springer, Berlin 1972.

Lippert, H.: Arterienvarietäten. Klinische Tabellen. Beilg. in Med. Klin. 18–32 (1967–1969).

Loose, K. E., R. J. van Dongen: Atlas of Angiography. Thieme, Stuttgart 1975.

Luisada, A. A.: Development and Structure of the Cardiovascular System. McGraw-Hill, New York 1961.

Maggio, E.: Microhemocirculation. Thomas, Springfield. Ill. 1965.

Mc Alpine, W. A.: Heart and Coronary Arteries. Springer, Berlin 1975.

Michels, N. A.: Blood Supply and Anatomy of Upper Abdominal Organs. Lippincott, Philadelphia 1956.

Nomura, T.: Atlas of Cerebral Angiography. Springer, Berlin 1970.

Puff, A.: Der funktionelle Bau der Herzkammern. In: Zwanglose Abhandlungen aus dem Gebiet der normalen und pathologischen Anatomie. H. 8, Ed. by *W. Bargmann, W. Doerr.* Thieme, Stuttgart 1960.

Rényi-Vámos, F.: Das innere Lymphgefäßsystem der Organe. Akadémiai Kiadó, Budapest 1960.

Routier, D.: Das Röntgenbild des Herzens. Thieme, Stuttgart 1963.

Rusznyák, I., M. Földi, G. Szabó: Lymphatics and Lymph Circulation. Pergamon Press, Oxford 1960.

Rusznyák, I., M. Földi, G. Szabó: Lymphologie, 2nd. Ed. Fischer, Stuttgart 1969.

Schoenmackers, J., H. Vieten: Atlas postmortaler Angiogramme. Thieme, Stuttgart 1954.

Staubesand, J.: Funktionelle Morphologie der Arterien, Venen und arteriovenösen Anastomosen. In: Angiologie, hrsg. von *M. Ratchow.* Thieme, Stuttgart 1959.

Stephens, R. B., D. L. Stilwell: Arteries and Veins of the Human Brain. Thomas, Springfield, Ill. 1969.

Yoffey, J. M., F. C. Courtice: Lymphatics, Lymph and Lymphomyeloid Complex. Academic Press, New York and London 1970.

Hematology and Immunology

Elves, M. W.: The Lymphocyte, 2nd Ed. Lloyd Luike, London 1972.

Goldstein, G., I. R. Mackay: The Human Thymus. Heinemann, London 1969.

Good, R. A., A. E. Gabrielsen: The Thymus in Immunobiology. Harper & Row, New York 1964.

Hayhoe, F. G., R. J. Flemans: Atlas der hämatologischen Zytologie. Springer, Berlin 1969.

Humphrey, J. H., R. G. White: Immunology for Medical Students. 3rd Ed. Blackwell, Oxford 1970.

Leder, L.-D.: Der Blutmonocyt. In: Experimentelle Medizin, Pathologie und Klinik, Vol. XXIII. Ed. by *F. Leuthardt, R. Schoen, H. Schwiegk, A. Studer, H. U. Zollinger.* Springer, Berlin 1967.

Lennert, K., D. Harms: Die Milz/The Spleen. Springer, Berlin 1970.

Lindenmann, J., J. Zielinski: Programmierte Immunologie. Thieme, Stuttgart 1971.

Mc Donald, G., A. Dodds, T. C. Cruickshank: Atlas of Haematology, 3rd Ed. Churchill Livingstone, Edinburgh 1970.

Mori, Y., K. Lennert: Electron Microscopic Atlas of Lymph Node Cytology and Pathology. Springer, Berlin 1969.

Rohr, K.: Das menschliche Knochenmark, 3rd. Ed., Thieme, Stuttgart 1960.

Rohr, K. R. Oechslin-Kütter, A. Uehlinger: Tabulae haematologicae, 4. Aufl. Thieme, Stuttgart 1966.

Roitt, I.: Essential Immunology. Blackwell, Oxford 1971.

Stuart, A. E.: The Reticuloendothelial System. Livingstone, Edinburgh 1970.

Thompson, R. B.: Textbook of Haematology, 4th Ed. Pitman, London 1975.

Tischendorf, F.: Die Milz. In: Handbuch der mikroskopischen Anatomie des Menschen, Vol. VI/6, Ed. by *W. Bargman.* Springer, Berlin 1969.

Vernon-Roberts, B.: The Macrophage. Cambridge University Press, Cambridge 1972.

Yoffey, J. M., F. C. Courtice: Lymphatics, Lymph and Lymphomyeloid Complex. Academic Press, New York 1970.

Respiratory System

Boyden, E. A.: Segmental Anatomy of the Lungs. McGraw Hill, New York 1965.

Emery, J.: The Anatomy of the Developing Lung. Heinemann, London 1969.

Fishman, A. P., H. H. Hecht: The Pulmonary Circulation and Interstitial Space. University of Chicago Press, Chicago 1969.

Frazer, R. G., J. A. P. Paré: Structure and Function of the Lung. Saunders, Philadelphia 1971.

v. Hayek, H.: The Human Lung. Haffner, New York 1960.

v. Hayek, H.: Die menschliche Lunge, 2nd. Ed. Springer, Berlin 1970.

Kubik, St.: Klinische Anatomie, Vol. III. Thieme, Stuttgart 1969.

Liebow, A. A., D. E. Smith: The Lung. Williams & Wilkins, Baltimore 1968.

Weibel, E. R.: Morphological basis of alveolar-capillary gas exchange. Physiol. Rev. 53 (1973) 419–495.

Weibel, E. R.: Morphometry of the Human Lung. Springer, Berlin 1963.

Endocrine System

Bachmann, R.: Die Nebenniere. In: Handbuch der mikroskopischen Anatomie des Menschen, Bd. VI/5, Ed. by *W. Bargmann.* Springer, Berlin 1954.

Bargmann, W.: Die funktionelle Morphologie des endokrinen Regulationssystems. In: Handbuch der allgemeinen Pathologie, Bd. VIII/1, Ed. by *H.-W. Altmann, F. Büchner, H. Cottier, E. Grundmann, G. Holle, E. Letterer, W. Masshoff, H. Meessen, F. Roulet, G.*

Seifert, G. Siebert, A. Studer. Springer, Berlin 1971.

Bargmann, W., B. Scharrer: Aspects of Neuroendocrinology. Springer, Berlin 1970.

Battaglia, G.: Ultrastructural Observations on the Biogenic Amines in the Carotid and Aortic-Abdominal Bodies of the Human Fetus. Z. Zellforsch. 99, 529–537 (1969).

Baumgarten, H. G.: Morphological basis of gastrointestinal motility: Structure and innervation of gastrointestinal tract. In *Bertacchini, G.:* Handbook of experimental Pharmacology, Vol. LIX/1: Mechanisms of Intestinal Secretion. Springer, Berlin 1982 (pp. 7–53).

Bloom, S. R., J. M. Polak: Gut Hormones, 2nd ed. Churchill-Livingstone, Edinburgh 1981.

Cantin, M.: Cell Biology of the Secretory Process. Karger, Basel 1984.

Capen, C. C., G. N. Rowland: Ultrastructural Evaluation of the Parathyroid Glands of Young Cats with Experimental Hyperparathyroidism. Z. Zellforsch. 90, 495–506, 1968.

Dhom, G.: Die Nebennierenrinde im Kindesalter. Springer, Berlin 1965.

Douglas, W. W.: Stimulus-Secretion Coupling in the Adrenal Medulla and the Neurohypophysis. In: Neurosecretion, Ed. *F. Stutinsky.* Springer, Berlin 1967.

Ferner, H.: Das Inselsystem des Pankreas. Thieme, Stuttgart 1952.

Fujita, T.: Gastro-Entero-Pancreatic Endocrine System: A Cell Biological Approach. Thieme, Stuttgart 1974.

Gaillard, P. J., R. V. Talmage, A. M. Budy: The Parathyroid Glands: Ultrastructure, Secretion and Function. University of Chicago Press, Chicago 1965.

Greep, R. O., R. V. Talmage: The Parathyroids. Thomas, Springfield, Ill. 1961.

Harris, G. W., B. J. Donovan: The Pituitary Gland. Vols. I–III. Butterworth, London 1966.

Hellerström, C., B. Hellman, B. Peterson, G. Alun: The Structure and Metabolism of Pancreatic Islets. Pergamon, New York 1964.

James, V. H. T.: The Endocrine Function of the Testis. Lewis, London 1971.

Krisch, B.: Immunocytochemistry of neuroendocrine systems (vasopressin, somatostatin, luliberin). Progr. Histochem. Cytochem. 13/2 (1980) 1–167.

Lupulescu, A., A. Petrovici: Ultrastructure of the Thyroid Gland. Karger, Basel 1968.

Neumann, K. H.: Die Morphokinetik der Schilddrüse. Fischer, Stuttgart 1963.

Pearse, A. G. E.: The diffuse neuroendocrine system and the APUD concept: related "endocrine" peptides in brain, intestine, putitary, placenta and anuran cutaneous glands. Med. Biol. 55 (1977) 115–125.

Vollrath, L.: The pineal organ. In *Oksche, A., L. Vollrath:* Handbuch der mikroskopischen Anatomie des Menschen, Vol. VI/7. Springer, Berlin 1981.

Watzka, M.: Die Paraganglien. In: Handbuch der mikroskopischen Anatomie des Menschen, Bd. VI/4, hrsg. von *W. v. Möllendorff.* Springer, Berlin 1943.

Winkler, J.: Zur Lage und Funktion der extramedullären chromaffinen Zellen. Z. Zellforsch. 96 (1969) 490–494.

Alimentary System

Beck, I. T., D. G. Sinclair: The Exocrine Pancreas. Churchill, London 1971.

Beck, K., W. Dischler, M. Helms, B. Kiani, K. Sickinger, R. Tenner: Atlas der Laparoskopie. Schattauer, Stuttgart 1968.

Becker, V.: Pankreas. In: Spezielle pathologische Anatomie, Bd. VI, ed. by *W. Doerr, G. Seifert, E. Uehlinger.* Springer, Berlin 1972.

Elias, H., J. C. Sherrick: Morphology of the Liver. Academic Press, New York 1969.

Födisch, H.: Feingewebliche Studien zur Orthologie und Pathologie der Papilla Vateri. Thieme, Stuttgart 1972.

Gall, E. A., F. K. Mostofi: The Liver. Krieger, Huntington/N. Y. 1980.

Kaufmann, P., W. Lierse, J. Stark, F. Stelzner: Die Muskelanordnung in der Speiseröhre (Mensch, Rhesusaffe, Kaninchen, Maus, Ratte, Seehund). Ergebn. Anat. Entwickl.-Gesch. 40 H. 3 1968.

McMinn, R. M. H., M. H. Hobdell: The Functional Anatomy of the Digestive System. Pitman Medical, London 1974.

Meyer, W.: Die Zahn-, Mund- und Kieferheilkunde, Bd. I. Urban & Schwarzenberg, München 1958.

Michels, N. A.: Blood Supply and Anatomy of the Upper Abdominal Organs. Lippincott, Philadelphia 1955.

Millat, B., J. P. Chevrel: The pylorus: an anatomical and physiological study. Anat. clin. 3 (1981) 161–175.

Ottenjann, R., K. Elster, S. Witte: Gastroenterologische Endoskopie, Biopsie und Zytologie. Thieme, Stuttgart 1970.

Popper, M., F. Schaffner: The Liver: Structure and Function. McGraw Hill, New York 1957.

Prévôt, R., M. A. Lassrich: Röntgendiagnostik des Magen-Darmkanals. Thieme, Stuttgart 1959.

de Reuck, A. V. S., M. P. Cameron: The Exocrine Pancreas, Normal and Abnormal Function. Ciba Foundation. Churchill, London 1962.

Schumacher, G. H.: Der maxillo-mandibuläre Apparat unter dem Einfluß formgestaltender Faktoren. Barth, Leipzig 1968.

Shiner, M.: Ultrastructure of the Small Intestinal Mucosa. Springer, Berlin 1982.

Sicher, H., E. L. Du Brul: Oral Anatomy. Mosby, St. Louis 1970.

Toner, P. G., K. E. Carr, G. M. Wyburn: The Digestive System. An Ultrastructural Atlas and Review. Butterworth, London 1971.

Wallraff, J.: Die Leber- und Gallengangsysteme, Gallenblase und Galle. In: Handbuch der mikroskopischen Anatomie des Menschen, Erg. zu Bd. V/2 Ed. *W. Bargmann.* Springer, Berlin 1969.

Urogenital System

Bargmann, W., K. Fleischhauer, A. Knoop: Über die Morphologie der Milchsekretion. Z. Zellforsch. 53: 545–568, 1960.

Beck, L.: Morphologie und Funktion der Muskulatur der weiblichen Harnröhre. Enke, Stuttgart 1969.

Bulmer, D.: The Functional Anatomy of the Urogenital System. Pitman Medical, London 1974.

Christensen, A. K., D. W. Fawcett: The Fine Structure of the Interstitial Cells of the Mouse Testis. Amer. J. Anat. 118: 551–571, 1966.

Dabelow, A.: Die Milchdrüse. In Handbuch der mikroskopischen Anatomie des Menschen, Erg. zu Vol. III/1, Ed. by *W. Bargmann.* Springer, Berlin 1957.

Dalton, A. J., F. Haguenau: Ultrastructure of the Kidney. Academic Press, New York 1967.

Ferner, H., A. Gisel, H. v. Hayek, W. Krause, Ch. Zaki: Die Anatomie der Harn- und Geschlechtsorgane. In: Handbuch der Urologie, Vol. I, Ed. by. *C. E. Alken, V. W. Dix, W. E. Goodwin, E. Wildbolz.* Springer, Berlin 1969.

Fourman, J., D. B. Moffat: The Blood Vessels of the Kidney. Blackwell, Oxford 1971.

Frangenheim, H., H.-T. Lindeman: Die Laparoskopie und die Kuldoskopie in der Gynäkologie, 3rd. Ed. Thieme, Stuttgart 1970.

Gil Vernet, S.: Morphology and Function of Vesico-Prostatico-Urethral Musculature. Canova, Treviso 1968.

Grosser, O., R. Ortmann: Grundriß der Entwicklungsgeschichte des Menschen, 7th Ed. Springer, Berlin 1970.

v. Hayek, H.: Die Entwicklung der Harn- und Geschlechtsorgane. In: Handbuch der Urologie, Vol. I, Ed. by C. E. Alken, V. W. Dix, W. E. Goodwin, E. Wildbolz. Springer, Berlin 1969.

Holstein, A.-F.: Morphologische Studien am Nebenhoden des Menschen. Thieme, Stuttgart 1969.

Holstein, A. F., E. C. Roosen-Runge: Atlas of the Spermatogenesis. Grosse, Berlin 1981.

Horstmann, E., H.-E. Stegner: Tube, Vagina und äußere weibliche Geschlechtsorgane. In: Handbuch der mikroskopischen Anatomie des Menschen, Erg. Vol. VII/1. Ed. by W. Bargmann. Springer, Berlin 1966.

Hutch, J. A.: Anatomy and Physiology of the Bladder, Trigone and Urethra. Butterworth, London 1972.

Hutch, J. A., O. S. Rambo: A Study of the Prostate, Prostatic Urethra and Urinary Sphincter System. J. Urol. 104: 443–452, 1970.

v. Krügelgen, A., B. Kuhlo, W. Kuhlo, Kl.-J. Otto: Die Gefäßarchitektur der Niere, 5. Aufl. Thieme, Stuttgart 1959.

Johnson, A. D., W. R. Gomes, N. L. Vandermark: The Testis. Vol. 1 Academic Press, New York and London 1970.

Langman, J.: Medical Embryology. 2nd Ed. Williams & Wilkins, Baltimore 1969.

Langman, J.: Medizinische Embryologie. 5. Aufl. Thieme, Stuttgart 1977.

Lenz, W.: Medizinische Genetik, 4th Ed. Thieme, Stuttgart 1978.

Martius, H., K. Droysen: Atlas der gynäkologischen Anatomie. Thieme, Stuttgart 1960.

Overzier, C.: Die Intersexualität. Thieme, Stuttgart 1961.

Pedersen, H.: Ultrastructure of the Ejaculated Human Spermatozoa. Z. Zellforsch. mikrosk. Anat. 94: 542–554, 1969.

Pernkopf, E., A. Pichler: Systematische und topographische Anatomie des weiblichen Beckens. In: Biologie und Pathologie des Weibes, Ed. by L. Seitz, A. J. Amreich. Urban & Schwarzenberg, Berlin 1953.

Potter, E. L.: Normal and Abnormal Development of the Kidney. Year Book Medical, Chicago 1972.

Ross, J. A., P. Edmond, I. S. Kirkland: Behaviour of the Human Ureter in Health and Disease. Churchill Livingstone, London and Edinburgh 1972.

Schmidt-Matthiesen, H.: Das normale menschliche Endometrium. Thieme, Stuttgart 1953.

Smout, C. F. V., F. Jacoby, E. W. Lilie: Gynaecological and Obstetrical Anatomy and Functional Histology. Arnold, London 1969.

Tanagho, E. A., R. C. B. Pugh: The Anatomy and Function of the Ureterovesical Junction. Brit. J. Urol. 35: 151–165, 1963.

Tanagho, E. A., F. H. Meyers, D. R. Smith: The Trigone: Anatomical and Physiological Considerations. I. In Relation to the Ureterovesical Junction. J. Urol., 100: 623–632, 1968.

Tonutti, E., O. Weller, E. Schuchardt, E. Heinke: Die männliche Keimdrüse. Thieme, Stuttgart 1960.

Ullrich, K. J.: Das Gegenstromsystem im Nierenmark. Schattauer, Stuttgart 1962.

Vogler, E., R. Herbst: Angiographie der Nieren. Thieme, Stuttgart 1958.

Watzka, M.: Weibliche Genitalorgane. In: Handbuch der mikroskopischen Anatomie des Menschen, Erg. zu Vol. VII/1, Ed. by W. Bargmann. Springer, Berlin 1957.

Woodruff, J. D., C. J. Pauerstein: The Fallopian Tube. Williams & Wilkins, Baltimore 1969.

Zuckerman, S.: The Ovary. Academic Press, New York 1962.

Skin

Breathnach, A. S.: An Atlas of the Ultrastructure of Human Skin. Churchill, London 1971.

Della Porta, G. O., O. Mühlbock: Structure and Control of the Melanocyte. Springer, Berlin 1966.

Horstmann, E.: Die Haut. In: Handbuch der mikroskopischen Anatomie des Menschen, Erg. zu Vol. III/1, Ed. by W. Bargmann. Springer, Berlin 1957.

Lyne, A. G., B. F. Short: Biology of Skin and Hair Growth. Angus and Robertson, Sydney 1965.

Montagna, W., R. A. Ellis, A. F. Silver, R. Billingham, F. Hu, R. A. Dobson: Advances in the Biology of Skin. Vols. I–IX, Pergamon Press, Oxford 1966–69.

Montagna, W., W. C. Lobitz: The Epidermis. Academic Press, New York 1964.

Montagna, W., P. F. Parakkal: The Structure and Function of Skin. 3rd. Ed. Academic Press, New York 1974.

Orfanos, C. E.: Feinstrukturelle Morphologie und Histologie der verhornenden Epidermis. Thieme, Stuttgart 1972.

Zelickson, A. S.: Ultrastructure of Normal and Abnormal Skin. Lea & Febiger, Philadelphia 1967.

Zelickson, A. S.: Ultrastructure of the Human Epidermis. In: Modern Trends in Dermatology. Ed. P. Borrie. Vol. 4, pp. 31–52. Butterworth, London 1971.

Index

Boldface page numbers indicate extensive coverage of the subject